Lecture Notes in Artificial Intelligence   10247

Subseries of Lecture Notes in Computer Science

More information about this series at http://www.springer.com/series/1244

Setsuya Kurahashi · Yuiko Ohta
Sachiyo Arai · Ken Satoh
Daisuke Bekki (Eds.)

New Frontiers in Artificial Intelligence

JSAI-isAI 2016 Workshops, LENLS
HAT-MASH, AI-Biz, JURISIN and SKL
Kanagawa, Japan, November 14–16, 2016
Revised Selected Papers

 Springer

Editors
Setsuya Kurahashi
Graduate School of Business Sciences
University of Tsukuba
Tokyo
Japan

Yuiko Ohta
Fujitsu Laboratories Ltd.
Kawasaki
Japan

Sachiyo Arai
Chiba University
Chiba
Japan

Ken Satoh
National Institute of Informatics
Tokyo
Japan

Daisuke Bekki
Ochanomizu University
Tokyo
Japan

ISSN 0302-9743   ISSN 1611-3349 (electronic)
Lecture Notes in Artificial Intelligence
ISBN 978-3-319-61571-4   ISBN 978-3-319-61572-1   (eBook)
DOI 10.1007/978-3-319-61572-1

Library of Congress Control Number: 2017936043

LNCS Sublibrary: SL7 – Artificial Intelligence

Printed on acid-free paper

This Springer imprint is published by Springer Nature
The registered company is Springer International Publishing AG
The registered company address is: Gewerbestrasse 11, 6330 Cham, Switzerland

Preface

JSAI-isAI (JSAI International Symposium on Artificial Intelligence) 2016 was the eighth international symposium on AI supported by the Japanese Society of Artificial Intelligence (JSAI). JSAI-isAI 2016 was successfully held during November 14–16 at Keio University in Kanagawa, Japan. In all, 177 people from 14 countries participated. The symposium took place after the JSAI SIG joint meeting. As the total number of participants for these two co-located events was about 990, it was the second-largest JSAI event in 2016 after the JSAI annual meeting.

JSAI-isAI 2016 included seven workshops, where 13 invited talks and 81 papers were presented. This volume, *"New Frontiers in Artificial Intelligence:JSAI-isAI 2016 Workshops,"* comprises the proceedings of JSAI-isAI 2016. From five (LENLS 13, HAT-MASH 2016, AI-Biz 2016, JURISIN 2016, and SKL 2016) out of seven workshops, 22 papers were carefully selected and revised according to the comments of the workshop Program Committees. About 34% of the total submissions were selected for inclusion in the conference proceedings.

LENLS 13 was the 13th event in the series, and it focused on the formal and theoretical aspects of natural language. LENLS (Logic and Engineering of Natural Language Semantics) is an annual international workshop recognized internationally in the formal syntax–semantics–pragmatics community. It brings together for discussion and interdisciplinary communication researchers working on formal theories of natural language syntax, semantics and pragmatics, (formal) philosophy, artificial intelligence, and computational linguistics.

HAT-MASH 2016 (Healthy Aging Tech Mashup Service, Data and People) was the second international workshop in the series; it bridges healthy aging and elderly care technology, information technology, and service engineering. The main objective of this workshop was to provide a forum to discuss important research questions and practical challenges in healthy aging and elderly care support to promote transdisciplinary approaches.

AI-Biz 2016 (Artificial Intelligence of and for Business) was the first workshop, in the series held to foster the concepts and techniques of business intelligence (BI) in artificial intelligence. BI should include such cutting-edge techniques as data science, agent-based modelling, complex adaptive systems, and IoT. The main purpose of this workshop is to provide a forum to discuss important research questions and practical challenges in BI, business informatics, data analysis and agent-based modelling so as to exchange the latest results and to join efforts in solving the common challenges.

JURISIN 2016 was the 10th international workshop on juris-informatics. Juris-informatics is a new research area that studies legal issues from the perspective of informatics. The purpose of this workshop was to discuss both the fundamental and practical issues among people from various backgrounds such as law, social science, information and intelligent technology, logic and philosophy, including the conventional "AI and law" area.

SKL 2016 (the Third International Workshop on Skill Science) aimed to internationalize the research on skill sciences through the meeting. Human skills involve well-attuned perception and fine motor control, often accompanied by thoughtful planning. The involvement of body, environment, and tools mediating them makes the study of skills unique among researches of human intelligence.

It is our great pleasure to be able to share some highlights of these fascinating workshops in this volume. We hope this book will introduce readers to the state-of-the-art research outcomes of JSAI-isAI 2016 and motivate them to participate in future JSAI-isAI events.

April 2017

Setsuya Kurahashi
Yuiko Ohta
Sachiyo Arai
Ken Satoh
Daisuke Bekki

Contents

LENLS 13

Proper Names in Interaction............................... 5
  Robin Cooper

An Analysis of Selectional Restrictions with Dependent Type Semantics.... 19
  Eriko Kinoshita, Koji Mineshima, and Daisuke Bekki

Non-canonical Coordination in the Transformational Approach........... 33
  Oleg Kiselyov

Rich Situated Attitudes 45
  Kristina Liefke and Mark Bowker

Reference and Pattern Recognition 62
  Youichi Matsusaka

Negotiating Epistemic Authority............................ 74
  Elin McCready and Grégoire Winterstein

Challenges in the Computational Implementation of Montagovian
Lexical Semantics...................................... 90
  Bruno Mery

Truth Conditionals and Use Conditionals: An Expressive Modal Analysis ... 108
  Lukas Rieser

On the Interpretation of Dependent Plural Anaphora
in a Dependently-Typed Setting 123
  Ribeka Tanaka, Koji Mineshima, and Daisuke Bekki

HAT-MASH 2016

Toward Sentiment Analysis in Elderly Care Facility 143
  Ken Fukuda, Satoshi Nishimura, Huizhi Liang, and Takuichi Nishimura

AI-Biz 2016

The Passenger Decision Making Mechanism of Self-service Kiosk
at the Airport... 159
  Keiichi Ueda and Setsuya Kurahashi

A Study of Crucial Factors for In-App Purchase of Game Software   176
  Meng-Ru Lin and Goutam Chakraborty

An Agent-Based Model for Evaluating Post-acquisition Integration
Strategies   188
  Jing Su, Mohsen Jafari Songhori, Takamasa Kikuchi,
  Masahiro Toriyama, and Takao Terano

Assessing Long-Term Care Resource Distribution in China
by Simulating Care-Seeking Behaviors   204
  Shuang Chang, Wei Yang, and Hiroshi Deguchi

The Research of Bankruptcies' Succession by Systemic Risk Index   220
  Morito Hashimoto and Setsuya Kurahashi

JURISIN 2016

Voluntary Manslaughter? A Case Study with Meta-Argumentation
with Supports   241
  Ryuta Arisaka and Ken Satoh

Argument-Based Logic Programming for Analogical Reasoning   253
  Teeradaj Racharak, Satoshi Tojo, Nguyen Duy Hung,
  and Prachya Boonkwan

Estimating Legal Document Structure by Considering Style Information
and Table of Contents   270
  Yoichi Hatsutori, Katsumasa Yoshikawa, and Haruki Imai

Legal Yes/No Question Answering System Using Case-Role Analysis   284
  Ryosuke Taniguchi and Yoshinobu Kano

Question Answering of Bar Exams by Paraphrasing and Legal
Text Analysis   299
  Mi-Young Kim, Ying Xu, Yao Lu, and Randy Goebel

SKL 2016

A Basic Study of Gaze Behavior Measurement Methodology for Drivers
in Autonomous Vehicles   317
  Rie Osawa, Shota Imafuku, and Susumu Shirayama

Toward a Mechanistic Account for Imitation Learning: An Analysis
of Pendulum Swing-Up   327
  Takuma Torii and Shohei Hidaka

Author Index   345

LENLS 13

Logic and Engineering of Natural Language Semantics

Alastair Butler

National Institute for Japanese Language and Linguistics, Tachikawa, Japan

1 The Workshop

On November 13–15, 2016, the Thirteenth International Workshop of Logic and Engineering of Natural Language Semantics (LENLS 13) took place. As an annual international workshop recognised internationally in the formal syntax-semantics-pragmatics community, LENLS has, since 2005, been bringing together for discussion and interdisciplinary communication researchers working on formal theories of natural language syntax, semantics and pragmatics, (formal) philosophy, artificial intelligence and computational linguistics.

On November 13th the workshop was located at the National Institute for Japanese Language and Linguistics, Tachikawa, and on subsequent days moved to the Raiousha Building, Keio University, as a workshop of the Eighth JSAI International Symposia on AI (JSAI-isAI 2011), organised by The Japan Society for Artificial Intelligence (JSAI).

The first day of LENLS 13 comprised an "Unshared Task" that asked participants to make use of datasets (FraCaS, MultiFraCaS and JSeM) as benchmarks for measuring and comparing the competence of syntax/semantic theories and computational processing systems. Submissions for the remaining days of LENLS came from topics in formal syntax, semantics and pragmatics, and related fields.

The first day had three twenty minute talks, three thirty minute talks, as well as three invited lectures that were each one hour in length. The invited speakers on this day were Robin Cooper (University of Gothenburg), who spoke about testing the FraCaS test suite, Tim O'Gorman (University of Colorado Boulder), who spoke about Abstract Meaning Representation performance on the Fracas test Suite, and Masaaki Nagata (NTT Communication Science Laboratories) who talked about how semantics can contribute to neural machine translation.

The following two days of LENLS 13 had fifteen submitted talks with the duration of thirty minutes each, and two invited lectures that were each one hour in length. Topics discussed by the submitted papers raised issues from Dynamic Semantics, Expressive meanings, Type Theoretic Semantics, language generation, syntactic analysis as well as fundamental themes from the philosophy of language. The invited speakers were Youichi Matsusaka (Tokyo Metropolitan University), who talked about a Metasemantic study of reference and pattern recognition, and again Robin Cooper (University of Gothenburg), who spoke about proper names in interaction.

For workshop participants a proceedings volume was made available containing 17 papers and 9 abstracts (five of which were for the invited talks), from which 9 papers were taken for the present volume.

2 Acknowledgements

Let me acknowledge some of those who helped with the workshop. The program committee and organisers, in addition to myself, were Daisuke Bekki, Elin McCready, Koji Mineshima, Richard Dietz, Yoshiki Mori, Yasuo Nakayama, David Y. Oshima, Katsuhiko Sano, Osamu Sawada, Wataru Uegaki, Katsuhiko Yabushita, Tomoyuki Yamada, Shunsuke Yatabe and Kei Yoshimoto. The organisers would like to thank the "Establishment of Knowledge-Intensive Structural Natural Language Processing and Construction of Knowledge Infrastructure" project, funded by JST CREST Programs "Advanced Core Technologies for Big Data Integration."

Proper Names in Interaction

Robin Cooper[✉]

University of Gothenburg, Gothenburg, Sweden
cooper@ling.gu.se

Abstract. Proper names in natural languages seem very simple from a linguistic point of view although getting their semantics correct turns out to be something of a challenge. In this paper we will suggest that there is an important advantage in viewing proper names as something like Saussurean signs incorporated in a formal theory which takes account of linguistic utterances as actions enabling linguistic agents to interact.

1 Introduction

In this paper we will consider some problems with considering proper names in natural languages as logically proper names, that is, considering them to be like constants which refer to one unique entity. We will argue that regarding proper names in terms of Saussurean signs in something like the version of them presented in Head Driven Phrase Structure Grammar (HPSG, Sag *et al.* 2003) although recast in terms of TTR (Type Theory with Records, Cooper 2012; Ginzburg 2012; Cooper and Ginzburg 2015; Cooper in prep) can help us to solve these problems. In particular we must take in account that proper names can be used in interaction and that the association of individuals with the phonology of names is something that can change not only during the course of a speaker's life, but also during the course of a dialogue, for example, when you are introduced to somebody you did not previously know.

In Sect. 2 we will argue that the notion of sign is important for the analysis of proper names and in Sect. 3 we will see that the signs related to proper names that an agent has access to can change during the course of a conversation. In Sect. 4 we will argue that in order to properly understand how proper names are working we need to have a model of how proper names relate to other cognitive resources like long term memory. We will then discuss (in Sect. 5) how this relates to the well-known Paderewski puzzle presented by Kripke (1979). Kripke's puzzle was posed as a puzzle about belief rather than as a puzzle about proper names as such and in Sect. 6 we will sketch how the view of proper names we have suggested relates to a theory of belief reports as seeking a match with an agent's long term memory. Finally, in Sect. 7 we will draw some conclusions from this about the nature of linguistic theory.

© Springer International Publishing AG 2017
S. Kurahashi et al. (Eds.): JSAI-isAI 2016, LNAI 10247, pp. 5–18, 2017.
DOI: 10.1007/978-3-319-61572-1_1

2   Proper Names and Signs

Consider a situation where Chris is standing on her own at a party, sipping a glass of warm sake. Somebody comes up to her with another person in tow and says (1).

(1)   Chris, I'd like to introduce you to my friend, Sam

Suppose that Chris is equipped with a grammar of proper names based on individual constants of the kind proposed by Montague (1973). Chris has never met the person she is being introduced to before but knows somebody else called Sam. How can Chris cope with this confusing situation? She replies with (2).

(2)   No, that's not Sam. I already know Sam and she is not this person.

This is because she associates exactly one proper name with the phonology "Sam" and she knows that it cannot be used to refer to the person to whom she is being introduced. Not surprisingly, nobody has ever really proposed that natural language proper names work in this way. We all know that more than one person can have the same name, that is, natural language proper names are not logically proper in this strict sense.

There is, however, a quick fix which seems apparently harmless and which preserves Montague's proposal and yet addresses this problem. This solution is to propose syntactically distinct proper names which just happen to be pronounced the same. For example, one can have names with silent subscripts. On this model, Chris' response to the situation might be different: realizing the seriousness of the situation, she quickly creates a new proper name and says (3).

(3)   Hi, Sam_2. I'm glad you could come.

There is a very important point to note about this, namely, that Chris' language has changed slightly during the course of the conversation. She has added a new word to the lexicon (albeit one with identical pronunciation to a word which she already had) and she has extended that part of the semantic universe of individuals she was aware of to include the person who can be referred to by that name. This is not a problem it seems to me. Proper names are an important, though comparatively simple, example of "language in flux" (Cooper and Kempson 2008), that is, the idea that our languages can change during the course of a conversation. This view of proper names is similar to one that has been defended by Ludlow (2014), referring back to earlier work by Larson and Ludlow (1993). There is, however, a slight problem with this approach: it means that two people cannot technically have the same name. The dialogue continues in (4).

(4)   Chris: Funny, I was just talking to a friend of mine who has
       a very similar name to you: Sam₁

Sam: Wow, cool. You mean we have the same name?

Chris (narrowing her eyes, and looking down her nose):
       Well, no, of course, not. That would be logically
       impossible since you are not identical with my friend.
       Your names are just pronounced the same.

This theory seems to commit us to a claim that technically two individuals cannot have the same name and that when we say in natural language that two individuals have the same name what we mean is that their names have the same phonology. One way to achieve this is perhaps to say that the extension of the English word *name* is the set of name phonologies, not the set of lexical items which are names.

Another way to approach this, which seems closer to what Ludlow (2014) is proposing and which might seem to avoid the problem with having the same name is to say that you use slightly different languages in different situations. This involves saying that you operate with slightly different micro-languages. In the case at point both micro-languages include the same lexical item *Sam* but it gets different denotations in the two different micro-languages. One potential problem with this is that you might expect Chris to be able to say (5).

(5)   Funny, I was just talking to a friend of mine who has a similar
       name to you in the language I use when I'm talking to her.

But this could be countered by saying that the English word *language* does not include micro-languages, which might at best in the vernacular be regarded as variants of the same language. There is, however, a deeper problem with this proposal. Consider the following situation. The name *Karin* is common in Sweden. I know at least three people with this name: my wife's sister, one of my graduate students and one of our neighbours. Consider the dialogue in (6) between me and Elisabet, my wife.

(6)   Elisabet: Karin called
       Robin: Karin?
       Elisabet: My sister

If I speak three variants of the language with different referents associated with *Karin* which of those variants am I speaking when I make that clarification request in (6)? Intuitively, it does not seem to be any of the three variants we have described but perhaps a fourth variant where my utterance of *Karin* refers to Elisabet's utterance of *Karin*. Note that my clarification could be replaced (perhaps a little pedantically) by (7a) (Ginzburg and Cooper 2014), but not naturally by anything like (7b).

(7)   a. Who did you refer to with your utterance of 'Karin'?
       b. Which language (variant) were you speaking when you said
          'Karin'?

I believe that these problems can be avoided if we construct our semantic analysis in terms of *signs*. Signs are pairings of speech events and semantic contents (de Saussure 1916; Sag *et al.* 2003, and much more). In terms of TTR this pairing can be represented as a record with two fields, one for a speech event and one for a semantic content, as in (8).

(8) $\begin{bmatrix} \text{sp-event} = e \\ \text{cont} \quad = k \end{bmatrix}$

Here e is an event of the phonological type "Karin" and k is Elisabet's sister. One type to which this record belongs is given in (9).

(9) $\begin{bmatrix} \text{sp-event}=e : \text{"Karin"} \\ \text{cont}=k \quad : Ind \end{bmatrix}$

A record of this type will have a 'sp-event'-field filled by the particular event, e, of the phonological type "Karin"[1] and a 'cont'-field filled by the particular individual, k, of type *Ind(ividual)*. In addition to (8), any record containing additional fields with other labels would be of this type. We will take such types of signs to be the currency of communication between agents. In making the utterance, e, of the word *Karin*, Elisabet intends me to recognize that she has produced a sign of type (9). However, the type I get from her utterance is not fully specified. I know that the speech event was of the phonological type "Karin" and I know that its content should be an individual, but I do not know exactly which individual it is. Thus the type I get from her utterance is (10).

(10) $\begin{bmatrix} \text{sp-event}=e : \text{"Karin"} \\ \text{cont} \quad : Ind \end{bmatrix}$

That is, the type I manage to compute from the speech event she created is *underspecified* with respect to content.

Underspecified sign types can also be used to model lexical resources in memory. Consider (11) where k is my wife's sister, kc is my graduate student and kg is our neighbour.

(11) a. $\begin{bmatrix} \text{sp-event} : \text{"Karin"} \\ \text{cont}=k \ : Ind \end{bmatrix}$

b. $\begin{bmatrix} \text{sp-event} : \text{"Karin"} \\ \text{cont}=kc : Ind \end{bmatrix}$

c. $\begin{bmatrix} \text{sp-event} : \text{"Karin"} \\ \text{cont}=kg : Ind \end{bmatrix}$

---

[1] Throughout this paper we will represent phonological types informally as the standard orthography enclosed in inverted commas.

In (11) it is the speech events which are underspecified. We know that the speech event must be of the phonological type "Karin" but we do not know exactly which speech event is involved. An agent with these resources can attempt to convey signs of these types by *creating* an event of type "Karin" (Cooper 2014).

This provides us with a simple and intuitive way of handling situations where more than one person has the same name. There is no need for unpronounced indices in a syntactic representation or for proposing that the same name refers to different people in different languages or variants of a language. We just have three distinct sign types which happen to share the same phonological type. We can take the name *Karin* to refer to the phonological type "Karin" and this opens the way for a straightforward and intuitive account of our intuition that these three individuals have the same name.

3   Creating Linguistic Resources

But what does it mean to have the linguistic knowledge that *Karin* is a proper name? It seems that it is not quite right to say that this involves knowing or believing one or more people to be called Karin, although this might be regarded as sufficient knowledge for knowing that *Karin* is a proper name. It seems that we can know that *Karin* is a proper name without actually having resources which link the phonological type "Karin" to any particular individual. Suppose, for example, that you do not know anybody called Karin but I tell you that many women in Sweden are called Karin and you take what I say to be true. It seems then that you would count *Karin* as a proper name without being able to link the name to any particular individual.

Suppose that $\text{Lex}_{\text{PropName}}$ is an operation on objects in our type theoretical universe which takes two arguments: a phonological type and an individual and returns a type defined so that $\text{Lex}_{\text{PropName}}(\text{"Karin"}, k)$ is (11a). Now we can say that we have the resource (12).

(12) $\lambda x{:}Ind$. $\text{Lex}_{\text{PropName}}(\text{"Karin"}, x)$

(12) is a function from individuals to types. Since it is a function which returns a type, it is also called a *dependent type* – which type you get depends on which individual you feed to the function. One could argue that a function like this represents "linguistic knowledge". That is, the "linguistic knowledge" that *Karin* is a proper name has to do with knowing that it can be used to name individuals. Knowing which individuals are called Karin is more like "world knowledge". I put the terms "linguistic knowledge" and "world knowledge" in scare quotes because I think that there is no way of defining a firm boundary between them. What I think is important here is that we are not required to know who is called Karin in order to know a language containing *Karin* as a proper name. Put another way, it is not a measure of my competence in English (or Swedish) that I know a number of people called Karin, though arguably a competent speaker of Swedish does know that *Karin* is a proper name, whether they know anybody with that name or not.

Note that the function in (12) is rather different from what we traditionally see in semantic treatments of natural languages. It is a function which creates a new linguistic resource which we can then use in the interpretation and generation of utterances. Our linguistic knowledge here does not so much involve a fact about a static language but a rule for generating a new linguistic resource we did not have before, a way of extending our language. Knowing what a proper name is, is knowing how they can be used when confronted with a new individual we have not encountered before. This is as we would expect if natural languages are in a state of flux (Cooper 2012).

4   Relating Linguistic Resources to Other Cognitive Resources

What does it mean to say that the type (11a) is, or models, a cognitive resource? In current work under development we have a proposal for how TTR types could in principle be represented as events on neural networks. Among other things this would involve a neural representation of k, my wife's sister. Clearly, k herself is not in my head. This is important if we want to say something about what happens when I manage to compute the underspecified type (10) after hearing Elisabet's utterance of *Karin*. What do I have to search for in order to specify this type and figure out who Elisabet might be referring to? Clearly not my wife's sister, but perhaps a representation of my wife's sister somewhere in memory.

In order to say something about how linguistic resources can be related to other cognitive resources we are going to complicate the notion of content of a proper name. We will introduce the notion of *parametric content*. A parametric content is a pair (a record with two fields) of a background type (a record type) and a function from records of that type to the non-parametric content, in this case an individual. In the case of *Karin* the background type we will use is (13).

$$(13) \quad \begin{bmatrix} \text{x} : Ind \\ \text{e} : \text{named(x, "Karin")} \end{bmatrix}$$

We think of records of this type as modelling situations in which there is an individual named Karin. A situation of this type is a requirement for establishing a successful reference for an utterance of *Karin*. The type (13) represents a kind of presupposition associated with utterances of the proper name. In the parametric content we will label this type with the label 'bg' (for "background") and the function from records of this type to an individual with 'fg' (for "foreground"). We have chosen this neutral terminology so as not to prejudge the issue of how what we are going to do relates to the notions of presupposition discussed in the semantics and pragmatics literature. The function we will use for the foreground is one that maps records of this background type to the individual which is in the 'x'-field of the record, as in (14). If r is a record with a field labelled with ℓ, then we use $r.\ell$ to represent what is in the 'ℓ'-field of r.

$$(14) \quad \lambda r: \begin{bmatrix} \text{x}:Ind \\ \text{e}:\text{named(x, "Karin")} \end{bmatrix} . \, r.\text{x}$$

Intuitively, this function maps situations containing an individual named Karin to that individual.

So now the underspecified content that I compute after having heard Elisabet utter *Karin* is (15).

$$
(15) \quad
\begin{bmatrix}
\text{sp-event}=e & & & : \text{``Karin''} \\[4pt]
\text{cont}=
\begin{bmatrix}
\text{bg}=
\begin{bmatrix}
\text{x:}Ind \\
\text{e:named(x, ``Karin'')}
\end{bmatrix} \\[10pt]
\text{fg}=\lambda r:
\begin{bmatrix}
\text{x:}Ind \\
\text{e:named(x, ``Karin'')}
\end{bmatrix} . r.\text{x}
\end{bmatrix}
& : &
\begin{bmatrix}
\text{bg:}RecType \\
\text{fg:}(bg{\rightarrow}Ind)
\end{bmatrix}
\end{bmatrix}
$$

Now I need to find a match for the background type elsewhere among my cognitive resources. The first place to look is on the dialogue gameboard (Ginzburg 2012) or conversational record. Among other things this keeps track of my view of what Ginzburg calls FACTS, Larsson (2002) calls shared commitments and what is standardly referred to in the literature as common ground (Stalnaker 2002). If there is a match there then I should use it. If there are several then I should use the most salient one. Otherwise, I should look for an appropriate match in long-term memory, and if I find one accommodate it to the shared commitments on the gameboard and otherwise I should use the background type to accommodate a new item to shared commitments. This algorithm can be represented diagrammatically as in (16).

(16)

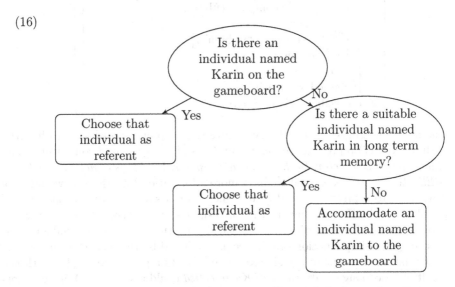

(16) is obviously a simplification of what is actually needed. One other place to look, for example, is in the visual scene apparent to the agent at the time of the utterance. This simple algorithm will be enough for us to consider for our present purposes, however.

Let us explore in more detail how this might work. We will model shared commitments as a type of situations that would be realized if what we have committed to in the dialogue were true. Whenever we update shared commitments we create a new type where the current type is embedded under the label 'prev' ("previous"). Suppose that T_{curr} is the current shared commitments and T_{new} is a type representing new information we wish to update it with. The result of the update will be: $[\text{prev}:T_{curr}] \wedge T_{new}$. Here '$\wedge$' represents the *merge* operation (Cooper and Ginzburg 2015; Cooper in prep). Basically if one of T_1, T_2 is not a record type then $T_1 \wedge T_2$ is identical with $T_1 \wedge T_2$, that is, the meet type formed from T_1 and T_2 whose witness condition is defined by: $a : T_1 \wedge T_2$ iff $a : T_1$ and $a : T_2$. If T_1, T_2 are both record types, then for labels they do not have in common, $T_1 \wedge T_2$ will contain both the fields with those labels from T_1 and T_2. For labels, ℓ, they do have in common, $T_1 \wedge T_2$ will contain a field labelled ℓ with the merge of the two types in that field in T_1 and T_2. Merge corresponds to the notion of unification in the feature-based grammar literature (Shieber 1986). (17) is an example of shared commitments where successive updates have been made, starting from an initial empty commitment *Rec*, the type of any record.

$$
(17) \quad
\begin{bmatrix}
\text{prev:} \begin{bmatrix}
\text{prev:} \begin{bmatrix}
\text{prev:} \begin{bmatrix}
\text{prev:}\textit{Rec} \\
\text{bg:} \begin{bmatrix} \text{x:}\textit{Ind} \\ \text{e:named(x, "Dudamel")} \end{bmatrix} \\
\text{fg:} \begin{bmatrix} \text{e:conductor}(\Uparrow \text{bg.x}) \end{bmatrix}
\end{bmatrix} \\
\text{bg:} \begin{bmatrix} \text{x:}\textit{Ind} \\ \text{e:named(x, "Beethoven")} \end{bmatrix} \\
\text{fg:} \begin{bmatrix} \text{e:composer}(\Uparrow \text{bg.x}) \end{bmatrix}
\end{bmatrix} \\
\text{bg:} \begin{bmatrix} \text{x:}\textit{Ind} \\ \text{e:named(x, "Uchida")} \end{bmatrix} \\
\text{fg:} \begin{bmatrix} \text{e:pianist}(\Uparrow \text{bg.x}) \end{bmatrix}
\end{bmatrix} \\
\text{bg:} \begin{bmatrix} \text{x:}\textit{Ind} \\ \text{e:named(x, "Karin")} \end{bmatrix} \\
\text{fg:} \begin{bmatrix} \text{e:Elisabet's\_sister}(\Uparrow \text{bg.x}) \end{bmatrix}
\end{bmatrix}
$$

The notation '$\Uparrow$bg.x' means that you have to go up one record level to find the path 'bg.x'. (17) represents successive updates of empty shared commitments by *Dudamel is a conductor, Beethoven is a composer, Uchida is a pianist* and *Karin is Elisabet's sister*. Note that we have more information here than we would have if we were to represent shared commitments by a successively shrinking set of possible worlds. Here, thanks to the 'prev'-labels, we can see the order in which various pieces of information were added during the course of the dialogue and we thus get a simple notion of salience as predicted by the recency of mention.[2]

Updating the view of the shared commitments represented in (17) with the result of processing an utterance of *Karin called* could result in (18), if we understand *Karin* to refer to Elisabet's sister. (We have abbreviated the type in (17) as '(17)'.)

---

[2] A complete theory of salience would, of course, need to take more than recency into account.

$$(18) \quad \begin{bmatrix} \text{prev} : (17) \\ \text{bg} \quad : \begin{bmatrix} \text{x=}\Uparrow\text{prev.bg.x} : \textit{Ind} \\ \text{e} \quad\quad\quad\quad\quad : \text{named(x, "Karin")} \end{bmatrix} \\ \text{fg} \quad : \begin{bmatrix} \text{e} : \text{call}(\Uparrow\text{bg.x}) \end{bmatrix} \end{bmatrix}$$

Note that we have linked the *Karin* of *Karin called* to the *Karin* of *Karin is Elisabet's sister* by using the manifest field $\begin{bmatrix} \text{x=}\Uparrow\text{prev.bg.x}:\textit{Ind} \end{bmatrix}$. This means that Karin remains most salient since she was involved in the most recent update to shared commitments.

In order to achieve this update we need to match the background type in the content in (15) in the shared commitments in (17), that is, we have to look for somebody named Karin in the shared commitments. Intuitively, it is obvious that the match should succeed in this case, but defining the notion of match in general terms involves some formal machinery. We need matching to succeed when the labels are different and even when the structure of the record type is different. Here we will explain the nature of the problem and its solution. A more rigorous characterization of matching is given in Cooper (in prep). Consider the type in (19).

$$(19) \quad T = \begin{bmatrix} \ell_1 : T_1 \\ \ell_2 : \begin{bmatrix} \ell_3 : T_2 \\ \ell_4 : T_3 \end{bmatrix} \end{bmatrix}$$

If T is non-empty then there are some a, b, c such that $a : T_1$, $b : T_2$ and $c : T_3$. Suppose we are looking for a match in (20).

$$(20) \quad T' = \begin{bmatrix} \ell_5 : \begin{bmatrix} \ell_6 : T_1 \\ \ell_7 : T_4 \\ \ell_8 : T_2 \end{bmatrix} \\ \ell_9 : T_3 \end{bmatrix}$$

If T' is non-empty then there are some a, b, c, d such that $a : T_1$, $b : T_2$, $c : T_3$ and $d : T_4$. Intuitively, T has a match in T' even though the labels and the structure are different. The match can be formally characterized using a flattening operator, φ, and relabelling. The effect on flattening on these two types is given in (21).

$$(21) \quad \text{a.} \quad \varphi(T) = \begin{bmatrix} \ell_1 \quad : T_1 \\ \ell_2.\ell_3 : T_2 \\ \ell_2.\ell_4 : T_3 \end{bmatrix}$$

$$\text{b.} \quad \varphi(T') = \begin{bmatrix} \ell_5.\ell_6 : T_1 \\ \ell_5.\ell_7 : T_4 \\ \ell_5.\ell_8 : T_2 \\ \ell_9 \quad : T_3 \end{bmatrix}$$

We can then characterize a relabelling, η, of $\varphi(T)$ as a map from labels in $\varphi(T)$ to other labels, in this case to other labels which are labels in $\varphi(T')$ as characterized in (22).

(22) $\ell_1 \rightarrow \ell_5.\ell_6, \ell_2.\ell_3 \rightarrow \ell_5.\ell_8, \ell_2.\ell_4 \rightarrow \ell_9$

It can now be seen that $\varphi(T')$ is a subtype of the relabelling of $\varphi(T)$ with η. In general we can say that there's a match for T_1 in T_2 just in case there is a relabelling, η, of $\varphi(T_1)$ such that $\varphi(T_2)$ is a subtype of $\varphi(T_1)$ relabelled by η.

Just as we represent shared commitments as a type representing the way the world would be if what has been committed to in the dialogue were true, so we will also represent long term memory as a type representing the way the world would be if the memory is correct. We will, however, not use the 'prev'-label to encode the order in which items have been introduced but rather give each item a permanent unique identifier label 'id$_i$', where i is a natural number. Thus a long term memory with the same information as (17) could be as in (23).

(23) $M = \begin{bmatrix} \text{id}_0 : Rec \\ \text{id}_1 : \begin{bmatrix} \text{x}:Ind \\ \text{e:named(x, "Dudamel")} \end{bmatrix} \\ \text{id}_2 : \begin{bmatrix} \text{e:conductor}(\Uparrow\text{id}_1.\text{x}) \end{bmatrix} \\ \text{id}_3 : \begin{bmatrix} \text{x}:Ind \\ \text{e:named(x, "Beethoven")} \end{bmatrix} \\ \text{id}_4 : \begin{bmatrix} \text{e:composer}(\Uparrow\text{id}_3.\text{x}) \end{bmatrix} \\ \text{id}_5 : \begin{bmatrix} \text{x}:Ind \\ \text{e:named(x, "Uchida")} \end{bmatrix} \\ \text{id}_6 : \begin{bmatrix} \text{x:pianist}(\Uparrow\text{id}_5.\text{x}) \end{bmatrix} \\ \text{id}_7 : \begin{bmatrix} \text{x}:Ind \\ \text{e:named(x, "Karin")} \end{bmatrix} \\ \text{id}_8 : \begin{bmatrix} \text{e:Elisabet's\_sister}(\Uparrow\text{id}_7.\text{x}) \end{bmatrix} \end{bmatrix}$

But now we know that the same technique for matching which we used in connection with the shared commitments will also work with long term memory even though the structure and labelling of the type is different.

5   Kripke's Paderewski and Interaction

Kripke (1979) introduces the well-known Paderewski example as "a puzzle about belief" (the title of his paper). I want to argue that this puzzle seems a lot less challenging when we take interaction and the organization of memory into account as well as the fact that it is common for different people to have the same name. The story goes likes this. Peter learns about a pianist named Paderewski. Then, as Kripke puts it, "Later, in a different circle, Peter learns of someone called 'Paderewski' who was a Polish nationalist leader and prime minister." The fact is that there was a pianist named Paderewski who was for a period prime minister of Poland. Peter, however, thinks there are two distinct people who happen to have the same name. As we will see below it is important that the information about Paderewski being prime minister is introduced to Peter in a different context, but this is not what Kripke concentrates on. For him the important question is *Does Peter believe that Paderewski is a pianist or not?* Thinking of him as the pianist he believes that he is and thinking of him as the prime minister he believes that he is not. Thus he appears to believe a contradiction while at the same time apparently behaving perfectly rationally.

Let us consider what Peter's long term memory might be like after he has learned these two separate facts. A possibility is given in (24).

$$
(24) \quad
\begin{bmatrix}
\dots \\
\mathrm{id}_i : \begin{bmatrix} \text{x:} Ind \\ \text{e:named(x, ``Paderewski'')} \end{bmatrix} \\
\mathrm{id}_j : \begin{bmatrix} \text{e:pianist}(\Uparrow\mathrm{id}_i.\text{x}) \end{bmatrix} \\
\mathrm{id}_k : \begin{bmatrix} \text{x:} Ind \\ \text{e:named(x, ``Paderewski'')} \end{bmatrix} \\
\mathrm{id}_l : \begin{bmatrix} \text{e:statesman}(\Uparrow\mathrm{id}_k.\text{x}) \end{bmatrix}
\end{bmatrix}
$$

This is a perfectly reasonable memory state, given that more than one person can have the same name. Ludlow (2014) claims that there is a fallacy in Kripke's argument because different microlanguages are involved and we cannot bring them together, but this doesn't seem to be right for the reasons we gave in connection with my attempt to process my wife's utterance of *Karin called*.

However, if Peter had learned about the statesman in the same dialogue as he learned about the pianist, he probably would have assumed that the same person was being talked about. Within a dialogue, proper names are "logical" (i.e. refer uniquely) unless it is explicit that this is not the case. It seems that the default assumption *within a single dialogue* is that proper names have unique referents unless explicitly stated otherwise or some kind of construction is used which indicates that the two occurrences of the name have different referents. Some examples are given in (25).

(25)   a. I know another person named Paderewski

b. Churchland and Churchland think that replacement of symbol manipulation computer-like devices... with connectionist machines hold (*sic*) great promise (Globus 1995, p. 21)

c. John E: I remember John as an inspiring professor when I was a student

John P: Well, I remember John as an extremely bright student

Other person: I didn't realize you'd known each other that long

d. A: John, I'd like to introduce you to my good friend John

B: Glad to meet you. Another John, eh?

In (25a) there is an explicit statement that more than one person named Paderewski is involved. In (25b) a conjunction of two proper names is used which normally indicates that two people are being referred to. In (25c), it is the speaker of the proper name who is the other person with the same name and it is not usual to refer to yourself with a proper name. In (25d), one occurrence of the proper name is a vocative and it is not usual to introduce people to themselves. In Kripke's story Peter was introduced to the two different guises of Paderewski in different dialogues and so it was natural for him to assume that they were two different people, given that pianists do not usually become prime ministers.

So what should we conclude that Peter believes about the single person Paderewski?

6   Belief as Matches to Long Term Memory

We have so far talked about two roles that types can play. They can be the contents of declarative sentences (used among other things to update shared commitments). The content of *Karin called* is a type of situation in which Karin called. It is true just in case there was such a situation. Types can also be used in the role of memory. A memory that Karin called is a type of situation in which Karin called. The memory is correct or true just in case there was such a situation.

Now we are going to suggest that types can also be the objects of attitude predicates. The type 'believe(a, T)' ("a believes T") is a non-empty type (i.e. is true) just in case there is a match between T and a's long term memory. This means then that a believes that Paderewski is a pianist if a's long-term memory is such that there's an individual named Paderewski who is a pianist. The labels might be different and the way the information is structured may be different, but the basic requirement on the world is the same. A complication with this is that when we report somebody's belief we normally mean something different to this, namely that the person whom we would call Paderewski is believed to be a pianist. That is, we may look for a person named Paderewski in other long term memories such as the speaker's or the audience's. This involves taking a *point of view* on what we believe to be the type representing Peter's long term memory. (For a technical development of the notion of point of view see Cooper in prep, Chap. 6.) Consider then the sentence in (26) which involves Peter believing a contradiction.

(26)   Peter believes that Paderewski is a pianist and not a pianist

(26) may be true if we are using our point of view on Peter's long term memory as a resource for *Paderewski*. According to our view there is a single person named Paderewski of whom Peter believes that he is both a pianist and not a pianist. It may also be false if we are using Peter's long term memory as a resource for *Paderewski* since in Peter's long term memory there is not a single person named Paderewski who is both a pianist and not a pianist. Our view is then that we cannot answer *simpliciter* whether (26) is true or not. It depends on the resources we are using for *Paderewski*.

7   Conclusion

In this paper we have tried to argue that proper names gain from being treated as signs in the sense of Saussure and HPSG. We have also argued that proper names are a good example of the need to treat language as a system in flux rather than the standard conception of them in semantic research as being like static formal languages. We suggested that the analysis of proper names gains from a dialogical analysis involving dialogue gameboards and other cognitive resources as well as the notion of accommodation. We have taken a brief look at Kripke's well-known Paderewski puzzle and suggested that what underlies it is a rather natural view of interaction and memory and we have sketched a view of belief which builds on this.

References

Cooper, R.: Type theory and semantics in flux. In: Kempson, R., Asher, N., Fernando, T. (eds.) Handbook of the Philosophy of Science. Philosophy of Linguistics, vol. 14, pp. 271–323. Elsevier BV (2012). General eds.: D.M. Gabbay, P. Thagard and J. Woods

Cooper, R.: How to do things with types. In: de Paiva, V., Neuper, W., Quaresma, P., Retoré, C., Moss, L.S., Saludes, J. (eds.) Joint Proceedings of the Second Workshop on Natural Language and Computer Science (NLCS 2014) and 1st International Workshop on Natural Language Services for Reasoners (NLSR 2014), 17–18 July 2014, Vienna, Austria, pp. 149–158. Center for Informatics and Systems of the University of Coimbra (2014)

Cooper, R.: Type theory and language: from perception to linguistic communication (in prep). Draft of book chapters available from https://sites.google.com/site/typetheorywithrecords/drafts

Cooper, R., Ginzburg, J.: Type theory with records for natural language semantics. In: Lappin, S., Fox, C. (eds.) The Handbook of Contemporary Semantic Theory, 2nd edn, pp. 375–407. Wiley-Blackwell, Hoboken (2015)

Cooper, R., Kempson, R. (eds.): Language in Flux: Dialogue Coordination, Language Variation, Change and Evolution (Communication, Mind and Language 1). College Publications, London (2008)

Ginzburg, J.: The Interactive Stance: Meaning for Conversation. Oxford University Press, Oxford (2012)

Ginzburg, J., Cooper, R.: Quotation via dialogical interaction. J. Logic Lang. Inform. **23**(3), 287–311 (2014)

Globus, G.G.: The Postmodern Brain (Advances in Con.sciousness Research 1). John Benjamins Publishing Company, Amsterdam (1995)

Kripke, S.: A puzzle about belief. In: Margalit, A. (ed.) Meaning and Use. Reidel, Kufstein (1979)

Larson, R.K., Ludlow, P.: Interpreted logical forms. Synthese **96**, 305–355 (1993)

Larsson, S.: Issue-based dialogue management. Ph.D. dissertation, University of Gothenburg (2002)

Ludlow, P.: Meaning Underdetermination and the Dynamic Lexicon. Oxford University Press, Oxford (2014)

Montague, R.: The proper treatment of quantification in ordinary english. In: Hintikka, J., Moravcsik, J., Suppes, P. (eds.) Approaches to Natural Language: Proceedings of the 1970 Stanford Workshop on Grammar and Semantics, pp. 247–270. D. Reidel Publishing Company, Dordrecht (1973)

Sag, I.A., Wasow, T., Bender, E.M.: Syntactic Theory: A Formal Introduction, 2nd edn. CSLI Publications, Stanford (2003)

de Saussure, F.: Cours de linguistique générale. Payot, Lausanne and Paris, edited by Charles Bally and Albert Séchehaye (1916)

Shieber, S.: An Introduction to Unification-Based Approaches to Grammar. CSLI Publications, Stanford (1986)

Stalnaker, R.: Common ground. Linguist. Philos. **25**, 701–721 (2002)

An Analysis of Selectional Restrictions with Dependent Type Semantics

Eriko Kinoshita[1]([✉]), Koji Mineshima[1,2], and Daisuke Bekki[1,2]

[1] Ochanomizu University, Bunkyō, Japan
kinoshita.eriko@is.ocha.ac.jp
[2] CREST, Japan Science and Technology Agency, Kawaguchi, Japan

Abstract. Predicates in natural languages impose selectional restrictions on their arguments. In this paper, we analyze selectional restrictions of predicates within the framework of Dependent Type Semantics, a framework of natural language semantics based on dependent type theory. We also introduce operators that shift the meanings of predicates and analyze two phenomena, coercion and copredication for logical polysemous nouns, that present challenges to simple analysis of selectional restrictions.

1 Introduction

Predicates in natural languages impose selectional restrictions on their arguments. For example, the transitive verb *marry* expects its subject and object to be expressions that denote humans. Thus, from the utterance of (1), we can infer that Bob and Ann are both human.

(1) Bob married Ann.

One potential way to explain this inference is to treat selectional restrictions of predicates as entailment. According to this analysis, the verb *marry* is assigned the meaning in (2a) and the whole sentence in (1) has the interpretation in (2b).

(2) a. $\lambda y \lambda x.\mathbf{marry}(x, y) \wedge \mathbf{human}(x) \wedge \mathbf{human}(y)$

   b. $\mathbf{marry}(bob, ann) \wedge \mathbf{human}(bob) \wedge \mathbf{human}(ann)$

A problem with this analysis is that it cannot handle the inference in (3).

(3) Bob didn't marry Ann. $\Rightarrow$ Bob and Ann are human.

From the negation of (1), one can also infer that Bob and Ann are human. If selectional restrictions of predicates were part of entailment, we would assign the interpretation (4) to the negative sentence in (3). This does not account for the inference in (3).

(4) $\neg(\mathbf{marry}(bob, ann) \wedge \mathbf{human}(bob) \wedge \mathbf{human}(ann))$

© Springer International Publishing AG 2017
S. Kurahashi et al. (Eds.): JSAI-isAI 2016, LNAI 10247, pp. 19–32, 2017.
DOI: 10.1007/978-3-319-61572-1_2

In general, the contents of selectional restrictions project out of the scope of negation, modals, and conditionals (Asher [2], Magidor [7]). This is a common feature of inferences known as *presupposition projection* (see, e.g., Beaver [3] for an overview).

The goal of this paper is to propose an analysis that treats selectional restrictions as presupposition within the framework of Dependent Type Semantics (DTS; Bekki [4], Bekki and Mineshima [5]). Using this framework, we also present a formal analysis of two lexical phenomena related to selectional restriction, namely, coercion and copredication for logical polysemy.

2   Selectional Restriction: Types vs. Predicates

Although the presuppositional analysis of selectional restriction goes back at least as far as McCawley [9], it seems fair to say that its precise formulation has been mostly neglected in the simply typed setting of standard formal semantics, where only e (entity) and t (truth-value) are taken as base types.

Recently, some proposals in the literature have suggested ways to handle selectional restrictions with extended type-theoretic frameworks (Asher [1], Luo [6], Retoré [11]). There are two possible approaches here. One is to represent selectional restrictions as *types*; as two examples, using **animate** and **human** as base types, one can assign a type **animate** → **prop** to the predicate **cry** and **human** → **human** → **prop** to the predicate **marry**. According to this approach, violation of a selectional restriction is to be treated as a type mismatch. One problem with this approach is the problem of *subtyping*. That is, to combine the predicate **cry** of type **animate** → **prop** with the term **john** of type **human**, one needs a subtyping relation **human** < **animate** and extra subtyping rules (cf. Luo [6], Retoré [11]). One drawback is that with additional subtyping rules, the resulting compositional semantics becomes complicated.

Alternatively, one can preserve the base type for entities and represent selectional restrictions as *predicates* over entities. This view seems to be underdeveloped, but it has the advantage that it can dispense with subtyping and preserve the clear, well-understood conception of syntax-semantics mapping. Our theory is based on this second approach.

3   Dependent Type Semantics

The main challenge here is how to provide a presuppositional analysis of selectional restrictions combined with the selectional-restriction-as-predicate view. We use DTS (Bekki [4], Bekki and Mineshima [5]) as a theoretical framework, which provides two crucial tools: *dependent types* (which are a generalization of simple types) and *underspecified terms*. DTS is a proof-theoretic semantics of natural language based on dependent type theory (Martin-Löf [8]). It characterizes the meaning of a sentence from the perspective of *inferences*.

DTS uses two kinds of dependent types.

(i) Π-type (dependent function type), written as $(x : A) \to B$, is a generalized form of a function type $A \to B$; a term of type $(x : A) \to B$ is a function f that takes a term a of type A and returns a term $f(a)$ of type $B(a)$.

(ii) Σ-type (dependent product type), written as $(x : A) \times B$ or $\begin{bmatrix} x : A \\ B \end{bmatrix}$, is a generalized form of a product type $A \times B$; a term of type $(x : A) \times B$ is a pair (t, u) such that t is of type A and u is of type $B(t)$. The projection operators π_1 and π_2 are defined in such a way that $\pi_1(t, u) = t$ and $\pi_2(t, u) = u$.

Under the so-called propositions-as-types principle (Martin-Löf [8]), types and propositions are identified; a term t having type A (i.e., $t : A$) serves as a *proof term* for the proposition A.

In the dependently typed setting, Π-type and Σ-type correspond to universal and existential quantifiers, respectively. For example, in DTS, the sentence in (5a) is given the semantic representation (SR) in (5b):

(5)   a.   Every man entered.

       b. $\left(u : \begin{bmatrix} x : \texttt{entity} \\ \textbf{man}(x) \end{bmatrix} \right) \to \textbf{enter}(\pi_1(u))$

The term u here has a Σ-type: it consists of a term (let it be x) having type `entity` and some proof term having type $\textbf{man}(x)$ that depends on x. The term $\pi_1(u)$ in $\textbf{enter}(\pi_1(u))$ picks up the entity that is the first component of u. In DTS, common nouns such as *man* are treated as predicates rather than as types. In other words, that a term x has a property *man* is represented as a proposition $\textbf{man}(x)$, rather than as a judgement $x : \textbf{man}$. See Bekki and Mineshima [5] for more discussions on the interpretation of common nouns in our framework.

For Π-types and Σ-types, we use the following formation rules (ΠF, ΣF), introduction rules (ΠI, ΣI), and elimination rules (ΠE, ΣE).

$$\frac{A : \mathsf{s_1} \quad B : \mathsf{s_2}}{(x : A) \to B : \mathsf{s_2}} \; (\Pi F), i \qquad \overset{\overline{x : A}^{\;(i)}}{\vdots}$$

$$\frac{A : \mathsf{type} \quad B : \mathsf{s}}{\begin{bmatrix} x : A \\ B(x) \end{bmatrix} : \mathsf{s}} \; (\Sigma F), i$$

$$\frac{(x : A) \to B : \mathsf{s} \quad M : B}{\lambda x.M : (x : A) \to B} \; (\Pi I), i \qquad \frac{M : A \quad N : B[M/x]}{(M, N) : \begin{bmatrix} x : A \\ B(x) \end{bmatrix}} \; (\Sigma I)$$

$$\frac{M : (x : A) \to B \quad N : A}{MN : B[N/x]} \; (\Pi E) \qquad \frac{M : \begin{bmatrix} x : A \\ B(x) \end{bmatrix}}{\pi_1(M) : A} \; (\Sigma E) \qquad \frac{M : \begin{bmatrix} x : A \\ B(x) \end{bmatrix}}{\pi_2(M) : B[\pi_1(M)/x]} \; (\Sigma E)$$

Here, s, $\mathsf{s_1}$ and $\mathsf{s_2}$ are kind or type (see Bekki and Mineshima [5] for more details).

DTS has an underspecified term @ to handle anaphora and presupposition. We use type annotation for underspecified terms; we write @ : A, where the underspecified term @ is annotated with its type A. By using underspecified terms, we can uniformly handle semantic phenomena that depend on the preceding contexts.

Presupposition and anaphora are resolved by constructing a proof term for @ : A with type checking and then replacing @ : A by the constructed term. Type checking ensures that an SR is well-formed (i.e., having type type). For underspecified terms, we use the rule

$$\frac{A : \mathsf{s} \quad A\ true}{(@ : A) : A}\ (@)$$

where $\mathsf{s} \in \{\mathsf{kind}, \mathsf{type}\}$. The judgement $A\ true$ triggers a proof search to construct a term having the type A in a given context. The constructed term is to be replaced with @ in the final representation. The annotated type A may contain another underspecified term, for which the type checking is triggered by the judgement $A : \mathsf{s}$ (e.g., $A : \mathsf{type}$) in the @ rule.

As an illustration, consider the sentence in (6a). For this sentence, one can compositionally derive the SR in (6b).[1]

(6)   a. He whistled.

b. $\mathbf{whistle}\left(\pi_1\left(@ : \begin{bmatrix} x : \mathtt{entity} \\ \mathbf{man}(x) \end{bmatrix}\right)\right)$

The SR (6b) contains an underspecified term @ annotated with the Σ-type corresponding to the proposition that there is an entity x such that x is a man. For the SR in (6b), the type checking runs as follows.

$$\cfrac{\cfrac{\cfrac{}{\mathtt{entity : type}}\ (CON)}{\begin{bmatrix} x : \mathtt{entity} \\ \mathbf{man}(x) \end{bmatrix} : \mathtt{type}}\ \cfrac{\cfrac{\cfrac{}{\mathtt{man : entity} \rightarrow \mathtt{type}}\ (CON) \quad \cfrac{}{x : \mathtt{entity}}\ (1)}{\mathbf{man}(x) : \mathtt{type}}\ (\Pi E)}{}\ (\Sigma F), 1}$$

$$\cfrac{\cfrac{}{\mathtt{whistle : entity} \rightarrow \mathtt{type}}\ (CON) \quad \cfrac{\cfrac{\left(@ : \begin{bmatrix} x : \mathtt{entity} \\ \mathbf{man}(x) \end{bmatrix}\right) : \begin{bmatrix} x : \mathtt{entity} \\ \mathbf{man}(x) \end{bmatrix}}{\pi_1\left(@ : \begin{bmatrix} x : \mathtt{entity} \\ \mathbf{man}(x) \end{bmatrix}\right) : \mathtt{entity}}\ (\Sigma E)}{}\ (\Pi E)}{\mathbf{whistle}\left(\pi_1\left(@ : \begin{bmatrix} x : \mathtt{entity} \\ \mathbf{man}(x) \end{bmatrix}\right)\right) : \mathtt{type}}$$

where the right branch above has $\begin{bmatrix} x : \mathtt{entity} \\ \mathbf{man}(x) \end{bmatrix}\ true$ derived by (@).

The application of the @ rule in this derivation triggers a proof search for the judgement:

$$\begin{bmatrix} x : \mathtt{entity} \\ \mathbf{man}(x) \end{bmatrix}\ true.$$

Assuming that we have $john : \mathtt{entity}$ and $t : \mathbf{man}(john)$ in the background global context, we can construct a term $(john, t)$ having the Σ-type in question, i.e., a type annotated for the underspecified term @. This term serves as an antecedent of the pronoun he. Replacing @ with the specific term $(john, t)$, the semantic representation in (6b) ends up with $\mathbf{whistle}(\pi_1(john, t))$, which reduces to $\mathbf{whistle}(john)$. In this way, we can derive the interpretation for the sentence containing a pronoun in (6a).

---

[1] See Bekki [4] for details on the compositional derivations of SRs in DTS.

4 Selectional Restriction in DTS

To handle selectional restrictions of predicates as presuppositions, we need to calculate whether selectional restrictions are satisfied at the stage of type checking. We propose that selectional restrictions of predicates are specified in the lexicon. For instance, we can define lexical entries of intransitive and transitive verbs as follows.

	syntax	semantic representation
cry	$S \backslash NP$	$\lambda x.\mathbf{cry}(x, @ : \mathbf{animate}(x))$
marry	$(S \backslash NP)/NP$	$\lambda y.\lambda x.\mathbf{marry}(y, @_i : \mathbf{human}(y))(x, @_j : \mathbf{human}(x))$

To be concrete, we use Combinatory Categorial Grammar (CCG; Steedman [12]) as a syntactic framework. The types of the predicates **cry** and **marry** in the above SRs are defined as follows.

$$\mathbf{cry} : \begin{bmatrix} x : \mathtt{entity} \\ \mathbf{animate}(x) \end{bmatrix} \to \mathsf{type}$$

$$\mathbf{marry} : \begin{bmatrix} y : \mathtt{entity} \\ \mathbf{human}(y) \end{bmatrix} \to \begin{bmatrix} x : \mathtt{entity} \\ \mathbf{human}(x) \end{bmatrix} \to \mathsf{type}$$

For example, the predicate **cry** takes a pair consisting of an entity x and a proof term for the proposition $\mathbf{animate}(x)$ as an argument and returns a type (as a proposition). In the lexical entry for the intransitive verb *cry*, the proof term for the proposition $\mathbf{animate}(x)$ is underspecified; given that there is an underspecified term $@ : \mathbf{animate}(x)$ in the SR, we have to prove $\mathbf{animate}(x)$ during the stage of type checking in order to ensure that the subject of *cry* is animate.

As an illustration, consider the sentence in (1). For this sentence, we can derive the following SR in a compositional way.

(7) $\mathbf{marry}(ann, @_1 : \mathbf{human}(ann))(bob, @_2 : \mathbf{human}(bob))$

The following is the compositional derivation of this SR.

$$
\cfrac{
  \cfrac{
    \cfrac{\text{married}}{(S\backslash NP)/NP} \quad : \lambda y.\lambda x.\mathbf{marry}(x, @_1 : \mathbf{human}(x))(y, @_2 : \mathbf{human}(y)) \qquad \cfrac{\text{Ann}}{NP} \quad : ann
  }{S \backslash NP \qquad : \lambda x.\mathbf{marry}(ann, @_1 : \mathbf{human}(ann))(x, @_2 : \mathbf{human}(x))} >
  \qquad \cfrac{\text{Bob}}{NP} \quad : bob
}{\mathbf{marry}(ann, @_1 : \mathbf{human}(ann))(bob, @_2 : \mathbf{human}(bob))} <
$$

Now, type checking to ensure that the SR in (7) is well-formed runs as follows.

$$
\cfrac{
\cfrac{
\text{marry} : \begin{bmatrix} x : \mathsf{e} \\ \mathbf{h}(x) \end{bmatrix} \to \begin{bmatrix} y : \mathsf{e} \\ \mathbf{h}(y) \end{bmatrix} \to \mathsf{type} \quad (Con) \qquad
\cfrac{\overline{a : \mathsf{e}}\ (Con) \quad (@_1 : \mathbf{h}(a)) : \mathbf{h}(a)}{(a, @_1 : \mathbf{h}(a)) : \begin{bmatrix} x : \mathsf{e} \\ \mathbf{h}(x) \end{bmatrix}}\ (\varSigma I)
}{
\text{marry}(a, @_1 : \mathbf{h}(a)) : \begin{bmatrix} y : \mathsf{e} \\ \mathbf{h}(y) \end{bmatrix} \to \mathsf{type}
}\ (\varPi E)
\qquad
\cfrac{\overline{b : \mathsf{e}}\ (Con) \quad (@_2 : \mathbf{h}(b)) : \mathbf{h}(b)}{(b, @_2 : \mathbf{h}(b)) : \begin{bmatrix} x : \mathsf{e} \\ \mathbf{h}(x) \end{bmatrix}}\ (\varSigma I)
}{
\text{marry}(a, @_1 : \mathbf{h}(a))(b, @_2 : \mathbf{h}(b)) : \mathsf{type}
}\ (\varPi E)
$$

Here, we abbreviate entity as e, **human** as **h**, *ann* as *a*, and *bob* as *b*. There are two open branches containing underspecified terms, $@_1$ and $@_2$, which show that we must search the preceding context to construct proof terms for **human**(*bob*) and **human**(*ann*). That is to say, for the semantic representation to be well-formed, it is presupposed that x and y, which are, respectively, the subject and the object of the verb *marry*, are both human. In this way, the selectional restriction of a predicate is derived as a presupposition.

Similarly, the SR of the negative sentence in (3) is given as follows.

(8) $\neg$**marry**$(ann, @_2 : \mathbf{human}(ann))(bob, @_1 : \mathbf{human}(bob))$

According to the formation rule of negation, A and $\neg A$ have the same well-formedness condition.

$$
\frac{A : \mathsf{type}}{\neg A : \mathsf{type}}\ (\neg F)
$$

That is, if we have $A : \mathsf{type}$, then we have $\neg A : \mathsf{type}$ as well. Therefore, the type checking for the negative SR in (8) ends up with the derivation that triggers a proof search in the same way as the type checking for the SR in (6b) given in Sect. 3. In this way, one can derive the inference pattern of presupposition projection out of the scope of negation. A similar explanation applies to the case of modals and conditionals.

Interestingly, a negative sentence of the form in (9) has two readings (cf. McCawley [9]).

(9)   The chair does not cry.

First, this sentence has a reading in which the selectional restriction projects out of the scope of negation, hence resulting in a violation of selectional restriction. In our terms, after composing the meaning of (9), one obtains the SR $\neg$**cry**$(chair, @ : \mathbf{animate}(chair))$; according to the formation rule of negation, the content of selectional restriction, that is, **animate**(*chair*), projects out of the scope of negation. Thus, for the SR to be well-formed, one needs to construct a proof term of **animate**(*chair*), which is not available in the standard context. Hence, it is predicted that under this reading, a violation of selectional restriction occurs in the sense that the derived SR is not well-formed.

Second, and more interesting, (9) can have a reading in which the selectional restriction does not project and is therefore interpreted inside the scope of negation. The presuppositional analysis correctly predicts this reading; by local accommodation, we can derive the SR $\neg(\mathbf{animate}(chair) \wedge \mathbf{cry}(chair))$

for (9). In this case, one does not have to construct a proof of **animate**(*chair*); hence, it is correctly predicted that under this reading, the utterance of (9) is meaningful and can be true. A detailed explanation of local accommodation in the framework of DTS is beyond the scope of this paper.

5 Coercion and Copredication for Logical Polysemy

5.1 Coercion

There are two phenomena that are not explained by a simple analysis of selectional restrictions of predicates. The first one is coercion (Nunberg [10]). For example, if we have a context in which there is a man who ate the omelet in a cafe, we can understand the meaning of (10a) as (10b).

(10)   a. The omelet escaped.
         b. The man who ate the omelet escaped.

To account for this phenomena, we define an operator, called *argument operator*, that transforms one predicate into another. The argument operators arg_1 for a one-place predicate and arg_2 for a two-place predicate are defined as follows.

$$arg_1 \equiv \lambda P.\lambda x.P\left(\pi_1\pi_1\left(@_5:\left(\left(@_4:\begin{bmatrix}z:\begin{bmatrix}x:\mathsf{e}\\@_2^{pr}(x)\end{bmatrix}\\\begin{bmatrix}x:\mathsf{e}\\@_1^{pr}(x)\end{bmatrix}\end{bmatrix}\rightarrow\begin{bmatrix}x:\mathsf{e}\\@_2^{pr}(x)\end{bmatrix}\right)\rightarrow\mathsf{type}\right)(x,(@_3^{pr}(x)))(z)\right)\right)$$

$$arg_2 \equiv$$
$$\lambda P.\lambda y.\lambda x.P\left(\pi_1\pi_1\left(@_7:\left(\left(@_6:\begin{bmatrix}z:\begin{bmatrix}x:\mathsf{e}\\@_3^{pr}(x)\end{bmatrix}\\\begin{bmatrix}x:\mathsf{e}\\@_1^{pr}(x)\end{bmatrix}\end{bmatrix}\rightarrow\begin{bmatrix}x:\mathsf{e}\\@_2^{pr}(x)\end{bmatrix}\rightarrow\begin{bmatrix}x:\mathsf{e}\\@_3^{pr}(x)\end{bmatrix}\right)\rightarrow\mathsf{type}\right)(y,(@_4:@_1^{pr}(y)))(x,(@_5:@_2^{pr}(x)))(z)\right)\right)(x)$$

Here an underspecified term $@_i^{pr}$ is an abbreviation for $@_i : \mathsf{e} \rightarrow \mathsf{type}$.

Let us first focus on the definition of the argument operator arg_1 for one-place predicates. In the definition of arg_1, the underspecified terms $@_1$ and $@_2$ in $@_1^{pr}$ and $@_2^{pr}$ are annotated with type $\mathsf{e} \rightarrow \mathsf{type}$; these are underspecified terms for properties. Intuitively, given a one-place predicate P and its argument x of type e, the argument operator arg_1 produces a new predicate P' that existentially introduces a new entity z having some relation R to x.

When one underspecified term appears inside a type annotated with another underspecified term, the inside term must be resolved first. Specifically, the underspecified terms contained in the argument operator arg_1 are resolved in the following way.

1. First, given the entity x (e.g., *the omelet* in (10a)), find a suitable property F (e.g., *edible*) holding for x. This property F replaces $@_1^{pr}$.
2. Second, if there is a proof term for the proposition that x has the property F (e.g., *the omelet is edible*), it replaces $@_3^{pr}$.

3. Also, a property G to be substituted for $@_2^{pr}$ is needed. The property G (e.g., *animate*) has to be chosen so that the newly introduced entity (the first element of the term z) satisfies G.
4. Next, find a relation R that is to be substituted for $@_4^{pr}$. In our example, a relation (e.g., *eat*) that has selectional restrictions specified by predicates **edible**(x) and **animate**(y) is needed. This relation R replaces $@_4$.
5. Finally, construct a term to be substituted for $@_5$. This is a tuple consisting of an entity z whose first element satisfies the property G and a proof term for the proposition that the relation R holds between x and z.

In this way, arg_1 transforms the predicate **escape** into a predicate whose argument is an animate entity that has the eating relation to the omelet.

Let us explain the derivation in more detail. To begin with, we can derive the SR of the sentence (10a) as follows.

$$\cfrac{\cfrac{\text{The omelet}}{NP}\; \cfrac{\cfrac{\text{escaped}}{S\backslash NP} \quad \cfrac{\epsilon}{(S\backslash NP)\backslash(S\backslash NP)}}{S\backslash NP} <}{S} <$$

$$: \lambda x.\textbf{escape}(x, @_5 : \textbf{animate}(x)) \qquad : arg_1$$

By unfolding the definition of arg_1, the sentence in (10a) is assigned the SR in (11).

(11)   **escape**$(Z_1, @_6 : \textbf{animate}(Z_1))$

Here, Z_1 abbreviates

$$\pi_1\pi_1\left(@_5 : \left[\begin{array}{l} z : \begin{bmatrix} x : \mathsf{e} \\ @_2^{pr}(x) \end{bmatrix} \\ (@_4 : \begin{bmatrix} x : \mathsf{e} \\ @_1^{pr}(x) \end{bmatrix} \to \begin{bmatrix} x : \mathsf{e} \\ @_2^{pr}(x) \end{bmatrix} \to \mathsf{type})(o, (@_3 : @_1^{pr}(o)))(z) \end{array}\right]\right).$$

Let us suppose that we have the following information in the global context $\mathcal{K}_1$:

$$\begin{aligned} \mathcal{K}_1 \equiv &\; \mathsf{type} : \mathsf{kind}, \; \mathsf{e} : \mathsf{type}, \\ &\; j : \mathsf{e}, \; o : \mathsf{e}, \\ &\; \textbf{animate} : \mathsf{e} \to \mathsf{type}, \; \textbf{edible} : \mathsf{e} \to \mathsf{type}, \\ &\; \textbf{eat} : \begin{bmatrix} y : \mathsf{e} \\ \textbf{edible}(y) \end{bmatrix} \to \begin{bmatrix} x : \mathsf{e} \\ \textbf{animate}(x) \end{bmatrix} \to \mathsf{type}, \\ &\; \textbf{escape} : \begin{bmatrix} x : \mathsf{e} \\ \textbf{animate}(x) \end{bmatrix} \to \mathsf{type}, \\ &\; p_1 : \textbf{animate}(j), \; p_2 : \textbf{edible}(o), \; p_3 : \textbf{eat}(o, p_2)(j, p_1). \end{aligned}$$

Now type checking is triggered to determine whether the SR (10) is well-formed. This is an example of nested presupposition, and underspecified terms are resolved outward from the most deeply embedded. Here, we focus on the step to find a relation R that is substituted for the following underspecified term:

$$@_4 : \begin{bmatrix} x : e \\ @_1^{pr}(x) \end{bmatrix} \rightarrow \begin{bmatrix} x : e \\ @_2^{pr}(x) \end{bmatrix} \rightarrow \text{type}.$$

The type checking tree for the relevant part looks as follows:

$$
\cfrac{
\cfrac{
\begin{matrix} \mathcal{D}_1 \\ \begin{bmatrix} x : e \\ @_1^{pr}(x) \end{bmatrix} : t \end{matrix}
\qquad
\cfrac{\begin{matrix} \mathcal{D}_2 \\ \begin{bmatrix} x : e \\ @_2^{pr}(x) \end{bmatrix} \rightarrow t : k \end{matrix}}{}\ (\Pi F)
}{
\begin{bmatrix} x : e \\ @_1^{pr}(x) \end{bmatrix} \rightarrow \begin{bmatrix} x : e \\ @_2^{pr}(x) \end{bmatrix} \rightarrow t : k
}
\qquad \vdots
}{
\left(@_4 : \begin{bmatrix} x : e \\ @_1^{pr}(x) \end{bmatrix} \rightarrow \begin{bmatrix} x : e \\ @_2^{pr}(x) \end{bmatrix} \rightarrow t\right) : \begin{bmatrix} x : e \\ @_1^{pr}(x) \end{bmatrix} \rightarrow \begin{bmatrix} x : e \\ @_2^{pr}(x) \end{bmatrix} \rightarrow t
}\ (@)
$$

where we use the abbreviations k for kind and t for type. The type checking for $\mathcal{D}_1$ runs as follows.

$$
\cfrac{
\cfrac{\ }{e : t}\ (CON)
\qquad
\cfrac{
\cfrac{\dfrac{\overline{e : t}\ (CON) \quad \overline{t : k}\ (CON)}{e \rightarrow t : k}\ (\Pi F) \qquad \dfrac{\vdots}{e \rightarrow t\ true}}{@_1^{pr} : e \rightarrow t}\ (@) \qquad \cfrac{\ }{x : e}\ (1)
}{
@_1^{pr}(x) : t
}\ (\Sigma F), 1
}{
\begin{bmatrix} x : e \\ @_1^{pr}(x) \end{bmatrix} : t
}\ (\Pi E)
$$

Similarly, the type checking for $\mathcal{D}_2$ runs as follows.

$$
\cfrac{
\cfrac{
\cfrac{\ }{e : t}\ (CON)
\qquad
\cfrac{
\cfrac{\dfrac{\overline{e : t}\ (CON) \quad \overline{t : k}\ (CON)}{e \rightarrow t : k}\ (\Pi F) \qquad \dfrac{\vdots}{e \rightarrow t\ true}}{@_2^{pr} : e \rightarrow t}\ (@) \qquad \cfrac{\ }{x : e}\ (1)
}{
@_2^{pr}(x) : t
}\ (\Sigma F), 1
}{
\begin{bmatrix} x : e \\ @_2^{pr}(x) \end{bmatrix} : t
}\ (\Pi E)
\qquad
\cfrac{\ }{t : k}\ (CON)
}{
\begin{bmatrix} x : e \\ @_2^{pr}(x) \end{bmatrix} \rightarrow t : k
}\ (\Pi F)
$$

The judgements $e \rightarrow t\ true$ in $\mathcal{D}_1$ and $\mathcal{D}_2$ trigger a proof search; given a suitable global context, we can find the antecedents **edible** of type $e \rightarrow t$ for $@_1^{sr}$, and **animate** of type $e \rightarrow t$ for $@_2^{sr}$. Replacing each underspecified term with its antecedent predicate, the above type checking tree is transformed as follows.

$$
\cfrac{
\cfrac{
\begin{matrix} \mathcal{D}_1 \\ \begin{bmatrix} x : e \\ \text{edible}(x) \end{bmatrix} : t \end{matrix}
\quad
\cfrac{\begin{matrix} \mathcal{D}_2 \\ \begin{bmatrix} x : e \\ \text{animate}(x) \end{bmatrix} \rightarrow t : k \end{matrix}}{}\ (\Pi F)
}{
\begin{bmatrix} x : e \\ \text{edible}(x) \end{bmatrix} \rightarrow \begin{bmatrix} x : e \\ \text{animate}(x) \end{bmatrix} \rightarrow t : k
}
\quad
\cfrac{\begin{bmatrix} x : e \\ \text{edible}(x) \end{bmatrix} \rightarrow \begin{bmatrix} x : e \\ \text{animate}(x) \end{bmatrix} \rightarrow t\ true}{\vdots}
}{
\left(@_4 : \begin{bmatrix} x : e \\ \text{edible}(x) \end{bmatrix} \rightarrow \begin{bmatrix} x : e \\ \text{animate}(x) \end{bmatrix} \rightarrow t\right) : \begin{bmatrix} x : e \\ \text{edible}(x) \end{bmatrix} \rightarrow \begin{bmatrix} x : e \\ \text{animate}(x) \end{bmatrix} \rightarrow t
}\ (@)
$$

Then, we can find an antecedent **eat** for $@_4$ that has a type

$$\begin{bmatrix} x : \mathsf{e} \\ \mathbf{edible}(x) \end{bmatrix} \rightarrow \begin{bmatrix} x : \mathsf{e} \\ \mathbf{animate}(x) \end{bmatrix} \rightarrow \mathsf{t}$$

in the context $\mathcal{K}_1$. In a similar way, we can find a proof term for other $@$-terms: p_2 for $@_3$, $((j, p_1), p_3)$ for $@_5$, and p_1 for $@_6$. By eliminating each $@$-term in (11) and reducing β-redexes, we obtain the SR **espace**(j, p_1) as a fully specified semantic representation for the sentence (10a).

5.2   Copredication for Logical Polysemy

The second phenomenon we consider is copredication of logically polysemous nouns. There are nouns having multiple meanings in natural language; the occurrences can be classified into accidental and logical polysemy (Asher [1]). For example, the noun *bank* in (12a) is accidentally polysemous, and the noun *book* in (12b) is logically polysemous.

(12)   a. # The bank is closed and is muddy.

   b. Mary memorized and burned the book.

The sentence (12b) shows that the logically polysemous noun *book* allows copredication, despite the fact that *memorize* and *burn* require different objects (i.e., informational objects and physical objects, respectively) as their object argument. To account for this fact, we can apply argument operators to the verbs *memorize* and *burn*, thereby avoiding the violation of selection restrictions.

   We introduce the logical polysemies of nouns as functions. For example, we assign the following functions to the noun *book*.

$$\mathbf{book_{infoOf}} : (x : \mathsf{e}) \rightarrow (\mathbf{book}(x) \rightarrow \begin{bmatrix} y : \mathsf{e} \\ \mathbf{infoOf}(x)(y) \end{bmatrix})$$

$$\mathbf{book_{phyObjOf}} : (x : \mathsf{e}) \rightarrow (\mathbf{book}(x) \rightarrow \begin{bmatrix} y : \mathsf{e} \\ \mathbf{phyObjOf}(x)(y) \end{bmatrix})$$

The function $\mathbf{book_{infoOf}}$ (resp., $\mathbf{book_{phyObjOf}}$) takes an entity x and a proof of $\mathbf{book}(x)$ and returns an entity y that is the informational aspect (resp., the physical aspect) of x.

   Now we can derive the SR of the sentence (12b) as follows.

$$
\dfrac{
\dfrac{
\dfrac{\underset{:MEM}{\overset{\text{memorized}}{S\backslash NP/NP}}\quad \underset{:arg_2}{\overset{\epsilon}{(S\backslash NP/NP)\backslash (S\backslash NP/NP)}}}{\underset{:arg_2(MEM)}{S\backslash NP/NP}}<\quad \underset{:\lambda p.\lambda q.\lambda y.\lambda x.\begin{bmatrix}p(y)(x)\\q(y)(x)\end{bmatrix}}{\overset{\text{and}}{CONJ}}\quad \dfrac{\underset{:BURN}{\overset{\text{burned}}{S\backslash NP/NP}}\quad \underset{:arg_2}{\overset{\epsilon}{(S\backslash NP/NP)\backslash (S\backslash NP/NP)}}}{\underset{:arg_2(BURN)}{S\backslash NP/NP}}<
}{\underset{:\lambda y.\lambda x.\begin{bmatrix}arg_2(MEM)(y)(x)\\arg_2(BURN)(y)(x)\end{bmatrix}}{S\backslash NP/NP}}\langle\Phi\rangle \quad \underset{:b}{\overset{\text{the book}}{NP}}
}{\underset{:\lambda x.\begin{bmatrix}arg_2(MEM)(b)(x)\\arg_2(BURN)(b)(x)\end{bmatrix}}{S\backslash NP}}>
$$

$$
\dfrac{\underset{:m}{\overset{\text{Mary}}{NP}}\quad \cdots}{\underset{:\begin{bmatrix}arg_2(MEM)(b)(m)\\arg_2(BURN)(b)(m)\end{bmatrix}}{S}}<
$$

where

$$
MEM \equiv \lambda y.\lambda x.\mathbf{memorize}(y, @_i : \begin{bmatrix}w:\mathsf{e}\\ \mathbf{infoOf}(w)(y)\end{bmatrix})(x, @_j : \mathbf{animate}(x)),
$$

and

$$
BURN \equiv \lambda y.\lambda x.\mathbf{burn}(y, @_i : \begin{bmatrix}w:\mathsf{e}\\ \mathbf{phyObjOf}(w)(y)\end{bmatrix})(x, @_j : \mathbf{animate}(x)).
$$

Thus, the sentence in (12b) is assigned the following SR.

$$
(13)\quad \begin{bmatrix}\mathbf{memorize}(Z_2, @_{15} : \begin{bmatrix}x:\mathsf{e}\\ \mathbf{infoOf}(x)(Z_2)\end{bmatrix})(m, @_{16} : \mathbf{animate}(m))\\[2mm] \mathbf{burn}(Z_3, @_{17} : \begin{bmatrix}x:\mathsf{e}\\ \mathbf{phyObjOf}(x)(Z_3)\end{bmatrix})(m, @_{18} : \mathbf{animate}(m))\end{bmatrix}
$$

where Z_2 abbreviates

$$
\pi_1\pi_1\left(@_7 : \begin{bmatrix}z:\begin{bmatrix}x:\mathsf{e}\\ @_2^{pr}(x)\end{bmatrix}\\ (@_6 : \begin{bmatrix}x:\mathsf{e}\\ @_1^{pr}(x)\end{bmatrix} \to \begin{bmatrix}x:\mathsf{e}\\ @_2^{pr}(x)\end{bmatrix} \to \begin{bmatrix}x:\mathsf{e}\\ @_3^{pr}(x)\end{bmatrix} \to \mathsf{t})(b,(@_4 : @_1^{pr}(b)))(m,(@_5 : @_2^{pr}(m)))(z)\end{bmatrix}\right),
$$

and Z_3 abbreviates

$$
\pi_1\pi_1\left(@_{14} : \begin{bmatrix}z:\begin{bmatrix}x:\mathsf{e}\\ @_9^{pr}(x)\end{bmatrix}\\ (@_{13} : \begin{bmatrix}x:\mathsf{e}\\ @_8^{pr}(x)\end{bmatrix} \to \begin{bmatrix}x:\mathsf{e}\\ @_9^{pr}(x)\end{bmatrix} \to \begin{bmatrix}x:\mathsf{e}\\ @_{10}^{pr}(x)\end{bmatrix} \to \mathsf{t})(b,(@_{11} : @_8^{pr}(b)))(m,(@_{12} : @_9^{pr}(m)))(z)\end{bmatrix}\right).
$$

Let us suppose that we have the following information in the global context $\mathcal{K}_2$:

$$
\mathcal{K}_2 \equiv \mathsf{t}:\mathsf{k},\ \mathsf{e}:\mathsf{t},
$$
$$
m:\mathsf{e},\ b:\mathsf{e},\ i_b:\mathsf{e},\ p_b:\mathsf{e},
$$
$$
\mathbf{animate}:\mathsf{e}\to\mathsf{t},\ \mathbf{book}:\mathsf{e}\to\mathsf{t},
$$
$$
\mathbf{infoOf}:\mathsf{e}\to\mathsf{e}\to\mathsf{t},\ \mathbf{phyObjOf}:\mathsf{e}\to\mathsf{e}\to\mathsf{t},
$$

$$\textbf{memorize} : \left[\begin{array}{l} y : \mathsf{e} \\ \left[\begin{array}{l} w : \mathsf{e} \\ \textbf{infoOf}(w)(y) \end{array} \right] \end{array} \right] \rightarrow \left[\begin{array}{l} x : \mathsf{e} \\ \textbf{animate}(x) \end{array} \right] \rightarrow \mathsf{t},$$

$$\textbf{burn} : \left[\begin{array}{l} y : \mathsf{e} \\ \left[\begin{array}{l} w : \mathsf{e} \\ \textbf{phyObjOf}(w)(y) \end{array} \right] \end{array} \right] \rightarrow \left[\begin{array}{l} x : \mathsf{e} \\ \textbf{animate}(x) \end{array} \right] \rightarrow \mathsf{t},$$

$$\textbf{book}_{\textbf{infoOf}} : (x : \mathsf{e}) \rightarrow (\textbf{book}(x) \rightarrow \left[\begin{array}{l} y : \mathsf{e} \\ \textbf{infoOf}(x)(y) \end{array} \right]),$$

$$\textbf{book}_{\textbf{phyObjOf}} : (x : \mathsf{e}) \rightarrow (\textbf{book}(x) \rightarrow \left[\begin{array}{l} y : \mathsf{e} \\ \textbf{phyObjOf}(x)(y) \end{array} \right]),$$

$$p_1 : \textbf{animate}(m), \ p_2 : \textbf{book}(o),$$

$$p_3 : \textbf{infoOf}(b)(i_b), \ p_4 : \textbf{phyObjOd}(b)(p_b).$$

Then we can find a proof term for each @-term in SR Z_2 as follows. Here $@_i \longmapsto T$ means that the underspecified term $@_i$ is replaced with a term T.

$@_1 \longmapsto \textbf{book},$

$@_2 \longmapsto \textbf{animate},$

$@_3 \longmapsto \textbf{infoOf}(b),$

$@_4 \longmapsto p_2,$

$@_5 \longmapsto p_1,$

$$@_6 \longmapsto \lambda y.\lambda x.\lambda z. \left[\begin{array}{l} u : \textbf{book}(y) \\ \textbf{book}_{\textbf{infoOf}}(y)(u) =_{\mathsf{e}} z \end{array} \right],$$

$$@_7 \longmapsto$$
$$((i_b, \ p_3), (\lambda y.\lambda x.\lambda z. \left[\begin{array}{l} u : \textbf{book}(y) \\ \textbf{book}_{\textbf{infoOf}}(y)(u) =_{\mathsf{e}} z \end{array} \right])(b, \ p_2)(m, \ p_1)(i_b, \ p_3)).$$

And we can also find a proof term for each @-term in SR Z_3:

$@_8 \longmapsto \textbf{book},$

$@_9 \longmapsto \textbf{animate},$

$@_{10} \longmapsto \textbf{phyObjOf}(b),$

$@_{11} \longmapsto p_2,$

$@_{12} \longmapsto p_1,$

$$@_{13} \longmapsto \lambda y.\lambda x.\lambda z. \left[\begin{array}{l} u : \textbf{book}(y) \\ \textbf{book}_{\textbf{phyObjOf}}(y)(u) =_{\mathsf{e}} z \end{array} \right],$$

$$@_{14} \longmapsto$$
$$((p_b, \ p_3), (\lambda y.\lambda x.\lambda z. \left[\begin{array}{l} u : \textbf{book}(y) \\ \textbf{book}_{\textbf{phyObjOf}}(y)(u) =_{\mathsf{e}} z \end{array} \right])(b, \ p_2)(m, \ p_1)(p_b, \ p_4)).$$

The rest of the underspecified terms can also be replaced with specific terms as follows.

$@_{15} \longmapsto p_3,$

$@_{16} \longmapsto p_1,$

$@_{17} \longmapsto p_4,$

$@_{18} \longmapsto p_1.$

By eliminating each @-term in (13) and reducing β-redexes, we obtain the following as a fully specified semantic representation for the sentence (12b).

$$(14) \quad \begin{bmatrix} \mathbf{memorize}(i_b, \ p_3)(m, \ p_1) \\ \mathbf{burn}(p_b, \ p_4)(m, \ p_1) \end{bmatrix}$$

6 Conclusion

In this paper, we proposed an analysis that treats the selectional restrictions of predicates as presuppositions. In addition, using argument operators, we gave a unified analysis of lexical phenomena that are not accounted for by simple analyses of selectional restrictions. Future work includes extending our analysis to phenomena such as metaphors, which Asher [1] opened up a way to analyze in type theoretical settings.

Acknowledgements. We thank the two anonymous reviewers for helpful comments and suggestions. We also thank the audience of LENLS13 for their valuable comments and discussions. This work was supported by CREST, Japan Science and Technology Agency.

References

1. Asher, N.: Lexical Meaning in Context: A Web of Words. Cambridge University Press, Cambridge (2011)
2. Asher, N.: Selectional restrictions, types and categories. J. Appl. Logic **12**(1), 75–87 (2014)
3. Beaver, D.I.: Presupposition and Assertion in Dynamic Semantics. Studies in Logic, Language and Information. CSLI Publications & FoLLI, Stanford (2001)
4. Bekki, D.: Representing anaphora with dependent types. In: Asher, N., Soloviev, S. (eds.) LACL 2014. LNCS, vol. 8535, pp. 14–29. Springer, Heidelberg (2014). doi:10.1007/978-3-662-43742-1_2
5. Bekki, D., Mineshima, K.: Context-passing and underspecification in dependent type semantics. In: Chatzikyriakidis, S., Luo, Z. (eds.) Modern Perspectives in Type-Theoretical Semantics. SLP, vol. 98, pp. 11–41. Springer, Cham (2017). doi:10.1007/978-3-319-50422-3_2
6. Luo, Z.: Formal semantics in modern type theories with coercive subtyping. Linguist. Philos. **35**(6), 491–513 (2012)
7. Magidor, O.: Categiry Mistakes. Oxford University Press, Oxford (2013)

8. Martin-Löf, P.: Intuitionistic Type Theory. Bibliopolis, Naples (1984)
9. McCawley, J.D.: Concerning the base component of a transformational grammar. Found. Lang. **4**(3), 243–269 (1968)
10. Nunberg, G.: Transfers of meaning. J. Semant. **12**(2), 109–132 (1995)
11. Retoré, C.: The montagovian generative lexicon lambda tyn: a type theoretical framework for natural language semantics. In: 19th International Conference on Types for Proofs and Programs (TYPES 2013), pp. 202–229 (2014)
12. Steedman, M.: Surface Structure and Interpretation. The MIT Press, Cambridge (1996)

Non-canonical Coordination
in the Transformational Approach

Oleg Kiselyov[✉]

Tohoku University, Sendai, Japan
oleg@okmij.org

Abstract. Recently introduced Transformational Semantics (TS) formalizes, restraints and makes rigorous the transformational approach epitomized by QR and Transformational Grammars: deriving a meaning (in the form of a logical formula or a logical form) by a series of transformations from a suitably abstract (tecto-) form of a sentence. TS generalizes various 'monad' or 'continuation-based' computational approaches, abstracting away irrelevant details (such as monads, etc.) while overcoming their rigidity and brittleness. Unlike QR, each transformation in TS is rigorously and precisely defined, typed, and deterministic. The restraints of TS and the sparsity of the choice points (in the order of applying the deterministic transformation steps) make it easier to derive negative predictions and control over-generation.

We apply TS to right-node raising (RNR), gapping and other instances of non-constituent coordination. Our analyses straightforwardly represent the intuition that coordinated phrases must in some sense be 'parallel', with a matching structure. Coordinated material is not necessarily constituent – even 'below the surface' – and we do not pretend it is. We answer the Kubota, Levine and Moot challenge (the KLM problem) of analyzing RNR and gapping without directional types, yet avoiding massive over-generation. We thus formalize the old idea of 'coordination reduction' and show how to make it work for generalized quantifiers.

1 Introduction

Non-canonical coordination – right-node raising (RNR) as in (1), argument-cluster coordination (2) and, in particular, gapping (3–7) – provides an unending stream of puzzles for the theory of semantics [8,10]:

(1) John likes and Mary hates Bill.
(2) John gave a present to Robin on Thursday and to Leslie on Friday.
(3) Mary liked Chicago and Bill Detroit.
(4) One gave me a book and the other a CD.
(5) Terry can go with me and Pat with you.
(6) Mrs. J can't live in Boston and Mr. J in LA.
(7) Pete wasn't called by Vanessa but rather John by Jesse.

© Springer International Publishing AG 2017
S. Kurahashi et al. (Eds.): JSAI-isAI 2016, LNAI 10247, pp. 33–44, 2017.
DOI: 10.1007/978-3-319-61572-1_3

With gapping, it is not just a simple verb that can "go missing", as in (3). It can be a complex phrase of a verb with arguments and complements – or, as in (4), a verb and an auxiliary verb. Interactions of coordination with scope-taking are particularly challenging: a competent theory needs to handle both narrow- and wide-scope reading of "a present" in (2) and the narrow- and wide-scope coordination in (6). In (7), negation somehow scopes over the first "coordinated structure" but not over the second.

Recently in [8,9], Kubota and Levin put forward new analyses of non-canonical coordination, applying hybrid categorial grammars they have been developing. In contrast, the analyses in [6] use plain old non-associative Lambek grammar. However, the main ideas of [6] are completely hidden behind thickets of complicated types and their interactions within a derivation. The intuition that coordinated structures must be parallel is thus lost in the details.

We present a new analysis of non-constituent coordination using the more intuitive and less round-about framework TS (formerly called AACG) [7], designed to take the 'hacking' out of tree-hacking. TS lets us talk about QR and other transformations towards some semantic form in a rigorous, formal, mostly deterministic way. We remind of TS in Sect. 2.

Our analyses re-expose ideas from the earlier approach of [6], but free them from the bondage of encoding. A notable feature of TS is the absence of directional types. We use it to answer the challenge posited by Kubota and Levin [10] and Moot (dubbed "the KLM problem" by Morrill): to analyze RNR within categorial-grammar–like formalisms without directional types, while avoiding massive over-generation.

One may categorize the various approaches to non-canonical coordination based on what exactly is being coordinated. Take (1), repeated below

(1) John likes and Mary hates Bill.

which will be our running example for a while. Are the complete sentences being coordinated behind the scene, as in "John likes Bill" and "Mary hates Bill" with "Bill" being later elided? Or perhaps sentences with holes are being coordinated, as in "John likes hyp_{obj}"? (as done in [6,8,9].) Or perhaps we regard "John likes" and "Mary hates" as constituents and coordinate as such (as in CCG). In this paper we give another answer: we analyze (1) as the coordination of the complete clause "Mary hates Bill" with the cluster "John" and "likes". The types of the cluster components and their order guide the transformation that picks the needed material from the clause "Mary hates Bill" to make the cluster the complete clause. The 'picking transformation' can be naturally supported within the existing setup of TS, using the same mechanism used in [7] to analyze quantification and inverse linking. The intuition of 'picking' is made precise and formal in Sect. 3.

The structure of the paper is as follows. Section 2 reminds TS, in a different, clearer presentation. We then describe our approach to coordination: transforming non-canonical one to the ordinary coordination of clauses. Section 4 discusses the related work that forms the context of our approach. The rigorous nature of TS makes it easier to carry analyses mechanically, by a computer. In fact, the analyses in the paper have been so programmed and executed. The implementation, in the form of a domain-specific language embedded in Haskell – 'the semantic calculator' – is publicly available at http://okmij.org/ftp/gengo/transformational-semantics/.

2 TS Background

Traditional Categorial Grammar approaches draw parallels between proof systems and grammars: grammaticality is identified with the existence of a derivation. It is rather challenging however to prove the absence of a derivation, and to overview the space of possible derivations in general.

TS (formerly, AACG) [7] in contrast pursues the computational approach, harking back to Transformational Generative Grammars [2] of 1960s: Rather than trying to *deduce* a derivation, it tries to *induce* the meaning (the logical formula) by applying a sequence of precisely and formally defined transformations to a suitably abstract form of a sentence. The latter abstracts away the case and the number agreement, declination, etc. The transformations are deterministic; the order of their applications is generally not. (There may still be dependencies between particular transformations imposing the order.) The transformations are partial: the failure is taken as ungrammaticality of the original sentence.

Formally, TS deals with term languages that represent typed finite trees. Each T-language is a set of well-typed terms built from typed constants (function symbols) c. Types are

$$\text{Base types} \quad v$$
$$\text{T-Types} \quad \sigma ::= v \mid \sigma \to \sigma$$

The set terms d is then inductively defined as: (i) each constant c of the type σ is a term; (ii) if c has the type $\sigma_1 \to \sigma$ and d is a term of type σ_1, then $c\,d$ is a term of type σ; (iii) nothing else is a term. The set of constants and their types is a (multi-sorted) algebraic signature; A T-language is hence a term language over the signature, which defines the language.

Table 1 shows three sample languages. T_S has the single base type string and numerous constants "John", "greet", "every", etc. of that type. It describes the surface, "phonetic", form of a sentence. The constant $- \cdots - :$ string $\to$ string $\to$ string (usually written as the infix operation) signifies string concatenation. The language T_A whose types are familiar categories represents the abstract form. T_L is the language of formulas of predicate logic, which describe the meaning of sentences. The (infinite) sets of *constants* $\mathsf{var_x}, \mathsf{var_y}, \ldots$ and the corresponding $\mathsf{U_x}, \ldots$ and $\mathsf{E_x}, \ldots$ represent (to be) bound variables and their binders. Unlike the conventional (lambda-bound) variables, they are not subject to substitution, α-conversion or capture-avoidance. T_L likewise has constants $x, y, z, \ldots$ of the type e and the corresponding sets of *constants* $\forall_x, \forall_y, \ldots, \exists_x, \exists_y, \ldots$ intended as binders.

As a way to introduce TS we show the quantification analysis of "John greeted every participant". The sample sentence in the language T_A has the form

cl john (argp greet (every$_\mathsf{x}$ participant))

Table 1. Signatures of various T-languages

υ		c
T_S	string	$\cdot$: string $\to$ string $\to$ string "John" , "greet" , "every" , . . . : string
T_A	S, NP, N, VP, PP, TV	John: NP participant: N greet: TV cl: $NP \to VP \to S$ argp: $TV \to NP \to VP$ ppadv: $VP \to PP \to VP$ every$_x$, every$_y$, a$_z$: $N \to NP$ var$_x$, var$_y$, . . . : NP U$_x$, U$_y$, . . . , E$_x$, E$_y$, . . . : $N \to S \to S$
T_L	e, t	conj , disj , . . . : $t \to t \to t$ john: e participant: $e \to t$ greet: $e \to t \to t$ $\forall_x, \exists_y$: $t \to t$ $x, y, z, \ldots$: e

to be referred to as *jgep*. The constant cl combines an NP and a VP into a clause. (Likewise, argp attaches an argument to a verb and ppadv attaches a prepositional phrase (PP) as a VP complement.) Quantifiers are uniquely labeled by x, y, z, etc. We assume it is the job of a parser to uniquely label the quantifiers in the abstract form.

Before taking on meaning we illustrate the recovering of the surface form of *jgep*, by applying the following 'phonetic' transformation $\mathcal{L}_{syn}$.

$$\begin{aligned}
\mathcal{L}_{syn}\ulcorner cl\ d_1\ d_2 \urcorner &\mapsto \mathcal{L}_{syn}\ulcorner d_1 \urcorner \cdot \mathcal{L}_{syn}\ulcorner d_2 \urcorner \\
\mathcal{L}_{syn}\ulcorner argp\ d_1\ d_2 \urcorner &\mapsto \mathcal{L}_{syn}\ulcorner d_1 \urcorner \cdot \mathcal{L}_{syn}\ulcorner d_2 \urcorner \\
\mathcal{L}_{syn}\ulcorner john \urcorner &\mapsto \text{"john"} \\
\mathcal{L}_{syn}\ulcorner every_x \urcorner &\mapsto \text{"every"} \\
\mathcal{L}_{syn}\ulcorner participant \urcorner &\mapsto \text{"participant"}
\end{aligned}$$

. . .

The rules are written in the form reminiscent of top-down tree transducers. The result $\mathcal{L}\ulcorner d \urcorner$ of transforming a term d is obtained by trying to match d against the pattern in the left-hand-side of every rule. The right-hand-side of the matching rule gives the result. If no matching rule is found, the transformation is not defined (i.e., 'fails'). The patterns may contain variables, which stand for the corresponding subterms. For example, in the first rule, d_1 and d_2 match the two children of a term whose head is cl. The occurrences of these variables in the right-hand side of the rule are replaced by the corresponding matching branches.

Intuitively, $\mathcal{L}_{sem}$ looks like a context-free-grammar of the sample sentence, with $jgep$ being its derivation tree.

The meaning is derived by applying a sequence of transformations to a T_A term. The transformation $\mathcal{L}_{Ux}$ gets rid of every$_x$, introducing var$_x$ and U$_x$ instead. This transformation is context-sensitive. Therefore, we first define context C – a term (tree) with a hole – as follows:

$$C = [] \mid \mathsf{cl}\ C\ d \mid \mathsf{cl}\ d\ C \mid \mathsf{argp}\ d\ C \mid \mathsf{ppadv}\ C\ d \mid \mathsf{ppadv}\ d\ C$$

where the meta-variable d stands for an arbitrary term. In words: a context is the bare hole $[]$, or a clause (the cl term) that contains a hole in the subject or the predicate, or a VP made of a transitive verb whose argument has a hole, or a complemented VP with the hole in the head or the complement, etc. We write $C[d]$ for the term obtained by plugging d into the hole of C. We further distinguish two subsets of contexts C_{cl} and C_{ncl}:

$$C_{cl} = \mathsf{cl}\ C_{ncl}\ d \mid \mathsf{cl}\ d\ C_{ncl}$$
$$C_{ncl} = [] \mid \mathsf{argp}\ d\ C_{ncl} \mid \mathsf{ppadv}\ C_{ncl}\ d \mid \mathsf{ppadv}\ d\ C_{ncl}$$

Intuitively, C_{cl} is the smallest context that has a hole within a clause.

The transformation $\mathcal{L}_{Ux}$ is then stated as follows:

$$\mathcal{L}_{Ux}\ulcorner C_{cl}[\mathsf{every}_x\ d_r]\urcorner \;\mapsto\; \mathsf{U}_x\ (\mathcal{L}_{Ux}\ulcorner d_r\urcorner)\ (\mathcal{L}_{Ux}\ulcorner C_{cl}[\mathsf{var}_x\ ]\urcorner)$$

We now use extended top-down tree transducers, whose patterns are 'deep', that is, contain matching expressions within arbitrary context. As before, whenever a pattern, e.g., $C_{cl}[\mathsf{every}_x\ d_r]$, matches the source term, it is replaced with $\mathsf{U}_x\ d_r\ C_{cl}[\mathsf{var}_x]$, and the transformation is re-applied to its subterms. That is, $C_{cl}[\mathsf{every}_x\ d_r]$ on the left hand-side of the rule matches a tree that contains, somewhere inside, a sub-expression of the form every$_x\ d_r$ (a branch headed by every$_x$). On the right-hand side of the rule, $C_{cl}[\mathsf{var}_x]$ is the same tree in which every$_x\ d_r$ subterm has been replaced with var$_x$. Unlike $\mathcal{L}_{syn}$ above, the $\mathcal{L}_{Ux}$ transformation does not look like a context-free grammar. It is context-sensitive. The other difference is the presence of a default rule: if $\mathcal{L}_{Ux}\ulcorner d\urcorner$ finds no match for d, $\mathcal{L}_{Ux}$ is repeated on sub-expressions of d. In particular, $\mathcal{L}_{Ux}\ulcorner c\urcorner$ is the constant c itself (unless there is an explicit rule for that particular c). For $\mathcal{L}_{syn}$, which translates from one language, T_A, to another, T_S, the default rule does not make sense.

Our example $jgep$ matches the left-hand side of $\mathcal{L}_{Ux}$ immediately: d_r matches participant and C_{cl} is john (argp greet $[]$), The result

$$(\mathsf{U}_x\ \mathsf{participant})\ (\mathsf{cl}\ \mathsf{john}\ (\mathsf{argp}\ \mathsf{greet}\ \mathsf{var}_x))$$

is in effect the Quantifier Raising (QR) of "every participant", but in a rigorous, deterministic way. The intent of the new constants should become clear: U$_x$ is to represent the raised quantifier, and var$_x$ its trace. Unlike QR, the raised quantifier (U$_x$ participant) lands not just on any suitable place. $\mathcal{L}_U$ puts it at the closest boundary marked by the clause-forming constant cl. $\mathcal{L}_U$ is type-preserving: it maps a well-typed term to also a well-typed term. Again unlike QR, we state the correctness properties such as type-preservation. The type preservation is the necessary condition for the correctness of the transformations.

To finally obtain the meaning we apply the transformation $\mathcal{L}_{sem}$:

$$
\begin{array}{rcl}
\mathcal{L}_{sem}\ulcorner \mathsf{cl}\ d_1\ d_2\urcorner &\mapsto& \mathcal{L}_{sem}\ulcorner d_2\urcorner\ \mathcal{L}_{sem}\ulcorner d_1\urcorner \\
\mathcal{L}_{sem}\ulcorner \mathsf{argp}\ d_1\ d_2\urcorner &\mapsto& \mathcal{L}_{sem}\ulcorner d_1\urcorner\ \mathcal{L}_{sem}\ulcorner d_2\urcorner \\
\mathcal{L}_{sem}\ulcorner \mathsf{U}_\mathsf{x}\ d_1\ d_2\urcorner &\mapsto& \forall_x\ \mathcal{L}_{sem}\ulcorner d_2\urcorner\ x \implies \mathcal{L}_{sem}\ulcorner d_2\urcorner \\
\mathcal{L}_{sem}\ulcorner \mathsf{var}_\mathsf{x}\ \urcorner &\mapsto& x \\
\mathcal{L}_{sem}\ulcorner \mathsf{john}\ \urcorner &\mapsto& \mathrm{john} \\
\mathcal{L}_{sem}\ulcorner \mathsf{participant}\ \urcorner &\mapsto& \mathrm{participant} \\
&\cdots&
\end{array}
$$

that produces the logical formula representing the term's meaning. The transformation replaces john, etc. with the corresponding logical constants and U_x with the universal quantifier. Since $\mathcal{L}_{sem}$ translates one language, T_A, into a different one, T_L, this transformation, like $\mathcal{L}_{syn}$, has no default rule. If the source term does not match the pattern of any $\mathcal{L}_{sem}$ rule, the transformation is undefined. In particular, applying $\mathcal{L}_{sem}$ to the original $jgep$ term straight away is not defined because there is no rule for every$_\mathsf{x}$. The failure means that $jgep$ cannot be given meaning – directly. However, $\mathcal{L}_{sem}\ulcorner\mathcal{L}_{Ux}\ulcorner jgep\urcorner\urcorner$ is well-defined, resulting in

$$\forall_x\ \mathrm{participant}\ x \implies (\mathrm{greet}\ x\ \mathrm{john})$$

3 Coordination in TS

We now apply TS to the analysis of (non-canonical) coordination. As a warm-up, we take the non-problematic "John tripped and fell," which is an example of the conventional VP coordination. We analyze it differently, however, as 'left-node raising' so to speak, to introduce the technique to be later used in right-node raising (RNR), argument cluster coordination (ACC) and gapping[1].

The abstract form of our example is

$$\mathsf{and}_{\mathsf{S,VP}}\ (\mathsf{cl}\ \mathsf{john}\ \mathsf{tripped})\ \mathsf{fell}$$

The new constant $\mathsf{and}_{\mathsf{S,VP}}$ has the type $S \to VP \to S$. As common, we assume a whole family of constants $\mathsf{and}_{\mathsf{x,y}}$ of different types. The constant $\mathsf{and}_{\mathsf{S,VP}}$ – like every$_\mathsf{x}$ in the example of the previous section – is not in the domain of $\mathcal{L}_{sem}$. Therefore, to be able to derive the logical formula, we have to transform it away. The following transformation $\mathcal{L}_a$ does that:

$$
\begin{array}{l}
\mathcal{L}_a\ulcorner \mathsf{and}_{\mathsf{S,VP}}\ (\mathsf{cl}\ d_{NP}\ d_{VP})\ d\urcorner \mapsto \\
\quad \mathrm{and}\ \mathcal{L}_a\ulcorner(\mathsf{cl}\ d_{NP}\ d_{VP})\urcorner\ \mathcal{L}_a\ulcorner(\mathsf{cl}\ d\ d_{VP})\urcorner
\end{array}
$$

---

[1] We may even analyze NP coordination as a sort of RNR: after all, "John and Mary left" can have the meaning of the conjunction of truth conditions of "John left" and "Mary left". Certainly, "John and Mary left" may also mean that "John and Mary", taken as a group, left. In the later case, the group can be referred as "they". Our analysis applies to the former (conjunction) case but not the latter. Hence we posit that 'and' is not only polytypic but also polysemic.

The rule again is written in the form of extended top-down tree transducers: when the source term matches the rule's pattern, it is replaced with the right-hand-side of the rule. Again, d with various subscripts are meta-variables that stand for arbitrary subterms (tree branches). Like $\mathcal{L}_{Ux}$, there is a default rule: a term that does not match the rule undergoes $\mathcal{L}_a$ on its subterms, if any. Applying $\mathcal{L}_a$ to our T_A term transforms it to

$$\text{and (cl john tripped) (cl john fell)}$$

where and is the ordinary coordination, of the type $S \to S \to S$, which can be given the meaning of propositional disjunction and which hence is in the domain of $\mathcal{L}_{sem}$. The result is straightforward to transform to a logical formula T_L.

3.1   RNR in TS

Our next example is the proper RNR: "John likes and Mary hates Bill", whose abstract form is

$$\text{and}_{(NP,TV),S} \ (\text{john}, \text{like}) \ (\text{cl mary (argp hate bill)})$$

We have added to T_A tuples (d, d) and tuple types (σ, σ). The constant $\text{and}_{(NP,TV),S}$ has the type $(NP, TV) \to S \to S$. Whereas (cl mary (argp hate bill)) is the complete sentence, (john, like) is certainly not. It is not even a constituent; it is just a sequence of words: a cluster. Since we added to T_A tuples and new constants, we may need to extend our earlier transformation rules, specifically, $\mathcal{L}_{syn}$ for transforming into the surface form of the sentence T_S:

$$\mathcal{L}_{syn}\ulcorner \text{and}_{(NP,TV),S} \ d_1 \ d_2 \urcorner \quad \mapsto \quad \mathcal{L}_{syn}\ulcorner d_1 \urcorner \cdot \text{"and"} \cdot \mathcal{L}_{syn}\ulcorner d_2 \urcorner$$
$$\mathcal{L}_{syn}\ulcorner (d_1, d_2) \urcorner \quad \mapsto \quad \mathcal{L}_{syn}\ulcorner d_1 \urcorner \cdot \mathcal{L}_{syn}\ulcorner d_2 \urcorner$$

Applying $\mathcal{L}_{syn}$ to our T_A clearly gives "John likes and Mary hates Bill". This 'phonetic' transformation is dull and uninteresting, in contrast to the higher-order phonetics of [8].

Let us derive the meaning, the T_L formula, from the same T_A term. Before we can apply $\mathcal{L}_{sem}$ we need to transform away $\text{and}_{(NP,TV),S}$, which is not in the domain of that transformation. We extend the $\mathcal{L}_a$ with a new clause:

$$\mathcal{L}_a\ulcorner \text{and}_{(NP,TV),S} \ (d_1, d_2) \ (\text{cl } d \ C[\text{argp } d_4 \ d_5]) \urcorner \quad \mapsto$$
$$\text{and } \mathcal{L}_a\ulcorner (\text{cl } d_1 \ (\text{argp } d_2 \ d_5)) \urcorner \ \mathcal{L}_a\ulcorner (\text{cl } d \ C[\text{argp } d_4 \ d_5]) \urcorner$$

where d_1, d, d_5 have to be of the type NP and d_2 and d_4 of the type TV. The transformation is context-sensitive and type-directed. It may be regarded as matching of (d_1, d_2) against the complete sentence (the second argument of $\text{and}_{(NP,TV),S}$). The matching is determined by the type of $\text{and}_{(NP,TV),S}$. The parallel structure of the coordination is clearly visible.

Analyses of RNR without directional types (e.g., using ACG) run into trouble of over-generating "*John likes Bill and Mary hates". Although we can write the abstract form for that sentence as well:

$$\text{and}_{S,(NP,TV)} \ (\text{cl john (argp like bill)}) \ (\text{mary}, \text{hate})$$

we do not provide the $\mathcal{L}_a$ rule with the constant $\mathsf{and}_{\mathsf{S},(\mathsf{NP},\mathsf{TV})}$. Since it remains uneliminated, $\mathcal{L}_{sem}$ cannot be applied and the meaning cannot be derived. In TS, transformations are partial and are not guaranteed to always succeed. The original sentence is considered ungrammatical then. We discuss the choice of transformable $\mathsf{and}_{\mathsf{XY}}$ constants in Sect. 3.4.

Let us consider another well-known troublesome example, due to P. Dekker:

(1)   *The mother of and John thinks that Mary left.

In categorial grammar approaches, 'the mother of' and 'John thinks that' may be given the same type, $(S/(N\backslash S))/N$. The two phrases may hence be coordinated, over-generating (1). In TS, 'the mother of' cannot be given any type at all (likewise, 'John thinks that' is not a constituent and has no type.) We can only treat 'the mother of' as a cluster, of the determiner, N and the proposition. We do provide the constant $\mathsf{and}_{(\mathsf{DET},\mathsf{N},\mathsf{POF}),\mathsf{S}}$ with the corresponding rule

$$\mathcal{L}_a \ulcorner \mathsf{and}_{(\mathsf{DET},\mathsf{N},\mathsf{POF}),\mathsf{S}} \ (d_1, d_2, \mathsf{of}) \ (C_{cl}[d_{det} \ (\mathsf{ppadj} \ d_n \ \mathsf{of} \ d_{np})]) \urcorner \quad \mapsto$$
$$\mathsf{and} \ \mathcal{L}_a \ulcorner (C_{cl}[d_1 \ (\mathsf{ppadj} \ d_2 \ \mathsf{of} \ d_{np})]) \urcorner \ \mathcal{L}_a \ulcorner (C_{cl}[d_{det} \ (\mathsf{ppadj} \ d_n \ \mathsf{of} \ d_{np})]) \urcorner$$

which can be used to analyze "The mother of, as well as the father of John died". The rule does not apply to the problematic (1) since there is no similar parallel structure of the of-headed PP.

3.2   Argument Cluster Coordination and Gapping

The same transformation idea also works for argument cluster coordination (ACC) and gappping. Take for example, "Mary liked Chicago and Bill Detroit", or, in the abstract form:

$$\mathsf{and}_{\mathsf{S},(\mathsf{NP},\mathsf{NP})} \ (\mathsf{cl} \ \mathsf{mary} \ (\mathsf{argp} \ \mathsf{liked} \ \mathsf{chicago})) \ (\mathsf{bill}, \mathsf{detroit})$$

The transformational rule for the constant $\mathsf{and}_{\mathsf{S},(\mathsf{NP},\mathsf{NP})}$ picks a suitable subterm that can relate two NPs from the left conjunct

$$\mathcal{L}_a \ulcorner \mathsf{and}_{\mathsf{S},(\mathsf{NP},\mathsf{NP})} \ (\mathsf{cl} \ d \ C[\mathsf{argp} \ d_4 \ d_5]) \ (d_1, d_2) \urcorner \quad \mapsto$$
$$\mathsf{and} \ \mathcal{L}_a \ulcorner (\mathsf{cl} \ d \ C[\mathsf{argp} \ d_4 \ d_5]) \urcorner \ \mathcal{L}_a \ulcorner (\mathsf{cl} \ d_1 \ C[(\mathsf{argp} \ d_4 \ d_2)]) \urcorner$$

It turns our T_A term to

$$\mathsf{and} \ (\mathsf{cl} \ \mathsf{mary} \ (\mathsf{argp} \ \mathsf{liked} \ \mathsf{chicago})) \ (\mathsf{cl} \ \mathsf{bill} \ (\mathsf{argp} \ \mathsf{liked} \ \mathsf{detroit}))$$

with the clear meaning. The examples (2) and (4) of Sect. 1 are dealt with similarly. One may observe that the analysis of gapping is nearly the same as that of VP coordination, used in the warm-up example.

3.3   Coordination and Scoping

The interaction of non-canonical coordination with quantification is not much different from that of the ordinary coordination of two clauses. For example, take (2) of Sect. 1, whose abstract form is

and$_{S,(PP,PP)}$
   (cl speaker (ppadv (ppadv (argp gave (a$_x$ present)) (to robin))(on thu)))
   (to leslie, on fri)
contains two components to be eliminated by transformations: and$_{S,(PP,PP)}$ and
the QNP (a$_x$ present). The latter is to be handled by $\mathcal{L}_E$, which is analogous
to $\mathcal{L}_U$ but for the existential quantifier. The transformations $\mathcal{L}_a$ and $\mathcal{L}_E$ can
be applied in either order, which corresponds to the wide- and narrow-scope–
readings of (2). The narrow scope happens when $\mathcal{L}_a$ goes first, producing

and
   (cl speaker (ppadv (ppadv (argp gave (a$_x$ present)) (to robin))(on thu)))
   (cl speaker (ppadv (ppadv (argp gave (a$_x$ present)) (to leslie))(on fri)))
The $\mathcal{L}_{Ex}$ transformation then gives

and
   (E$_x$ present (cl speaker (ppadv (ppadv (argp gave var$_x$) (to robin))(on thu))))
   (E$_x$ present (cl speaker (ppadv (ppadv (argp gave var$_x$) (to leslie))(on fri))))
whose meaning is the conjunction of two existentially quantified formulas.

   If $\mathcal{L}_{Ex}$ is applied first to the original sentence, we get

and$_{S,(PP,PP)}$
   (E$_x$ present (cl speaker (ppadv (ppadv (argp gave var$_x$) (to robin))(on thu))))
   (to leslie, on fri)
Strictly speaking, the rule analogous to $\mathcal{L}_a$ from Sect. 3.2 does not apply since the
first conjunct now has the form E$_x$ d_r (cl d_1 d_2) rather than the bare (cl d_1 d_2).
We have to hence generalize the rule to

$$\mathcal{L}_a \ulcorner \text{and}_{S,(PP,PP)} \; C_{ncl} \, [(\text{cl } d \; C[\text{ppadv (ppadv } d_h \; d_4) \; d_5])] \; (d_1, d_2) \urcorner \quad \longmapsto$$
$$C_{ncl}[\text{and } \mathcal{L}_a \ulcorner(\text{cl } d \; C[\text{ppadv (ppadv } d_h \; d_4) \; d_5])\urcorner$$
$$\mathcal{L}_a \ulcorner(\text{cl } d \; C[\text{ppadv (ppadv } d_h \; d_1) \; d_2])\urcorner]$$

effectively pulling out the context C_{ncl} – the sequence of U$_x$ d and E$_x$ d quantifiers and their restrictors – and coordinating underneath. The coordination thus
receives narrow scope. Such pulling of the context may seem ad hoc; however,
it is this general form of $\mathcal{L}_a$ rules that gives the mechanism to account for the
anomalous scope of negation in (7) of Sect. 1, repeated below.

(7)   Pete wasn't called by Vanessa but rather John by Jesse.

The transformation involving the contrasting coordinating particle such as 'but
rather' gets a chance to examine C_{ncl} and determine if there is a negation to
contrast with:

$$\mathcal{L}_a \ulcorner \text{rather}_{S,(NP,PP)} \; (\text{Neg } (\text{cl } d \; C[\text{ppadv } d_4 \; d_5])) \; (d_1, d_2)\urcorner \quad \longmapsto$$
$$\text{and } (\text{Neg } \mathcal{L}_a \ulcorner(\text{cl } d \; C[\text{ppadv } d_4 \; d_5])\urcorner) \; \mathcal{L}_a \ulcorner(\text{cl } d_1 \; (\text{ppadv } d_4 \; d_2))\urcorner$$

where Neg is the constant analogous to U$_x$.

3.4   Discussion

We have presented the uniform analysis of both the canonical and non-canonical
coordination, reducing the variety of coordination (VP, RNR, ACC, Gapping)
to the choice of the coordinating constants and$_{S,x}$ or and$_{x,S}$ that adjoin material

(often just a cluster of words) to a sentence. The transformation rules driven by the constants pick the pieces from the sentence to complete the material to a clause. We have thus provided a uniform *mechanism* of coordination. The corresponding policy is embodied in the coordinator constants like and hence lexicalized.

There remains a question of a general principle/pattern that governs the choice of the constants. For example, the fact that in English the coordinated sentence appears on the right for RNR but on the left for ACC and Gapping boils down to the presence of and$_{(NP,TV),S}$ and and$_{S,(NP,NP)}$ and the absence of and$_{S,(NP,TV)}$ and and$_{(NP,NP),S}$. In contrast, one may say that this fact 'falls out' as a consequence of like-category coordination analyses in directional categorial grammars. One may also say that the like-category coordination is itself a postulate, which does not come from any general principle, but does have significant empirical justification. Like any empirical principle, it has exceptions: unlike-category coordination, e.g., "John saw the facts and that Mary had been right". Also, the like-category coordination leads to overgeneration, as we saw in the Dekker's example in Sect. 3.1.

Since our TS approach is still new, we have not yet accumulated enough empirical data to discern patterns and formulate postulates that underlie the presence of coordination constants for some types and their absence for others. For now, we leave the question open.

4   Related Work

Our transformational approach is rooted in Transformational Generative Grammars [2,3], later carried into Minimalism [4]. Our abstract form T_A is similar to the spell-out of Minimalism. However, whereas the spell-out is near culmination of a syntactic derivation for Minimalists, for us, it is just the beginning. We are not interested in how structure is created through a sequence of Merges from lexical selections. Rather, we consider our abstract form as given (by a parser) and investigate its transformations into a semantic form. Our transformations are hence all covert.

Closely related to TS is the work of Butler [1], who also obtains a semantic representation as a result of a transformation from a parsed tree. Unlike us, he has applied his approach to a wealth of empirical data in many languages and has truly achieved wide coverage. His transformations are rather complex and coarse, doing many things at once, and not typed. One may view TS as an attempt to re-engineer and understand Butler's approach and decompose his transformations into elementary steps.

We are grateful to the anonymous reviewer for pointing out the analysis of ACC and Gapping in [14].

(1) The interpretation of an elliptical construction is obtained by uniformly substituting its immediate constituents into some immediately preceding structure, and computing the interpretation of the results. [14, p. 162, (119)]

We indeed share the underlying idea of picking and substituting of 'immediate constituents' into the coordinated material (understood at some level as an elliptical construction). The proposal of [14] remained rather informal; the present paper may be seen as an attempt to formalize the idea, as well as to extend it to scope phenomena.

There have been other attempts to solve the KLM problem without directional types (within the ACG-like formalisms). Kanazawa [5] proposes 'regular constraints' to prevent over-generation (which recall structural constraints in Government and Binding). This amounts however to duplication of lexical entries. The approach [13] reins in the over-generation using subtyping. Either proposal can be classified as 'proof search' rather than computational like TS; in case of [13] with no guarantees that the proof search ever terminates (and, as the authors admitted, no good way to characterize the space of available derivations and detect over-generation).

5   Conclusions

We have demonstrated the transformational analyses of RNR and Gapping. The analyses make precise various eliding schemas, demanding type preservation. The asymmetry of the type of $\mathsf{and}_{(\mathsf{NP},\mathsf{TV}),\mathsf{S}}$ and similar constants is what lets us answer the Kubota, Levine and Moot challenge: how to prevent over-generation in analyses of RNR and gapping without directional types.

The idiosyncrasies of coordination are distilled to the ad hoc choice of constants $\mathsf{and}_{\mathsf{XY}}$. There are transformations for some types XY but not for the others. There may be a pattern there. Collecting the arbitrariness in one place might make the pattern easier to find. Being able to handle the entire ellipsis part of the FraCaS corpus seems the natural first step in searching for that pattern.

It is interesting to consider interpreting the "sequence of words" as a discontinuous sentence in the sense of Morrill et al. [12].

Another future work task is to apply TS to more complicated scoping phenomena including 'same', 'different', 'the total of' – as well as to various wh-movement phenomena.

Acknowledgments. I am very grateful to Leo Tingchen Hsu for numerous perceptive and stimulating discussions. I thank an anonymous reviewer for many very insightful and helpful comments. Numerous discussions with Yusuke Kubota, Bob Levine, Alastair Butler, Greg Kobele and the participants of the workshop "New Landscapes in Theoretical Computational Linguistics" (Ohio State University, October 14–16, 2016) are gratefully acknowledged.

References

1. Butler, A.: Linguistic Expressions and Semantic Processing - A Practical Approach. Springer, Cham (2015)
2. Chomsky, N.: Aspects of a Theory of Syntax. MIT Press, Cambridge (1965)

3. Chomsky, N.: Lectures on Government and Binding. Foris, Dordrecht (1981)
4. Chomsky, N.: The Minimalist Program. The MIT Press, Cambridge (1995)
5. Kanazawa, M.: Syntactic features for regular constraints and an approximation of directional slashes in abstract categorial grammars. In: Kubota and Levine [11], pp. 34–70
6. Kiselyov, O.: Canonical constituents and non-canonical coordination. In: Murata, T., Mineshima, K., Bekki, D. (eds.) JSAI-isAI 2014. LNCS, vol. 9067, pp. 99–113. Springer, Heidelberg (2015). doi:10.1007/978-3-662-48119-6_8
7. Kiselyov, O.: Applicative abstract categorial grammars in full swing. In: Proceedings of LENLS 12, November 2015. http://dx.doi.org/10.1007/978-3-319-50953-2_6
8. Kubota, Y., Levine, R.: Gapping as like-category coordination. In: Béchet, D., Dikovsky, A. (eds.) LACL 2012. LNCS, vol. 7351, pp. 135–150. Springer, Heidelberg (2012). doi:10.1007/978-3-642-31262-5_9
9. Kubota, Y., Levine, R.: Gapping as hypothetical reasoning. In: Natural Language and Linguistic Theory (2014, to appear). http://ling.auf.net/lingbuzz/002123
10. Kubota, Y., Levine, R.: Against ellipsis: arguments for the direct licensing of 'non-canonical' coordinations. Linguist. Philos. **38**(6), 521–576 (2015)
11. Kubota, Y., Levine, R. (eds.): Proceedings for ESSLLI 2015 Workshop 'Empirical Advances in Categorial Grammar'. University of Tsukuba and Ohio State University (2015)
12. Morrill, G., Valentín, O., Fadda, M.: The displacement calculus. J. Logic Lang. Inform. **20**(1), 1–48 (2011)
13. Pollard, C., Worth, C.: Coordination in linear categorial grammar with phenogrammatical subtyping. In: Kubota and Levine [11], pp. 162–182
14. Sag, I.A., Gazdar, G., Wasow, T., Weisler, S.: Coordination and how to distinguish categories. Nat. Lang. Linguist. Theory **3**(2), 117–171 (1985)

Rich Situated Attitudes

Kristina Liefke[(✉)] and Mark Bowker

Ludwig-Maximilians-University Munich, Munich Center for Mathematical
Philosophy, Faculty 10, Geschwister-Scholl-Platz 1, 80539 Munich, Germany
{Kristina.Liefke,Mark.Bowker}@lrz.uni-muenchen.de

Abstract. We outline a novel theory of natural language meaning, *Rich
Situated Semantics* [RSS], on which the content of sentential utterances
is semantically *rich* and informationally *situated*. In virtue of its situatedness, an utterance's rich situated content varies with the informational
situation of the cognitive agent interpreting the utterance. In virtue of its
richness, this content contains information beyond the utterance's lexically encoded information. The agent-dependence of rich situated content
solves a number of problems in semantics and the philosophy of language
(cf. [14,20,25]). In particular, since RSS varies the granularity of utterance contents with the interpreting agent's informational situation, it
solves the problem of finding suitably fine- or coarse-grained objects for
the content of propositional attitudes. In virtue of this variation, a layman will reason with more propositions than an expert.

Keywords: Information-sensitivity · Interpreter-dependence · Propositional attitude contents · Rich semantic content · Situated semantics

1   Introduction

The same utterance of a (non-indexical) sentence has a different meaning to
different interpreting agents. This is due to the fact that different agents have
different information about the sentence's subject matter, which is used in the
utterance's agent-specific interpretation: Depending on the agent's background
knowledge, the utterance of (1) in a particular context will be interpreted as an
informationally rich proposition (e.g. as a proposition which contains the information that *the inhabitant of Gobbler's Knob* is a groundhog/that Punxsutawney
Phil *is a member of the largest existing marmot species*) or as an informationally
poorer proposition which does not contain this additional information.

(1) Punxsutawney Phil is a groundhog.

We would like to thank two anonymous referees for LENLS 13, Robin Cooper, Sebas-
tian Löbner, Markus Werning, and Dietmar Zaefferer for their valuable comments
and suggestions. The research for this paper is supported by the German Research
Foundation (via K. Liefke's grant LI 2562/1-1) and by LMU Munich's Institutional
Strategy LMUexcellent within the framework of the German Excellence Initiative.

S. Kurahashi et al. (Eds.): JSAI-isAI 2016, LNAI 10247, pp. 45–61, 2017.
DOI: 10.1007/978-3-319-61572-1_4

Most formal theories of semantic natural language content (e.g. [9,27,28,32]) restrict the content of sentential utterances to the utterances' lexical information (for (1): to the information that the referent of the name *Punxsutawney Phil* is a groundhog), and delegate all other available information about the utterance's subject matter to areas like pragmatics or psychology. However, Moltmann [26] (cf. [7]) has observed that they thus seriously underspecify the content of propositional attitudes. We observe that, as a result, these theories are unable to explain why an inference is valid for some agents, but invalid for others.

This paper solves the above problem by complementing the traditional notions of utterance content with a new kind of semantic content, *rich situated content*. The latter includes non-lexically encoded information that is available to the interpreter of the utterance at the time of the interpretation. Below, we first sketch our new theory of linguistic meaning, called *Rich Situated Semantics* (in Sect. 2). We then present the rigid granularity problem for the content of propositional attitudes (in Sect. 3) and show how Rich Situated Semantics solves this problem (in Sect. 4). Section 5 answers a salient objection to our solution to the rigid granularity problem. The paper closes by identifying other intensional phenomena that lend themselves to a rich situated semantic treatment (in Sect. 6).

2   Rich Situated Semantics

Rich Situated Semantics [hereafter, RSS] (cf. [19,20]) is a new theory of natural language meaning on which the content of (utterances[1] of) declarative sentences is semantically *rich* and informationally *situated*. In virtue of its situatedness, the rich situated content of a sentence varies with the informational situation of the cognitive agent interpreting the sentence. In virtue of its richness, this content contains information beyond the sentence's lexically encoded information. Rich situated content is thus a special form of descriptive content.

Below, we first illustrate the richness and situatedness of sentential content and identify a number of theories from linguistics, philosophy, cognitive and computer science that suggest this richness and/or situatedness (in Sect. 2.1). We then specify the RSS-interpretation of sentences (in Sect. 2.2) and identify some notable consequences of this interpretation with respect to linguistic entailment and equivalence (in Sect. 2.3). The section closes with a definition of truth for Rich Situated Semantics (in Sect. 2.4).

---

[1] We hereafter sometimes use the expression 'content of a sentence' (or 'sentential content'), instead of 'content of *an utterance of* a sentence'. This is merely a terminological shortcut. The reader is asked to keep in mind that sentences are uttered by a speaker (with certain background information) in a spatiotemporal and communicative situation, and are directed at an addressee (with a certain, likely different, background information). The relevance of the addressee's information for the interpretation of the utterance is the central topic of this paper.

2.1 Illustration and Inspiration for RSS

To familiarize the reader with the core idea of RSS, we introduce rich linguistic contents by means of an example: Consider the interpretation of (1) by three agents, viz. Alf, Bea, and Chris. Assume that, *re* Punxsutawney Phil (hereafter, 'Phil'), these agents have the following information:

Alf: Phil lives in Gobbler's Knob.

Bea: Phil is celebrated each February 2nd.

Chris: Phil lives in Gobbler's Knob; Phil is celebrated each February 2nd.

Since rich situated content includes the interpreter's information about the sentence's subject matter (here: Phil), (1) is interpreted by Alf as (1.i), by Bea as (1.ii), and by Chris as (1.iii):

(1) i. Phil is a groundhog who/and lives in Gobbler's Knob.

ii. Phil is a groundhog who is celebrated each February 2nd.

iii. Phil is a groundhog who lives in Gobbler's Knob and is celebrated each February 2nd.

The non-identity of the rich contents of (1) at Alf's, Bea's, and Chris' informational situation witnesses the *situatedness* of linguistic content in RSS. The greater informativeness of the rich content of (1) at any of the above situations in comparison to the sentence's traditional, possible-worlds content (which only contains the sentence's lexical information) witnesses the *richness* of linguistic content in RSS. In particular, the rich content of (1) at Alf's informational situation contains the information that Phil lives in Gobbler's Knob, which is not contained in the sentence's the lexical information.

The *situatedness* of linguistic content is inspired by work in situation semantics (cf. [3, 15]), semantic contextualism (cf. [13, 17]), relativism (cf. [6, 22]), and dynamic semantics (cf. [11, 42]). Situation semantics assumes that sentences are uttered in and their utterances evaluated with respect to partial possible worlds (i.e. situations). Contextualism and relativism assume, respectively, that the same sentence can have a different content in different contexts and that the truth-value or the content of a sentence vary with the context of assessment. Dynamic semantics suggests the situatedness of linguistic content by interpreting sentences as *state transitions*, i.e. as functions from information states to the result of updating these states with the sentence's lexical information.

Rich linguistic content is found in Fregean theories of belief content (cf. [5, 10]), in semantic descriptivism and generalized quantifier theory (cf. [2, 39]), and in frame semantics (cf. [1, 21]). Fregean theories of belief content assume that any adequate representation of belief contents involves the modes of presentation of the individuals and properties the beliefs are about. Descriptivism and generalized quantifier theory assume that proper names are interpreted analogously to definite NPs, i.e. as sets of properties of individuals. Frame semantics represents utterance contents by rich recursive feature structures that account for the content of mental concepts.

2.2   The RSS-Interpretation of Sentential Utterances

To capture the *situatedness* of rich linguistic content, RSS interprets sentences as *functions from interpreters' informational situations* to the sentences' rich contents *at these situations* (i.e. to the sentences' *situated contents*). These functions are objects of type $s\alpha$, where α is the type for situated sentential contents.

The *richness* of situated sentential contents is captured via (characteristic functions of) partial sets of situations (s.t. $\alpha := st$).[2] Such sets are familiar from the representation of sentential contents in generalizations of possible world semantics, including some versions of situation semantics (e.g. [15, 28]). However, the set of situations that serves as the content of a sentence in RSS is generally much smaller than the set of situations that serves as the content of this sentence in situational generalizations of possible world semantics. This is due to the fact that – in addition to being restricted to situations in which the sentence is true – the RSS-set is further restricted to situations which contain the interpreting agent's information about the sentence's subject matter. For example, while (1)-as-received-by-Alf is interpreted as (2) in situational possible world semantics, it is interpreted as (3) in RSS. Below, i is a variable over situations, as reflected in the superscript s. The formulas $groundhog(phil)(i)$ and $livesinGK(phil)(i)$ assert that Phil is a groundhog in i and that Phil lives in Gobbler's Knob in i.

$$\lambda i^s [groundhog(phil)(i)] \tag{2}$$
$$\lambda i^s [groundhog(phil)(i) \wedge livesinGK(phil)(i)] \tag{3}$$

To capture the informational imperfection of cognitive agents, we identify situations with *partial* (i.e. informationally incomplete) spatio-temporal parts of worlds[3] in which the parts' individual inhabitants may fail to have some of the properties which they have at the relevant world-part. Situations in rich situated semantics are thus "partial specifications of some of the entities in the universe with [their] properties" [23, p. 614]. They are obtained from worlds by reducing the information about the world's inhabitants to the information available to the agent at the given point in time. As a result, situations are agent- and time-specific: the same agent may be in different informational situations at different points in time.

We assume that situated sentential contents are *partially* (or *selectively*) *rich*, i.e. that they contain – next to the sentence's lexical information – all and *only* information *about the sentence's subject matter* that is available to the interpreter of the sentence at the time of the interpretation. As a result, RSS interprets any sentence p as a function from informational situations i to

---

[2] One can increase the granularity of situated sentential contents by analyzing them instead as semantically primitive (i.e. non-analyzable) propositions (cf. [8, 29, 32, 41]). The development of hyperfine-grained RSS is left for another occasion.

[3] The inclusion of *impossible* worlds or situations (cf. [12, 35]) captures the possibility of agents' misinformedness or false belief. For reasons of space, the consideration of impossible worlds or situations is left for future research.

sets of situations whose members contain the lexical information of p together with all information from i which regards some individual about which p carries information. For convenience, sentences that carry information about some individual a will hereafter be called *aboutness-relevant with respect to a*, or *a-relevant*. Sentences that carry information about the same individuals are called *aboutness-identical (w.r.t. these individuals)*. The RSS-interpretation of a sentence p is given in (4).

$$\lambda i^s \lambda j^s [p^{st}(j) \wedge \forall q^{st}([q(i) \wedge \exists x^e (abt(x)(q) \wedge abt(x)(p))] \twoheadrightarrow q(j))] \qquad (4)$$

In (4), x is a variable over individuals. The formula $\varphi^t \twoheadrightarrow \psi^t$ asserts that ψ contains the information of φ (i.e. that ψ is less partial/better defined than φ), s.t. ψ is true if φ is true and is false if φ is false (cf. [28, pp. 50, 47]). $\varphi \twoheadrightarrow \psi$ is defined as $((\varphi \wedge \psi) \vee ((\varphi \vee \psi) \wedge *)) = \varphi$, where $*$ is the neither-true-nor-false formula. The introduction of $\twoheadrightarrow$ is made necessary by our association of t with the set of truth-*combinations*, by the resulting existence of two different orderings on the type-t domain (i.e. a truth- and an approximation-ordering), and by the reference of the material conditional to the 'wrong' ordering for our purposes (i.e. to the truth-ordering; on this ordering, ψ is true if φ is true, *but φ is false if ψ is false*).

The formula $abt(x^e)(q^{st})$ asserts that q carries information about the referent of x. The behavior of abt is governed by a variant of the axioms from [30, p. 129] (cf. [18, pp. 120–121]). These axioms include the aboutness-relevance (with respect to an individual) of atomic formulas that contain the designating constant for the individual as a constituent, the closure of aboutness-relevant formulas under non-contradictory conjunction,[4] and the closure of aboutness-relevant formulas under disjunction (given that both disjuncts contain information about the subject matter).

To better understand the interpretation of sentences in Rich Situated Semantics, consider the rich content of (1) at Alf's current informational situation, σ_{alf} (in (5)). We assume for simplicity that σ_{alf} only contains the information that Phil lives in Gobbler's Knob (cf. Sect. 2.1) and that Bea has red hair. Since only the first-mentioned informational item of σ_{alf} regards Phil, (1) will be RSS-interpreted at σ_{alf} as (3).

$$\lambda i [groundhog(phil)(i) \wedge \forall q([q(\sigma_{alf}) \wedge \exists x^e (abt(x)(q) \wedge abt(x)(p))] \twoheadrightarrow q(i))] \quad (5)$$
$$\equiv \lambda i [groundhog(phil)(i) \wedge \forall q([q(\sigma_{alf}) \wedge abt(phil)(q)] \twoheadrightarrow q(i))]$$

We next identify a concrete candidate for the set of situations described by (5): Assume a universe consisting of four situations, σ_{alf}, σ_1, σ_2, and σ_3 and two individuals: Phil (abbreviated p) and Bea (abbreviated b). We assume that Phil

---

[4] To avoid the inclusion of information that does not regard the subject matter, we demand (*contra* Perry) that *both* conjuncts be aboutness-relevant. This also avoids the problem of obtaining aboutness-'relevant' conjunctions by combining an aboutness-irrelevant sentence with a trivially aboutness-relevant *verum*, or with *falsum*.

lives in Gobbler's Knob (Kp) in σ_{alf}, σ_1, and σ_2, that Bea has red hair (Rb) in σ_{alf} and σ_2, and that Phil is a groundhog (Gp) in σ_1, σ_2, and σ_3 (cf. Fig. 1).

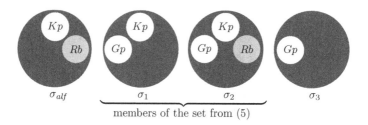

$$\sigma_{alf} \qquad \underbrace{\qquad \sigma_1 \qquad \qquad \sigma_2 \qquad} \qquad \sigma_3$$

members of the set from (5)

Fig. 1. The rich content-at-σ_{alf} of (1).

Then, since the lexical information of (1) (i.e. Phil is a groundhog) and the Phil-relevant information from σ_{alf} (i.e. Phil lives is Gobbler's Knob) are included only in σ_1 and σ_2 (and in none of the other situations), the rich situated content of (1) at σ_{alf} is represented by the set $\{\sigma_1, \sigma_2\}$ (underbraced in Fig. 1).

2.3   Consequences of RSS

The RSS-interpretation of sentential utterances has a number of important consequences for the individuation of situated sentential contents. In particular, since RSS updates the available information about a sentence's subject matter with the sentence's lexical information, it identifies the rich contents of sentences at situations whose information about the sentence's subject matter differs only with respect to the inclusion of the sentence's lexical information. Consider the interpretation of (1) at Len's informational situation in which Phil is a groundhog and lives in Gobbler's Knob. (We assume that this situation does not contain any other information about Phil, s.t., as regards Phil, it is identical to σ_1): at this situation, (1) has the same rich content (i.e. $\{\sigma_1, \sigma_2\}$) as at σ_{alf}.

Note that, although (1) has the same rich content at Alf's and at Len's informational situation, its utterance has a different *effect* on Alf's than on Len's situation: while (1)-as-received-by-Alf updates Alf's information about Phil with the information that Phil is a groundhog (s.t. Alf's information is extended to the information from σ_2), it leaves Len's informational situation unchanged. The updating effect of (1) on Alf's Phil-specific information is witnessed by the fact that (the information associated with) the rich content of (1) at Alf's informational situation (i.e. $\{\sigma_1, \sigma_2\}$) is properly contained in (the information associated with) the rich content of (6) at Alf's situation (i.e. $\{\sigma_{alf}, \sigma_1, \sigma_2\}$).

(6)   Punxsutawney Phil lives in Gobbler's Knob.

As a result of the richness of situated contents, RSS further identifies the contents of different aboutness-identical sentences at situations which contain the sentences' lexical information. Consider the interpretation of (1) and (6) at σ_1: since this situation already contains the lexical information of (1) and (6), these sentences have the same rich content (i.e. $\{\sigma_1, \sigma_2\}$) at this situation.

We will see in Sect. 4 that the identification-at-a-situation of the rich contents of different aboutness-identical sentences solves the problem of finding suitably fine- or coarse-grained objects for the content of propositional attitudes. This problem is described in Sect. 3.

2.4 Truth-Evaluation in RSS

We have described situated sentential contents as the results of updating the available information about the sentence's subject matter with the sentence's lexical information. As a result of this description, situated sentential contents in RSS contain much more information than sentential contents in situational generalizations of possible world semantics. However, much of this information is irrelevant for the sentences' evaluation. For example, it does not (or should not) matter for the truth of (1) whether Phil lives in Gobbler's Knob. Since *non*-situated sentential contents (type $s(st)$) do not have the 'right' type for truth-evaluable objects (i.e. they do not yield a truth-value when applied to a world), we need to provide a custom truth-evaluation procedure for sentences in RSS.

To evaluate the truth of a sentence in Rich Situated Semantics, we check whether the world of evaluation w is a member of the union of the sentence's rich contents at all informational situations. The resulting truth-definition is given below. In this definition, we use denotation brackets, $[\![\cdot]\!]$, as a notational device for rich non-situated sentential contents (i.e. type-$s(st)$):

Definition 1 (Truth at a world). *In Rich Situated Semantics, a sentence p is true at a world w iff $w \in \bigcup_{\sigma^s} [\![p]\!](\sigma)$, where $[\![p]\!](\sigma)$ is the rich interpretation of p at the situation σ.*

By taking the union of the rich contents, $[\![p]\!](\sigma)$, of p for each situation σ, we obtain the set of situations in which p is true. This set is a situational generalization of the classical Lewisian proposition denoted by p.

The rationale behind the above strategy is as follows: since we assume the existence of a situation for every consistent combination of information (including the 'empty' combination; cf. [31,32]), the members of the above union will never share *more* than the lexical information of p (plus p's presuppositions). Since we identify the result of updating a situation's information via incompatible information with the empty set of situations[5], the members of this union will never share *less* than the lexical information of p. In particular, situations which contain the information that not-p will not contribute their information to the above union.

---

[5] This is due to the fact that the available information about the sentence's subject matter at these situations will include the sentence's lexical information. Since we have excluded impossible situations from our considerations (cf. fn. 3), no situation contains both an item of information and its complement.

Notably, unions of rich situated contents provide an easy way of retrieving the traditional notion of (lexical, 'poor') content. This notion is required for the explanation of a number of phenomena, including the use of sentences to state facts (cf. (7)), to give reasons (cf. (8)), and to express shared belief (cf. (9)). For example, Eve may utter (1) to communicate the fact that Phil is a groundhog (rather than some other fact she knows about Phil) (cf. (7)) or to give a reason for Phil's long teeth (cf. (8)). Many other sentences which receive an identical interpretation at Eve's informational situation would not serve this purpose.

(7)   Eve asserted that Phil was a groundhog.

(8)   Since Phil is a groundhog, his teeth never stop growing.

(9)   Len and Eve believe that Phil is a groundhog.

We next turn to the rigid granularity problem for the content of propositional attitudes. This problem lies in the fact that most theories of linguistic content assume a single, uniform level of granularity for belief contents. As a result of this assumption, these theories cannot explain why an inference is valid for some epistemic agents (given their background knowledge), but invalid for others.

3   The Rigid Granularity Problem

To avoid predicting agents' logical omniscience, many theories of formal semantics (e.g. [9, 28, 32, 41]) assume hyperfine-grained sentential contents that have stricter identity-conditions than sets of possible worlds. The level of granularity of these contents is chosen in accordance with speakers' intuitions about synonymy (cf. [32, p. 553]). Since most speakers judge the contents of many intensionally equivalent sentences (e.g. of (1) and (10)) to be non-identical, hyperfine-grained semantics distinguish the contents of these sentences.[6]

(10)   Punxsutawney Phil is a member of the largest existing marmot species.

The success of these semantics is hampered by the fact that the above identity- (or *non*-identity-)judgements are not shared by all speakers for all sentence-pairs. This is due to the fact that speakers' judgements about sentential synonymy are influenced by their background information about the sentences' subject matter. Depending on their informational situation, speakers will thus identify or distinguish the contents of the same sentences. Consider the case of (1) and (10): since she is familiar with the different properties of groundhogs, a groundhog expert (e.g. Eve in (11)) will identify the contents of (1) and (10). Since he is unaware of this fact, a groundhog layman (e.g. Len in (12)) will treat the contents of (1) and (10) as distinct. Any reasoner who is familiar with Eve and Len's level of groundhog expertise (s.t. (s)he knows that (1) and (10)

---

[6] The identification of these sentences' traditional contents in possible world semantics is due to the fact that, in the actual world at the current time (cf. the adjective *existing* in (10)), groundhogs *are* the largest marmot species.

have the same rich content at Eve's, but different rich contents at Len's informational situation), will conclude (11b) from (11a), but not (12b) from (12a). Since hyperfine-grained semantics assume the same level of granularity of content for *all* agents interpreting a sentence, they cannot distinguish between the validity of these inferences.

(11) a. Eve knows that Phil is a groundhog. **T**

      b. Eve knows that Phil is a member of the largest existing marmot species. **T**

(12) a. Len knows that Phil is a groundhog. **T**

      b. Len knows that Phil is a member of the largest existing marmot species. **F**

In particular, since hyperfine-grained semantics distinguish the contents of (1) and (10), they will counterintuitively predict the invalidity of (11). Since traditional (coarse-grained) possible world semantics identifies the content of (1) and (10), it will counterintuitively predict the validity of (12).

4 Rich Situated Attitudes

Rich Situated Semantics solves the above problem by varying the granularity of sentential contents with the informational situation of the sentence's interpreter. This is possible since RSS identifies the contents of different aboutness-identical sentences at situations which contain the sentences' lexical information.

4.1 Solving the Rigid Granularity Problem

Since, as we will hereafter assume, Eve's informational situation, σ_{eve}, contains the lexical information of (1), (6), and (10), RSS *identifies* the contents of (1) and (10) at this situation (i.e. (14)). Since Len's situation does *not* contain the lexical information of (10) (s.t. (10) is interpreted as an update on Len's information about Phil), RSS *distinguishes* the contents of (1) (i.e. (3)) and (10) (i.e. (14)) at Len's informational situation. With respect to the relevant subject domain, a layman will thus reason with *more* sentential contents than an expert.

$$\lambda i^s [groundhog(phil)(i) \wedge largestmarmot(phil)(i)] \tag{13}$$

$$\lambda i^s ([groundhog(phil)(i) \wedge largestmarmot(phil)(i)] \wedge livesinGK(phil)(i)) \tag{14}$$

The variation of sentences' semantic granularity with the epistemic agent's informational situation explains the intuitive validity of the inference from (11) and the intuitive invalidity of the inference from (12). However, this explanation presupposes the reasoner's familiarity with Eve and Len's level of expertise about Phil (cf. Sect. 3). Reasoners who are *not* familiar with the two agents' levels of subject expertise (s.t. they are, in particular, unaware of Eve's identical interpretation of (1) and (10)) will not be able to make the inference from (11).[7]

---

[7] The ability of (11b) to extend the reasoner's knowledge depends on this unfamiliarity.

To capture the dependence of (11) on the reasoner's awareness of the agent's subject expertise, we stipulate the following: when they occur in the complement of epistemic verbs like *know*, sentences are interpreted as sets of situations whose members only encode the agent's information about the sentence's subject matter *of whose availability to the agent the reasoner is aware*.[8] For the occurrence of (1) from (11a), this set is specified in (15). There, r is a variable for the reasoner. The formula $aware(r)(q)(\sigma)$ asserts the reasoner's awareness that σ includes the information of q.

$$\lambda i\,[groundhog\,(phil)(i) \wedge \forall q([\boldsymbol{aware\,(r)(q)(\sigma_{eve})} \wedge abt\,(phil)(q)] \twoheadrightarrow q(i))] \quad (15)$$

We illustrate the reasoner-dependence of epistemic inferences by means of an example: Compare the interpretation of (11a) and (11b) by two reasoners, Dan and Fred, who have different degrees of familiarity with Eve's information about Phil. In particular, Dan knows that, in σ_{eve}, Phil is a groundhog, belongs to the largest existing marmot species, and lives in Gobbler's Knob. Fred only knows that Phil is a groundhog in this situation. The complements of the occurrences of *know* from (11a) and (11b) are then interpreted as (14) by Dan and as (2) (cf. (11a)) and (13) (cf. (11b)) by Fred. Since only Dan is, thus, aware of Eve's identification of the rich contents of (1) and (10), only he can make the inference from (11).

Notably, the inference from (11) can also be made solely on the basis of Dan's awareness of Eve's *general expertise* about Phil, which does not require Dan's familiarity with the particular content of Eve's informational situation. This expertise entails the inclusion-in-σ_{eve} of all Phil-relevant information that is true at the actual world, @. The resulting interpretation of (1) from (11a) is given in (16).

$$\lambda i\,[groundhog\,(phil)(i) \wedge \forall q([\boldsymbol{aware\,(r)(q)(@)} \wedge abt\,(phil)(q)] \twoheadrightarrow q(i))] \quad (16)$$

Our previous considerations suggest the distinction between two types of validity, relative to an agent's informational situation. The types are defined below:

Definition 2 (Situational validity). *An inference is* valid relative to the informational situation σ of some specific reasoner (or is valid-at-σ) *iff the rich content at σ of the inference's premise(s) is a subset of the rich content at σ of the inference's conclusion.*

Definition 3 (Validity simpliciter). *An inference is* valid simpliciter *iff, at all informational situations σ, the rich content at σ of the inference's premise(s) is a subset of the rich content at σ of the inference's conclusion.*

---

[8] Admittedly, the reasoner may wrongly assume that Len also knows (10). This assumption explains why the reasoner may still make the inference from (12). It can be captured by replacing the reasoner's required *awareness* of the inclusion of a particular item of information in the agent's information state by the reasoner's *belief* about this inclusion (which does not entail the factivity of this inclusion).

The condition from Definition 3 corresponds to requiring the entailment of the traditional, possible worlds-interpretation of the conclusion by the traditional, possible worlds-interpretation of the premise(s). The different types of validity are illustrated respectively by (11) and (17):

(17)   a. Eve knows that Phil is a groundhog and lives in Gobbler's Knob. **T**
         b. Eve knows that Phil is a groundhog. · **T**

Since the situated interpretation of (11a) does not entail the situated interpretation of (11b) at some situations (e.g. at σ_1), (11) is not valid *simpliciter*.

4.2   Consequences of Situating Attitudes

As a result of its rich situated interpretation of epistemic complements, RSS also predicts the validity of inferences between epistemic reports like (18), whose complements are not intensionally equivalent.

(18)   a. Eve knows that Phil is a groundhog.   **T**
         b. Eve knows that Phil lives in Gobbler's Knob. **T**

The validity of these inferences may be justified by the reasoner's familiarity with the epistemic agent's level of subject expertise: a reasoner (e.g. Dan) who is aware of the agent's degree of informedness about the interpreted sentence's subject matter will follow the agent in identifying his/her situated interpretation of the complements of *know* from (18a) and (18b). However, intuitively, inferences like (18) have a different kind of validity from inferences like (11).

To block inferences of the form of (18), we modify the content of the epistemically embedded occurrence of (1) from (15) to the set of situations whose members only encode *the information contained in the complement's lexical information* of whose availability to Eve the reasoner is aware. This modification restricts the set of validly substitutable complements of epistemic verbs like *know* to CPs that are classically entailed[9] by the CP. For the complement of *know* from (11a), this is achieved by (19). There, the variable w ranges over possible worlds.

$$\lambda i \forall q \left(\left[(\forall w \, [\textbf{\textit{groundhog}} \, (\textbf{\textit{phil}})(\textbf{\textit{w}}) \rightarrow \textbf{\textit{q}}(\textbf{\textit{w}})] \wedge aware \, (r)(q)(\sigma_{eve}) \right) \wedge \right. \qquad (19)$$
$$\left. abt \, (phil)(q) \right] \twoheadrightarrow q(i))$$

Consider Dan's interpretation of the complements from (18a) and (18b). Following (19), these complements are interpreted as (13) (cf. (18a)) and (20) (cf. (18b)). Since the set of situations denoted by (13) is not contained in the set of situations denoted by (20), the inference from (18) is no longer valid on this interpretation.

$$\lambda i^s \, [livesinGK \, (phil)(i)] \qquad (20)$$

The interpretation from (19) is in line with the understanding of propositional knowledge as focusing on *a particular item* of the agent's subject-relevant

---

[9] Entailment is here defined in terms of (subset) inclusion of sets of possible worlds.

information (at a given time), rather than as surveying *all* of his or her information (at this time). It differs from most attitude treatments by extending propositional knowledge to the *union* of the sentence's lexical information and the available aboutness-relevant information of its traditional entailments.

Our previous considerations may have made it seem as if our interpretation of epistemic complements was only an *ad hoc* move to prevent counterintuitive inferences of the form of (18). This is not the case: Since different verbs have differently strict requirements on the substitution of their complements (with verbs like *remember* even allowing the replacement by other than the classically entailed complements), the same sentence requires a differently fine-grained interpretation in different contexts. This observation calls for a 'modular' approach to granularity, which varies the granularity of sentential interpretations with the sentence's embedding context. By assuming interpretations with the granularity of (5) as the default case, and allowing different verbs to reduce (cf. (15)) or relatively increase the level of granularity (cf. (19)), RSS provides such modularity.

Consider the substitution properties of the complements of the verbs *say verbatim*, *know*, and *remember*: while *know* allows the substitution of its complement by sentences with the same subject matter to which the complement is traditionally equivalent (cf. the intuitive support for (11)), *say verbatim* does *not* allow such a substitution (cf. the intuitive support against (21), below). In contrast to the class of 'substitutable' complements of the verb *know*, the class of substitutable complements of the verb *remember* extends *beyond* the complement's traditional equivalents. The substitution-generality of the complement of *remember* is witnessed by the intuitive support for the inference from (22).[10]

(21) a. <u>Eve said verbatim that Phil was a groundhog.</u>   **T**
        b. Eve said verbatim that Phil was a woodchuck.   **F**

(22) a. <u>Dan remembered that Phil was nibbling at a dandelion.</u> **T**
        b. Dan remembered that Phil was endearing.   **T**

In addition to a modular account of complement restriction (above), RSS also enables a modular account of granularity that is determined by *non*-linguistic context. This account explains the observation that the same sentence requires a differently fine- (or coarse-)grained interpretation in different *communicative* contexts. Consider the complement of the verb *say verbatim* from (21a): This verb typically does not allow the substitution of its complement by any other sentence. The described ban on substitution even extends to pairs of classically equivalent sentences which receive an identical RSS-interpretation at the epistemic agent's current information state: in court, a witness' utterance of (21b) – instead of the original (21a) – will be counted against her and may even be punishable. However, these and other substitutions seem admissible in cases in which

---

[10] This inference assumes that the complements of the two occurrences of *remember* from (22) describe the same remembered situation. The intuitive validity of this type of inference is discussed in detail in [20, Sects. 4, 5].

less is at stake. These include less formal social contexts, like friends gossipping about Eve.

The interpretation of the complements of epistemic verbs from (15) and (19) suggests the possibility of providing a modular account of contextually determined granularity. Because of the semantic effect of pragmatic factors (here: the respective social context and its associated level of formality), this account would involve reference to some version of pragmatic enrichment (cf. [4,34]). We leave the detailed development of this account for another occasion.

5   Objections and Answers

We have shown in the preceding section that RSS solves the rigid granularity problem for the content of propositional attitudes. However, there exists a widely-used – arguably simpler and more salient – alternative solution. This solution lies in the assumption of a hyperfine-grained semantics that distinguishes the contents of intensionally equivalent sentences and in the introduction of an additional premise stating the epistemic agent's awareness of the co-intensionality (equivalence) of the two complements. For (11), such a premise is given in (11b)'.

(11)'   a. Eve knows that Phil is a groundhog.   **T**
         b. Eve knows that Phil is a member of the largest existing marmot
             species *iff* he is a groundhog.   **T**

         c. Eve knows that Phil is a member of the largest existing marmot
             species.   **T**

Premise (11b)' can even be replaced by the more general premise (11)'':

(11)''   b. Eve knows that groundhogs are the largest existing marmot species,
             and *vice versa*.

The introduction of either of the above premises serves the same role as the *rich* interpretation of the two complements in RSS: it connects the premise (i.e. (11a)) with the conclusion by asserting the obtaining of an equivalence relation between the complements of the two occurrences of the verb *know*. Since RSS identifies the rich contents of the two complements at Eve's informational situation (s.t. the complements are also equivalent in RSS), it does not require the introduction of an additional premise establishing this equivalence.

Its initial appeal notwithstanding, the above strategy lacks three desirable features of our RSS-account of the rigid granularity problem. These include the possibility of enabling inferences of the form of (11) without reference to a specific item of the agent's knowledge (i), the easy generalizability to granularity problems involving verbs from other verb classes (ii), and the provision of a modular account of (linguistic or non-linguistic) contextually determined granularity (iii). Feature (iii) has been discussed in some detail at the end of Sect. 4.2. Features (i) and (ii) are discussed below:

Ad (i): Premises (11b)' and (11b)'' specify the particular item of the agent's knowledge that enables the inference from (11a) to (11b). However, reasoning often proceeds more holistically through association (cf. [40]). In particular, to make the inference from (11), the reasoner does not need to identify a specific inference pattern that ensures formal validity. Instead, it suffices for him to know that, at σ_{eve}, the complements from (11a) and (11b) have the same rich content, such that they allow mutual substitution. This is achieved via the reasoner's awareness that σ_{eve} includes both the lexical information of (1) and (10) (cf. (15)) or through his awareness of Eve's general expertise about Phil (cf. (16)).

Ad (ii): Our previous considerations have focused on the complements of epistemic verbs like *know*. However, variants of the rigid granularity problem also arise for the contents of other attitudes, including perceptual attitudes (e.g. *see*), emotional attitudes (e.g. *fear*), and evaluative attitudes (e.g. *admire*). In contrast to premises like (11b)′ that specify epistemic attitudes, premises that specify perceptual, emotional, or evaluative attitudes are semantically deviant (cf. (23b))

(23) a. Eve saw/feared/admired that Len (would) pet Phil.   T

      b. #Eve saw/feared/admired that Len (would) pet the best-known member of the largest existing marmot species *iff* he (would) pet Phil.   ?

      c. Eve saw/feared/admired that Len (would) pet the best-known member of the largest existing marmot species   T

The deviance of (23b) can be removed by replacing the occurrence of *saw* (or of *feared* or *admire*) by the verb *know* (in (23b)′).

(23)′ b. Eve knew that Len (would) pet the best-known member of the largest existing marmot species *iff* he (would) pet Phil.

However, the resulting premise presupposes a connection between knowledge and perception (or between knowledge and evaluations or the emotions) that is not made explicit in everyday reasoning or in RSS.

The above is not to question the validity of the inference from (23a) and (23b)′ to (23c). Rather, it observes the difficulty of explaining this validity in semantic theories which exclude the agent's non-lexical information from the relevant notion of linguistic content. This difficulty originates in a lack of correspondence between the contents of perception (or of emotion or evaluation) and of knowledge in these theories. In particular, in these theories, the occurrences of the sentence *Len pets Phil* from (23a) and (23b)′ are interpreted, respectively, as subsets of the set of Len-petting-Phil situations that are consistent with Eve's current *perceptual* (or emotional, or evaluative) situation (cf. (23a)) and as the set of Len-petting-Phil situations that are consistent with Eve's current *epistemic, informational* situation (cf. (23b)′). The absence of an explicit relation between these two sets impedes the inference from (23a) and (23b)′ to (23c).

One could try to avoid the above problem by replacing (23b)′ instead by the premise (23b)″.

(23)″ b. Eve saw/feared/admired that Len (would) pet the best-known member of the largest existing marmot species *iff* she saw/feared/admired that Len (would) pet Phil.

This replacement yields a non-deviant sentence that ensures the validity of the inference. However, since any justification of (23b)″ will, again, need to establish a connection between knowledge and perception (or evaluation, or the emotions), it suffers from the same problem as the replacement of (23b) by (23b)′.

Rich Situated Semantics allows the inference from (23) by identifying the rich contents of the complements of *saw* from (23a) and (23c) at Eve's informational situation (cf. (5), (15)). Since this identification establishes a strong semantic connection between the two sentences, it does not require the introduction of an additional premise making this connection, or a specification of the relation between perception and knowledge.

6   Other Applications of RSS

We have shown above that Rich Situated Semantics solves the rigid granularity problem for the content of propositional attitudes. Our presentation of RSS suggests that this semantics can also be used to explain several other intensional phenomena. In particular, RSS helps solve some familiar problems of intensionality that have recently resurfaced in the philosophy of language. These include the cognitive accessibility problem for propositions (cf. [14,25]), the problem of rational illogical belief (cf. [36,38]), and the substitution problem for the objects and contents of propositional attitudes (cf. [24,33]). Respectively, these problems regard the difficulty of most mainstream theories of linguistic content to explain how communicative agents can grasp abstract propositions, how rational agents can jointly believe superficially contradictory propositions,[11] and how the *contents* of propositional attitudes (as denoted by the CP complements of epistemic verbs) differ from the *objects* of these attitudes (as denoted by the complements' nominalizations of the form *the proposition that* ___).

RSS solves these problems by incorporating the interpreting agents' information about the sentences' subject matter into the content of these sentences. This information corresponds to the agents' *mode of presentation* of the subject matter (cf. [5,10,37]). In RSS, an object's mode of presentation is represented by the set of situations (type st) in which the object has the properties that the agent associates with it.[12] Since rich situated sentential contents depend on the information of the sentence's interpreting agent, they are cognitively accessible. The ability of agents to interpret different occurrences of the same NP w.r.t. their informational situations at different times further explains the possibility of rational illogical belief. The non-substitutability of CPs by their NP nominalizations in many contexts is explained by the situated (rich) interpretation of embedded CPs and the non-situated (poor) interpretation of their nominalizations.

The RSS-solution to the substitution problem is presented in [20, Sect. 5.3]. The detailed description of a rich situated solution to the remaining problems is left as a project for future work.

---

[11] These include Pierre's simultaneous belief that London is pretty and that London is not pretty (cf. [16]).

[12] This contrasts with the standard formal semantic representation of modes of presentation as sets of the object's properties (type $(et)t$) (cf. [2,27]).

References

1. Barsalou, L.W.: Frames, concepts, and conceptual fields. In: Lehrer, A., Kittay, E.F. (eds.) Frames, Fields, and Contrasts: New Essays in Semantic and Lexical Organization. Erlbaum Associate, Erlbaum (1992)
2. Barwise, J., Cooper, R.: Generalized quantifiers and natural language. Linguist. Philos. **4**(2), 159–219 (1981)
3. Barwise, J., Perry, J.: Situations and Attitudes. MIT Press, Cambridge (1983)
4. Carston, R.: Explicit communication and 'free' pragmatic enrichment. In: Soria, B., Romero, E. (eds.) Explicit Communication: Robyn Carston's Pragmatics. Palgrave Macmillan, Basingstoke (2010)
5. Chalmers, D.: The representational character of experience. In: Leiter, B. (ed.) The Future for Philosophy. Oxford UP, Oxford (2004)
6. Egan, A., Hawthorne, J., Weatherson, B.: Epistemic modals in context. In: Preyer, G., Peter, G. (eds.) Contextualism in Philosophy. Oxford UP, Oxford (2005)
7. Fara, D.G.: Specifying desires. Noüs **47**(2), 250–272 (2013)
8. Fox, C., Lappin, S., Pollard, C.: A higher-order fine-grained logic for intensional semantics. In: Proceedings of the 7th International Symposium on Logic and Language (2002)
9. Fox, C., Lappin, S.: Foundations of Intensional Semantics. Blackwell, Hoboken (2005)
10. Frege, G.: Über Sinn und Bedeutung. Gottlob Frege: Kleine Schriften. Edition Classic Verlag Dr. Müller (2006)
11. Groenendijk, J., Stokhof, M.: Dynamic predicate logic. Linguist. Philos. **14**(1), 39–100 (1991)
12. Hintikka, J.: Impossible possible worlds vindicated. J. Philos. Logic **4**(4), 475–484 (1975)
13. Kaplan, D.: Demonstratives: an essay on the semantics, logic, metaphysics, and epistemology of demonstratives and other indexicals. In: Almog, J., Perry, J., Wettstein, H. (eds.) Themes from Kaplan. Oxford UP, Oxford (1989)
14. King, J., Soames, S., Speaks, J.: New Thinking about Propositions. Oxford UP, Oxford (2014)
15. Kratzer, A.: An investigation into the lumps of thought. Linguist. Philos. **12**(5), 607–653 (1989)
16. Kripke, S.: A puzzle about belief. In: Margalit, A. (ed.) Meaning and Use: papers presented at the Second Jerusalem Philosophy Encounter. Reidel, Kufstein (1979)
17. Lewis, D.: Index, context, and content. In: Kanger, S., Öhman, S. (eds.) Philosophy and Grammar. Synthese Library, vol. 143. Springer, Netherlands, Dordrecht (1981). doi:10.1007/978-94-009-9012-8_6
18. Liefke, K.: A single-type semantics for natural language. Doctoral dissertation, Center for Logic and Philosophy of Science, Tilburg (2014)
19. Liefke, K.: Towards an account of rich situated semantic content. In: Gamerschlag, T., et al. (eds.) Abstracts of Cognitive Structures: Linguistic, Philosophical, and Psychological Perspectives, Düsseldorf, 15–17 September 2016. http://cognitive-structures.phil.hhu.de/wp-content/uploads/2016/09/Updated-program-brochure-14-09.pdf
20. Liefke, K., Werning, M.: Evidence for single-type semantics - an alternative to e/t-based dual-type semantics (submitted)
21. Löbner, S.: Evidence for frames from human language. In: Gamerschlag, T., Gerland, D., Osswald, R., Petersen, W. (eds.) Frames and Concept Types. SLP, vol. 94, pp. 23–67. Springer, Cham (2014). doi:10.1007/978-3-319-01541-5_2

22. MacFarlane, J.: Epistemic modals are assessment-sensitive. In: Egan, A., Weatherson, B. (eds.) Epistemic Modality. Oxford UP, Oxford (2011)
23. Moltmann, F.: Part structures in situations: the semantics of individual and whole. Linguist. Philos. **28**(5), 599–641 (2005)
24. Moltmann, F.: Abstract Objects and the Semantics of Natural Language. Oxford UP, Oxford (2013)
25. Moltmann, F.: Propositions, attitudinal objects, and the distinction between actions and products. Can. J. Philos. **43**(5–6), 679–701 (2013)
26. Moltmann, F.: Cognitive products and the semantics of attitude verbs and deontic modals. In: Moltmann, F., Textor, M. (eds.) Act-Based Conceptions of Propositional Content: Contemporary and Historical Perspectives. Oxford UP, Oxford (forthcoming)
27. Montague, R.: The proper treatment of quantification in ordinary English. In: Thomason, R.H. (ed.) Formal Philosophy: selected papers of Richard Montague. Yale UP, New Haven (1974)
28. Muskens, R.: Meaning and Partiality. CSLI Publications, Stanford (1995)
29. Muskens, R.: Sense and the computation of reference. Linguist. Philos. **28**(4), 473–504 (2005)
30. Perry, J.: Possible worlds and subject matter. In: Sturé, A. (ed.) Possible Worlds in Humanities, Arts and Sciences. Walter de Gruyter, Berlin (1989)
31. Pollard, C.: Hyperintensions. J. Logic Comput. **18**(2), 257–282 (2008)
32. Pollard, C.: Agnostic hyperintensional semantics. Synthese **192**, 535–562 (2015)
33. Prior, A.: Objects of Thought. Clarendon Press, Wotton-under-Edge (1971)
34. Recanati, F.: Pragmatic enrichment. In: Russell, G., Fara, D.G. (eds.) The Routledge Companoin to Philosophy of Language. Routledge, Abingdon (2012)
35. Rantala, V.: Quantified modal logic: non-normal worlds and propositional attitudes. Stud. Logica. **41**(1), 41–65 (1982)
36. Salmon, N.: The resilience of illogical belief. Noûs **40**(2), 369–375 (2006)
37. Schiffer, S.: The mode-of-presentation problem. In: Anderson, C.A., Owens, J. (eds.) Propositional Attitudes: The Role of Content in Logic, Language, and Mind. CSLI Lecture Notes, vol. 20. CSLI Publications, Stanford (1990)
38. Schiffer, S.: Belief ascription. J. Philos. **89**(10), 499–521 (1992)
39. Searle, J.: Proper names. Mind **67**(266), 166–173 (1958)
40. Stanovic, K.E.: Who is Rational?: Studies of Individual Differences in Reasoning. Routledge Psychology Press, Abingdon (2011)
41. Thomason, R.H.: A model theory for the propositional attitudes. Linguist. Philos. **4**(1), 47–70 (1980)
42. Veltman, F.: Defaults in update semantics. J. Philos. Logic **25**(3), 221–261 (1996)

Reference and Pattern Recognition

Youichi Matsusaka[✉]

Tokyo Metropolitan University, Tokyo, Japan
matsusak@tmu.ac.jp

1 Introduction

In a well known passage in *Naming and Necessity*, Kripke expressed a pessimistic attitude toward a philosophical theory of reference. He said, of the view he attributed to Frege and Russell, that:

> It really is a nice theory. The only defect I think it has is probably common to all philosophical theories. It's wrong. You may suspect me of proposing another theory in its place; but I hope not, because I'm sure it's wrong too if it is a theory. (Kripke (1972), p. 64.)

Kripke presented his own account based on causal chains of name passing as a *picture* rather than a theory. I am not sure why Kripke said this, but after a long hesitation I myself came to have a similar opinion: There may not be a necessary and sufficient condition for referencehood. Whether there is a set of necessary and sufficient conditions for reference or not, what I am going to put forward is not a theory of reference. In fact, it may not be even a picture, but just part of a whole picture about reference. I believe, however, that the part is important, and has been unnoticed or at least unstressed in the philosophical literature.

I find it useful to begin by regarding *concepts*, rather than names, as the primary bearers of reference. In this, I side with Kaplan (1968), Evans (1973), Millikan (2000), and recent "mental file" theorists who seem to be gaining in numbers.[1] In the end, however, I will argue that names themselves play important roles in sustaining their reference. Names are not just "parasitic" on concepts.

I believe that a person has, for each object she knows well, a concept that is devoted to collecting and storing information about the object. Let us call such a concept an *"object-centered concept"*. (You might want to call it a *"de re* concept" instead.) Then, I have an object-centered concept devoted to my father, to my mother, and to many other objects as well. (When I talk about "concepts" below, I mean object-centered ones unless otherwise is indicated.) I would like to claim that my concept for a friend of mine *refers to* him, and that my use of his name somehow *exploits* the intentional character of the concepts people have of him.

How does a concept have reference? (If you are a mental file theorist, you might want to frame this question in terms of how a mental file manages to be a

---

[1] See Recanati (2013) and authors cited therein.

© Springer International Publishing AG 2017
S. Kurahashi et al. (Eds.): JSAI-isAI 2016, LNAI 10247, pp. 62–73, 2017.
DOI: 10.1007/978-3-319-61572-1_5

file *of* a particular individual). You can find varieties of answers that have been given to this question: Descriptive fit (classical descriptivism), Causal source (Kaplan (1968), Evans (1973)), Biological function (Millikan (2000)), Indexicality (Recanati (2013)), etc. I will not repeat, or initiate, the criticisms of these views here. In fact, to me they all seem natural, and true to some extent. I wish to give my own view below, but it is partly intended to provide a broad framework in which the insights of the previously given accounts can be properly understood.

Central to my account is a certain notion of the *stability* of a concept, which I will call *intentional stability*. I will explain its basic ideas in the next four paragraphs:

The function of an object-centered concept is to collect and store information about the object it is devoted to. (I am using the term "information" informally.) But how is it possible? The question of how you *generate* a concept for a particular individual is an interesting and important one, but I will not go into this question in this paper. (This question is sometimes referred to as that of "clustering" in the literature on machine learning.) I assume that you already have a set of concepts for the objects with which you have had experiential contact, and that your task is just to *revise* them on the basis of the experience you are having now.

In order to tackle this task, you have to decide whether the current experience is an experience *of* one of those objects about which you already possess concepts, and if the answer is yes, which one. Let us call this an *identification task*. If you decide, for example, that what you see now is a certain object you know well, you may revise the concept of the object on the basis of the visual information you are receiving. But how is an identification task is achieved? How do you recognize that what you see now is the face of your husband? There are no "magic recipes" for this. An overwhelmingly natural answer is that you tackle the task by utilizing the body of information *already* stored in the concepts you have. In the following, I assume that this is basically the correct view. Then, if you somehow decide that what you now see is something you know well, you may *update* the object-centered concept of that thing utilizing the visual information you are receiving.

Since we are fallible, we will misidentify an object from time to time. As a result, a concept may be "contaminated" with information coming from other objects. Such "contamination" could be serious. If worse comes to worst, your concept of your friend may be so contaminated with information from other objects that it is no longer centered around the friend, or even that it is centered around some other object. However, in most, favorable, cases, we can expect that as you come to know objects better, namely, as information in your concepts of the objects becomes richer, you will be able to perform identification tasks better, thereby enriching your concepts further and diminishing the risk of misidentification further. In this sense, a successful system of concepts has a *self-reinforcing nature*.

So far I have allowed myself to freely talk about "correct" or "incorrect" identification in describing the kind of conceptual revision I am interested in. In other words, I have *presupposed* the notion of a concept's denoting, or being *of*, an object. However, we can reverse the order of explanation here. Suppose that a subject keeps updating his concepts in a fixed environment. If they show a self-reinforcing nature in the above explained sense, we can expect that they will strengthen their tendency to collect information from certain objects. If a concept has this tendency, namely, a tendency to end up with a concept whose information gathering activities are concerned with a single object, let us say that the concept is *intentionally stable* with respect to that object (in the given environment). I believe this notion of intentional stability gives us a good starting point for discussing reference.

The purpose of what follows is to pursue this idea in the simplest setting. Since intentional stability is concerned with the dynamics of a concept's revision, I wish to say a bit more about concept revision in the next section.

2   Concept Revision and Pattern Recognition

Central to the notion of intentional stability is the dynamic process of concept revision. Unfortunately, we do not quite understand the nature of human concepts yet. In order to illustrate my ideas, I wish to use a particular model of concept called "exemplar theory",[2] or a somewhat simplified version of it.[3]

I choose this model mainly because of its simplicity. I believe the main arguments can be couched in terms of many other proposals about concept acquisition as well, e.g., prototype theory or neural networks ("deep learning").

According to this theory, a concept is simply a set of exemplars. Exemplars are, roughly, memories of objects. They typically take the form of perceptual images of the objects, or their summaries. A subject's concept of John consists in her memories of what *she judged* were encounters with John in her past

---

[2] See Medin and Schaffer (1991) and Nosofsky (2011). Exemplar theory has also close connection to the k-NN (k-nearest neighbor) method in machine learning. In both traditions, an exemplar is represented as a point in a multidimensional space, whose each dimension corresponds to a feature, or its mathematical transformation, of objects. The distance defined between two points in the space represents the similarity between exemplars. I should note that both exemplar theory and the k-NN method are usually applied to *classification* tasks, i.e., tasks of *classifying* things into classes, rather than to identification tasks. However, once one accepts the idea that a subject retains a number of exemplars for a single object, it is straightforward to extend the methods to the case of identification. In fact, Nosofsy (1988) suggests that we do retain multiple exemplars for a single object. See also Murphy (2002), pp. 58–60, for a discussion of Nosofsky's result.

[3] One important omission is the role of *contexts* in identification tasks. Undoubtedly we appeal to various contextual factors in tackling an identification task. You can recognize your colleague more easily and decisively in your office than on a beach in Okinawa, and the place, in this case, plays a crucial role. I this paper, however, I ignore any such effects of contexts just for the sake of simplicity.

experience. If an exemplar actually comes from John, let us call it a *John-exemplar*. The matter would be simpler if we could take a subject's concept of John to simply consist of John-exemplars, namely, exemplars obtained by the subject's actual encounter with John. But the subject may have made mistakes in identification tasks. Accordingly, some of the exemplars in the subject's concept of John would not be her memories of John.

According to this model, one tackles an identification task by accessing already-known exemplars of objects. If the perceptual appearance of a person in front of you shows stronger overall *similarities* to exemplars in your John-concept than to exemplars in other concepts, and the degree of the total sum of its similarities to the exemplars in your John-concept is high enough, then you will judge him to be John. Note that your John-concept does not have to have an exemplar that exactly *matches* the person in front of you. All you need is that its exemplars are similar enough to him.

Exemplar theory provides a simple model of how we manage to do *pattern recognition*. In fact, according to the simple model under consideration, identification is just a matter of pattern recognition. However, for this simple model to have even a limited initial plausibility, we need to make a number of assumptions about the nature of objects to be identified and its environment. For example, the ways an object perceptually appears to a subject must form a pattern that is distinguishable from the patterns other objects show to the subject. Moreover, the kind of process envisaged here works well only when the relevant objects weakly retain their properties over time. They can change, but not so radically and quickly. If one wants to deal with identification of butterflies over time, for example, one will need to appeal to some extra mechanism to handle critical transition points. Considerations of a case like this might lead one to severely limit the role of pattern recognition in explaining identification tasks, or even to dismiss it at all. Admittedly, the ways a human forms and revises his concepts seem extremely complex. Philosophers and psychologists have emphasized the roles of the theory and knowledge he has of an object, and the inferences he makes from and to his beliefs about the object. These factors do seem to be relevant. I still believe, however, it is much closer to the truth to say that in the vast majority of cases in which we identify our friends and colleagues in every-day occasions, what we rely on is a kind of pre-rational, non-inferential exercise of our ability to recognize patterns than to say that it is a kind of theoretical inference or reasoning.

Using exemplar theory, one can theorize about concept revision in varieties of ways, but the simplest one will be the model in which revision is done by just adding new exemplars. You will add the current perceptual information to your John-concept if you judge the person in front of you to be John.[4]

---

[4] The reader should not be misled by my use of a proper name in describing a concept. At this stage, I am assuming that proper names play no roles in identification-tasks or concept revision.

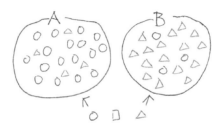

This simple model of what concepts are and how they are revised gives us a starting point for talking about reference. For the moment, let us confine ourselves to concepts that are formed purely perceptually. Suppose that you need to distinguish only two persons, John and Paul, and that they are perceptually distinguishable but not so clearly as to completely eliminate the possibility of misidentification. Suppose also that you have two concepts devoted to John and Paul.

We are assuming that both your John-concept and Paul-concept are inhabited by exemplars coming from objects. When you encounter an object, your John-concept will gain a new member if what you perceive now shows sufficiently strong overall similarities to the exemplars in your John-concept than to your Paul-concept, and similarly for your Paul-concept. No new member will be added to either concept if what you perceive is not similar enough to the exemplars in either concepts. We can think of this process as a kind of *game* between two concepts. In fact, with some simplifying assumptions, we can think of this as an instance of *population dynamics* widely studied in evolutionary game theory.

In this paper I cannot go into evolutionary game theory itself, but I can sketch the results of some typical situations. If you have two concepts, A and B, and they both comprise John-exemplars (50%) and Paul-exemplars (50%), then it is hard to predict the course of events that will await these concepts. However, if you start with the following:

A: John-exemplars (80%), Paul-exemplars (20%)
B: John-exemplars (20%), Paul-exemplars (80%)

we can predict that in each concept the majority will overwhelm the minority in the long run, due to the fact that the majority in one concept has more power to attract a new exemplar than the minority in the other. In this case, the John-exemplars in Concept A have more power to attract a new John-exemplar than the John-exemplars in Concept B, for the new John-exemplar is expected to show stronger overall similarities to exemplars in A than those in B. As a result, Concept A will stabilize as a John-concept (i.e., a concept whose exemplars are overwhelmingly dominated by John-exemplars) and concept B as a Paul-concept.

3   The Reference of a Concept

I think it reasonable to regard a concept, if it has stabilized as a John-concept in an environment involving John, as *referring* to John. But if a concept has not yet stabilized, but is intentionally stable, i.e., has a disposition to stabilize as a concept of a certain object, one might still find it natural to count it as referring to the object. A child who can distinguish cows from other animals *only roughly* may well be regarded as possessing a concept designating cows already. Or, one might want to say that such a chid has only a *premature* concept of cow, which does not quite designate cows. I believe intentional stability gives a good approximation of the notion of reference. But one can imagine a variety of cases, and I do not know where to draw a line. In fact, I do not know whether there is a line to be drawn at all.

Before discussing the reference of proper names, I wish to make a few remarks on the notion of intentional stability. Intentional stability is concerned with the dynamics of a concept's revision; it sees whether there is an object around which the concept's activities will center in the long run. In the first place, I wish to emphasize its *causal*, or *historical*, nature. Revision requires identification, and identification requires information collected *from* objects in the past. The notion of an exemplar *coming from* an object plays a central role in this account. In this sense, I take this to be a version of causal or historical account. On the other hand, it is also concerned with what a given system of concepts is *disposed to do* in a given environment. It looks at the future, not only the past, of the concepts. Thus, this account is *both* causal and dispositional. This seems a natural view to me; after all, concepts are "metal glue" (Murphy 2002) that ties our past experiences to the present and future ones.

I wish to emphasize its appeal to "descriptional" or "qualitative" aspect of this account. It uses the notion of *similarity* between exemplars, which are based on subjective representations of *qualitative features* of objects. In this sense, it might be regarded as a descendant of the "cluster theory of descriptions". In this account, though, qualities are required as the basis of pattern recognition. They do not have to be "true of" an object. This, I believe, makes its marked difference from the traditional versions of descriptivism.

I should also stress the environment-relative nature of this account. Whether a given concept of yours is intentionally stable with respect to a certain object depends on what other objects exist in your environment and how they appear to you. One can illustrate the point with "mistakes" made by a machine for recognizing handwritten digits:

$$4{\to}6 \quad 3{\to}5 \quad 8{\to}2 \quad 2{\to}1 \quad 5{\to}3 \quad 4{\to}8$$

$$9{\to}4 \quad 8{\to}0 \quad 7{\to}8 \quad 5{\to}3 \quad 8{\to}7 \quad 0{\to}6$$

From Murphy (2012), p. 570.

The leftmost example on the bottom shows that it is supposed to be an instance of "9" but the machine recognized it as a "4"; the rightmost one on the bottom is an instance of "0" mistakenly recognized by the machine to be an instance of "6", and so on. These examples show that successful pattern recognition requires that target objects be perceptually distinct. If people's hand-writings are *always* like these, *we* may not be able to form stable concepts, or conceptions,[5] of handwritten digits. This kind of example also shows that it is important to have a sufficient number of concepts. In this case, it shows that you had better have 10 concepts to deal with digits.

The account of reference presented here presupposes that the causal and the dispositional aspect of a concept are in harmony. I believe such cases are the home of our notion of reference. However, philosophers have spent enormous energy discussing cases where they *diverge*. Perhaps, your concept of Paul consists of exemplars coming from Paul on the earth, but now you are condemned, unbeknownst to you, to apply it to Paul on the twin-earth. Does your concept refer to Paul, twin-Paul, neither, or both? My account is intended to tell us nothing about such cases. This is another sense in which my account is just part of a picture. Again, however, I am not sure whether there should be a definite answer to such a question. Taxonomists do not know exactly what the fish is (Yoon 2010), but marine scientists generally do not miss the definition of "fish" in doing research about fish. Should philosophers, or linguists, really miss an exact theory of reference?

4   Names

I believe perceptual concepts discussed so far form the basis of children's learning of proper names. In many cases, a child learns a name *as* the name of this or that object of which she has a perceptual concept. Then she will begin to use the name to ask questions, express her desires, and insist her opinions about the object. A name will give her new ways of gaining information about the object, and new ways of revising the concept.

As I said, the ways a human forms or revises his concepts are extremely complex. Nonetheless, I believe *pattern recognition* plays a role in linguistically mediated ways of shaping concepts. In listening to others or reading a text, we often come across a piece of information intended to be *of* a particular object. If you are interested in whether the object talked about is one of those objects you know well, you are already working on an identification task.

But how do you know whether a linguistically given piece of information is intended by the speaker to be about an object you know well? A use of a name in the discourse will be a strong clue, but it is not decisive: two things can have the same name, and one thing can have two different names. Descriptive fit or unfit can also be a clue, but they are not decisive either: your concept may contain a false description, or the present utterance may be misdescribing the object. After all, *nothing* seem to be decisive.

Although none of these factors may be decisive individually, they may jointly give you a sufficient clue to accessing the *similarities* between the item talked about and the things you know.

---

[5] Note that I am using the term "concept" in a broader sense here than in other parts of the text. I hope what I mean here is clear enough.

I could not find a very apt example of a proper name, but here is an example of a natural kind term from Aristotle's *History of Animals*:

> The bonasos occurs in Paeonia on the Messapian mountain. Its size is that of a bull. Its form is generally like an ox except that it has a mane down to the shoulder like a horse. ... Their voice is like an ox; their horns are crooked, curved towards each other and not useful for defence ... blue. On being hit, ... it defends itself by kicking and by voiding dung at them, throwing it up to four rods [7 m] from itself; it uses this easily and frequently, and it scorches the hounds' coats so that they rub off (630a).

Here an animal called "bonasos" is the topic. The name is unknown to us, and the descriptions given do not quite fit any animal we know today. But scholars of Aristotle think that this animal is the *bison*. No animals we know defend themselves by throwing poisonous dung at the enemies, so perhaps the bison in the ancient time was a bit different from the modern one. Or—and this is more likely—perhaps this part was just a misdescription, due to Aristotle himself, or to someone else in the chains of communication through which these pieces of information have been transmitted to him. A misdescription or not, they could still recognize it as a talk about the bison. And I believe that the *overall similarity* it showed to what they knew about the bison played a crucial role in this recognition.

Incidentally, I also wish to add that no historical chain of a single name is involved here; the name "bonosos" had ceased to be in use before it reached the scholars.

I claimed that the notion of intentional stability, applied to perceptually formed concepts, gives a useful approximation of the notion of reference. The central idea was that a perceptual concept can be said to be *of* a certain object even when it is "contaminated" with exemplars from other sources, if they are destined to remain minorities so that they do not significantly affect the activities of the concept.

However, most of our concepts are not purely perceptual. My concepts of my close friends in part consist of perceptual information, and in part information I gained from conversations with people, including themselves. My concept of Socrates entirely consists of information I obtained from reading books and hearing what others say about him.

In order to apply the notion of intentional stability to such concepts, I need to make clear:

1. What are exemplars of objects in such concepts?

and

2. What is it for such concepts to be intentionally stable with respect to a certain object?

Let us say that a *linguistic exemplar* is that piece of information extracted from a discourse which is *intended* by the speaker to be *of* an object—an object of which the speaker has a concept. This notion—the notion of an utterance, or information, being *intended to be of a thing the speaker already has a concept of*—is central to my account.

I believe this notion is at least hinted when Kaplan says, in explaining Donnellan's ideas, that having something *in mind* can "guide" one's utterance, or that having something *in mind* can be "transmitted" from one person to another. (Kaplan 2012, p. 129.) If someone already has a concept of a particular object, and part of his speech is "guided" by the concept so that it is to be *of* the object, you can get a linguistic exemplar of the object by understanding his speech. This notion of *being a linguistic exemplar of* an object is partly causal: it requires someone who is already in possession of a concept of the object. As is the case with perceptual concepts, it also has a qualitative aspect: it is used in assessing similarity.

In the previous section, we saw that perceptual concepts require perceptual distinctness of their target objects for their stability. Likewise, in order for you to be able to form concepts of objects from what people say, your *linguistic environment* must satisfy some requirements, namely, linguistic exemplars of single object in a given community must be largely in agreement.

This means that, in a given community, what people *say* about a single object should be coherent enough to allow successful pattern recognition. The Hesperus-Phosphorus case shows that what people say about an object—Venus, in this case—need not be *completely* uniform, but if, say, people should randomly choose what they say about a given object, you would not be able to form a concept of the object from what they say.

There is also the requirement that linguistic exemplars of distinct objects in a given community must be dissimilar enough from each other. This too is needed for successful pattern recognition. I wish to emphasize that the use of proper names themselves contributes greatly to this requirement. Russell notoriously suggested that the name "Caesar" *abbreviates* the description "the person called *Caesar*". Without committing to his version of descriptivism, we can save Russell's insight by regarding the property of being referred to by "Caesar" as part of linguistic exemplars.[6] In some cases, this property alone may be enough to allow one to recognize who is being talked about, but as we noted, it may not be decisive in other cases.

On this account, properties stored in a given concept are used as the basis of pattern recognition. They need not be "true of" any object. In extreme cases, because of groundless rumors widespread in your community, your concept of a certain object may consist of properties most of which do not fit it. It can still

---

[6] After Russell, the idea that names themselves play certain roles in explaining the "cognitive significance" of sentences involving names has been taken up by a number of distinguished authors. See, for example, Burge (1973), Kaplan (1989), Kaplan (1990), and Perry (2012). My proposal is just an addition to this series of attempts. In this article, I cannot discuss the individuation of *names* or *words* themselves—the main topic of Kaplan (1990). I hope I can deal with this important issue in a future work.

be a concept *of* the object, for those false properties can help you to recognize others' intention, and to collect linguistic exemplars *of* the object. Although I am not sure how far such extreme cases can go—Can you imagine that your concept of Socrates turned out to be a concept of a refrigerator?—my account certainly does not require that most of what we believe about an object must be true of it. In this respect, it sides with Kripke's criticism of classical descriptivism.

However, this account does require that information about an object disseminated in a community form a *pattern*—a pattern recognizable by the speakers in the community. This, I believe, makes a marked difference from Kripke's picture. Here is what he says:

> An initial 'baptism' takes place. Here the object may be named by ostension, or the reference of the name may be fixed by a description. When the name is 'passed from link to link', the receiver of the name must, I think, intend when he learns it to use it with the same reference as the man from whom he heard it. (Kripke (1972), p. 96.)

Presumably, according to this picture, my use of "Socrates" refers to Socrates because it leads to his "initial baptism" by following the long chain through which the name has been passed to me. Here Kripke talks only of the passage of a name, but not of *information* or *opinions* conveyed by using the name. However, if I am right that, in order for me to be able to form a concept of Socrates from hearing others, what people say about him must form a pattern, they need to largely coordinate with respect to the *opinions* they have about him. It is hard to believe that the kind of causal links Kripke appeals to provides us with the required coordination.

To explain my skepticism about Kripke's account, let me give you a simple exercise in the theory of stochastic processes. Imagine a sort of "broken telephone" game. A player is supposed to transmit a certain piece of information about Socrates to the next player. The message to be transmitted is one of two messages: "Socrates had a snub nose" and "Socrates had a hooked nose". Suppose that from one player to the next, the message will be transmitted correctly with probability 0.99, but with probability 0.01, for whatever reason, the hearer will misunderstand the message so that the other message will be transmitted. If you start the game with the message "Socrates had a snub nose", what is the probability that the 50th player will receive the message correctly? What about the 100th player?

The answer is 0.685 (50th) and 0.567 (100th).[7] In fact, the "fixed point" of this process is 0.5 and it converges to the fixed point rather quickly. In the long run, no matter which message you start with, players will receive one of these messages with just a 50% chance. Admittedly, this is too simplified a model

---

[7] See, for example, Feller (1970), p. 432 for a general formula calculating these probabilities. In this case, the matrix of transition probabilities P is $\begin{pmatrix} 0.99 & 0.01 \\ 0.01 & 0.09 \end{pmatrix}$, and $P^{49} = \begin{pmatrix} 0.685 & 0.314 \\ 0.314 & 0.685 \end{pmatrix}$, and $P^{99} = \begin{pmatrix} 0.567 & 0.432 \\ 0.432 & 0.567 \end{pmatrix}$.

of actual communications, but I hope the moral is clear. Chains of message-transmissions are not very reliable.

I wish to hasten to add that I am not claiming that Kripke's causal chains do not exist, or that they are unimportant. I am just suggesting that something else is needed to explain why we still talk of Socrates in a sufficiently coherent and uniform way, hence still talk of him at all.

I can think of two possible explanations.

One is that people have talked to each other in such a manner as to make their opinions converge. Our daily conversations are not just transmissions of information. We sometimes refuse to accept what others say, try to persuade that they are wrong, and try to reach a conclusion with which participants in the conversation can agree. This kind of mutual correction may have kept what people say about Socrates roughly in harmony.

The other possibility is the existence of *texts*. Kripke's picture makes our use of the name "Socrates" look as if it is a very distant descendant of its use in ancient Greece. Although it *could* be true of "Socrates", it need not be. We have *texts* written by Plato and others, and we can consult them when we have disagreements about Socrates' doings. The texts themselves may contain a number of mistaken descriptions about him, but the existence of such "sacred texts" certainly seems to have prevented people's opinions about Socrates from diverging in hopeless ways.

Presumably, in actual cases, both these factors play roles in sustaining our use of a name, but in the case of the name of a person who died long ago like Socrates, the latter factor—the existence of "sacred texts"—seems to play a more important role.[8]

References

Burge, T.: Reference and proper names. J. Philos. **70**, 425–439 (1973)

---

[8] This is basically a mere transcription of what I read out at the meeting of LENLS 13 at Keio University in November 2016. In March 2017, I had an opportunity to present the material at Reed College in Portland, Oregon. I would like to thank the participants of both these occasions—to name a few, George Bealer, Robin Cooper, Troy Cross, Paul Hovda, and William Tasheck—for helpful comments and questions. Originally my plan was to revise and augment the arguments given above partly in response to the comments and questions, but soon it became clear that such a revision would require far more space and time than is allowed for in this post proceedings. I hope in a longer version of the views presented here I can do justice to them.

I wish to thank Naoya Fujikawa, Koji Minehima, Masahide Yotsu, and other members of the study group of the philosophy of language at Tokyo Metropolitan University for helping me to shape my ideas in earlier stages. I also wish to thank Jih-wen Lin for his technical assistance on questions about game theory. Special thanks should go to David Kaplan and Paul Hovda. They read earlier drafts of this paper and gave me invaluable comments and questions, as well as suggestions about how best to conventionalize my English.

Evans, G.: Causal theory of names. Aristot. Soc. Suppl. Vol. **47**, 187–208 (1973)

Feller, W.: An Introduction to Probability Theory and Its Applications, vol. 1, 3rd edn. Wiley, New York (1970)

Kaplan, D.: Quantifying in. Synthese **19**, 178–214 (1968)

Kaplan, D.: Afterthoughts. In: Almog, J., Perry, J., Wettstein, H. (eds.) Themes from Kaplan. Oxford University Press, Oxford (1989)

Kaplan, D.: Words. Aristot. Soc. Suppl. Vol. **64**, 93–119 (1990)

Kaplan, D.: An idea of Donnellan. In: Almog, J., Leonardi, P. (eds.) Having in Mind: The Philosophy of Keith Donnellan. Oxford University Press, Oxford (2012)

Kripke, S.: Naming and Necessity. Oxford University Press, Oxford (1972)

Medin, D., Schaffer, M.: Context theory of classification learning. Psychol. Rev. **85**(3), 207–238 (1978)

Millikan, R.G.: On Clear and Confused Ideas: An Essay about Substance Concepts. Cambridge University Press, Cambridge (2000)

Murphy, G.L.: The Big Book of Concepts. The MIT Press, Cambridge (2002)

Murphy, K.P.: Machine Learning: A Probabilistic Perspective. The MIT Press, Cambridge (2012)

Nosofsky, R.: Similarity, frequency, and category representations. J. Exp. Psychol. Learn. Mem. Cogn. **14**, 54–65 (1988)

Nosofsky, R.: The generalized context model: an exemplar model of classification. In: Pothos, E., Wills, A. (eds.) Formal Approaches in Categorization. Cambridge University Press, Cambridge (2011)

Perry, J.: Reference and Reflexivity, 2nd edn. CSLI Publications, Stanford (2012)

Recanati, F.: Mental Files. Oxford University Press, Oxford (2013)

Yoon, C.K.: Naming Nature: The Clash Between Instinct and Science. W. W. Norton & Co Inc., New York (2010)

Negotiating Epistemic Authority

Elin McCready[1(✉)] and Grégoire Winterstein[2]

[1] Department of English, Aoyama Gakuin University, Shibuya, Japan
mccready@cl.aoyama.ac.jp
[2] Department of Linguistics and Modern Language Studies,
The Education University of Hong Kong, Tai Po, Hong Kong
gregoire@eduhk.hk

Abstract. Why do we trust what other people say, and form beliefs on the basis of their speech? One answer: they are taken to have *epistemic authority*. Intuitively this means that the other person (or institution, or group) is taken to be authoritative in what they say, at least with respect to a particular domain. Here, we want to claim that there are (at least) two varieties of epistemic authority, one based on reliability and one on assuming (nonepistemic) authority. We claim that both are subject to linguistic negotiation. This paper begins by reviewing McCready's (2015) theory of reliability, and then turns to strategies for attempting to assume epistemic authority, focusing on those involving the use of not-at-issue content. We then show the results of two experiments which test the interaction of stereotypes about gender with epistemic authority, and how this is mediated by language use, focusing on the case of gendered pronouns. Finally, the results are explored for Bayesian views of argumentation and analyzed within McCready's Reliability Dynamic Logic.

1 Introduction

Why do we trust what other people say, and form beliefs on the basis of their speech? One answer: they are taken to have *epistemic authority*. Intuitively this means that the other person (or institution, or group) is taken to be authoritative in what they say, at least with respect to a particular domain. Here, we want to claim that there are (at least) two varieties of epistemic authority, one based on reliability and one on assuming (nonepistemic) authority. We claim that both are subject to linguistic negotiation. This paper begins by reviewing McCready's (2015) theory of reliability, and then turns to strategies for attempting to assume epistemic authority, focusing on those involving the use of not-at-issue content. We then show the results of two experiments which test the interaction of stereotypes about gender with epistemic authority, and how

The authors would like to acknowledge the support of JSPS Kiban C Grant #16K02640, which partially supported this research, and to thank Heather Burnett, Christopher Davis, Michael Erlewine, Regine Lai, Zoe Luk, Makoto Kanazawa, Lukas Rieser, Henriette de Swart, Christopher Tancredi, and audiences at LENLS13, the Education University of Hong Kong, Keio University and the University of Paris 7.

S. Kurahashi et al. (Eds.): JSAI-isAI 2016, LNAI 10247, pp. 74–89, 2017.
DOI: 10.1007/978-3-319-61572-1_6

this is mediated by language use, focusing on the case of gendered pronouns. The first experiment concerns English and the second Cantonese. Finally, the results are explored for Bayesian views of argumentation.

2   Passive Assumption of Authority

One way to be authoritative, in the sense of having one's speech consistently believed, is to be a speaker who is judged reliable with respect to speaking truth. If one is judged reliable, one is likely to have a kind of epistemic authority, in the sense that the things one says are likely to be believed. Here, reputation is key given that belief is a form of cooperation; it is known, for the game-theoretic case that the use of reputation in strategizing in repeated Prisoner's Dilemma [18,19] yields extremely good results, and is therefore likely to be evolutionarily stable.

One way to model reputation with respect to reliability is given by [13], which we will briefly summarize. On this theory, reputations can be derived in part from *histories*, defined as sequences of objects $act \in A, A$ the set of possible actions for a given agent in a given (repeated) game. These objects are records of an agent's actions in past repetitions of the game. Game histories are n-tuples of sequences of records representing the history of the agent's actions at each decision point. For the case of communication, these are of course histories of speech acts. A player's reputation in a game is derived from his history in that game. A player's reputation with respect to some choice is defined as his propensity, based on past performance, to make a particular move at that point in the game. Such propensities are computed from frequencies of this or that move in the history. Specifically, the propensity of player a to play a move m in a game g at move i is: the proportion of the total number of game repetitions that the player chose the action m at choice point i.

$$F_{H_a^{g,n}}(move) = \frac{card(\{act \in H_a^{g,n} | act = move\})}{card(H_a^{g,n})}$$

Always, $0 \leq F_{H_a^{g,n}}(move) \leq 1$, so the above number can be viewed as a probability: in effect, the information that the game participants have about a's likelihood of choosing move m.

An agent's propensity to play a strategy is a real number in $[0,1]$. This fact supports a scalar view of propensities, and indeed of cooperation itself: an agent has a propensity for using strategy σ iff the frequency exceeds the contextual standard for having that propensity [9]. Thus,

$$Prop(a, \sigma) \text{ iff } F_{H_a^g}\sigma \succ s,$$

where s is the contextual standard for propensity-having. These propensities can also be used to decide whether to assign someone epistemic authority with respect to some claim. In the context of the repeated prisoner's dilemma, [18, 19] make use of reputations and find that there are optimal strategies, given an index of reliability (here in $[0,1]$, but for them in the range 1–5), involving trust if, for example, a has a propensity for reliability (where this for them amounts to setting some arbitrary number above which cooperation is dictated), or if $\sum_{Coop(\sigma)} F_{H_a^g} \sigma$ is above some threshold (not necessarily s) (for the sum of frequencies of all a's cooperative strategies), or if the other agent's reliability index is higher than the choice maker's. Since such strategies are public, the other agent has an incentive to maintain her R-rating high: i.e. to genuinely be reliable. Any of the above seem reasonable bases to choose to accept someone's epistemic authority, or not.

The above must be combined with other information about reliability. This is so because of the need to decide whether to give someone epistemic authority even in the first communication, before any kind of history is available. This decision corresponds closely to the distinction between Humean and Reidian views on trust in testimony [13,15]. One way to model the Reidian view, on which decisions about trust aren't made automatically but rather on the basis of some metric, is that of [4], who takes speakers to make judgements about people's epistemic authority based on stereotypical information about factors like their gender, race, occupation, and personal grooming. This seems sensible: one might be more likely to believe a clean-shaven man in a suit about his having had his wallet stolen and needing money for the train than the same statement made by a homeless woman carrying a bottle in a brown paper bag (depending of course on one's other beliefs). This heuristic gives a first guess about reliability which can then be modified by interaction.

All this can be embedded in a more general model of information change; [13] proposes a new flavor of dynamic semantics for this purpose [6]. The basic idea is to virtually always update with content acquired from any source, but only 'conditionally.' To make this work, information states σ are complex and consist of possibly many substates. Each IS is a set of worlds (simplification), ordered with a 'plausibility ranking' reflecting epistemic preferences on states. Each substate is indexed by an index $j \in \mathsf{Source} \cup \mathcal{A}$. Here Source is the set of evidence sources and $\mathcal{A}$ the set of agents, which are constrained to only hold indices which the epistemic agent has had experience with. This set is ordered by a total ordering $\preceq_a$, where $i \prec j$ iff $P(Rel(i)) < P(Rel(j))$, when $P(Rel(i))$ is the probability that source i yields reliable information.

Updates are of the form $E_i \varphi$, for E_i an operator indicating source in i-type evidence. A sentence $E_i \varphi$ always induces update of state σ_i. Some cases are indeterminate cases, such as the use of direct evidentials in some languages that have them, where it may not be clear what the source is: visual, auditory, ... In such cases, all possible substates are updated. But in the testimonial case, states indexed with agentive sources a are updated. So, at the level of substates, update with φ always takes place when φ is observed—but this is *not* the same as coming to believe φ at a global level. Global beliefs are defined on the global state σ_T resulting from unifying all substates σ_i. This unification is done via a merge operation ($\cap$): all substate content survives when non-contradictory, but in case of conflict, information from higher-ranked sources trumps lower-ranked source-indexed information. Thus the global state almost never exhibits conflicts; it only will if two sources are precisely equally ranked, which is unlikely given the range of real values, but can be explicitly banned by enforcing a version of Lewis's Limit Assumption, here for sources rather than worlds [10].

More formally, global information states σ consist of sets of elements (substates) of the form $\sigma_i = \langle X, \leq_a \rangle$ where $X \subseteq W$ (the set of states). The substates are plausibility frames in the sense of [1,2]: multi-agent Kripke frames $\langle X, R_a \rangle_{a \in \mathcal{A}}$, where the accessibility relations R_a are called 'plausibility orders', written $\leq_a$, and assumed to be locally connected preorders. This simplifies a bit: sometimes the substates can be more complex, in particular in the case of testimonial agents, as the substates associated with them also have a similar structure. Total information states are written σ_T, and are of the form $\langle X, \leq_a \rangle$ for $X \in \wp(W)$. They are derived by recursively merging all plausibility relations found in $\sigma_i \in \sigma$ via a lexicographic merge operation, which respects priority ordering; so an agent's beliefs thus are derived on the basis of the most reliable source, and so on down the source hierarchy. From this, we get resolution in cases of conflicting sources.

Update in this system follows the $[.]_{\Uparrow}$ of [1,2], defined as follows.

$\sigma[\varphi]_{\Uparrow} = \sigma'$, where $S' = S$ and $s \leq'_a t$ iff either (i) $s \not\in \varphi$ and $t \in s(a) \cap \varphi$, or (ii) $s \leq_a t$.

This definition thus leaves the set of states the same, but upgrades those states which satisfy φ above those which don't, otherwise leaving the relative plausibilities untouched. Using this operation ensures that substates will be comparable without recourse to revision.

Support and entailment are defined as follows. A total information state $\langle X, \leq_a \rangle$ is said to *support* a proposition φ, $\sigma \models \varphi$, iff $\{s \in X | s \in best_a(s(a))\} \subseteq \phi$, where $best_a \phi := \{s \in \phi | t \leq_a s \text{ for all } t \in \phi\}$.[1] The definition of entailment is the standard fixed-point dynamic one modulo the use of $[.]_{\Uparrow}$, as defined above (with ';' dynamic conjunction as usual):

$$\phi_1, \ldots, \phi_n \models_\sigma \psi \text{ iff } \sigma[\phi_1]; \ldots; [\phi_n] = \sigma[\phi_1]; \ldots; [\phi_n]; [\psi].$$

Evidential update is defined via the following clause, which ensures that only the substate corresponding to the information source is updated, and all others are left alone.

$$\sigma[\mathsf{E}_i \varphi] = \sigma' \text{ where, for all } \sigma_j \in \sigma, \begin{cases} \sigma'_j = \sigma_j[\varphi] & \text{if } i = j \\ \sigma'_j = \sigma_j & \text{if } i \neq j \end{cases}$$

For an example, suppose agent a learns $\varphi = $ 'It is raining' from evidence source b (agent b). Then: $\sigma' = \sigma$ except that $\sigma'_b \in \sigma' = \sigma_b[\varphi]$, by the definition of evidential update.

---

[1] Note that this is essentially identical to the definition of belief in [2].

Thus: in all cases, the result of evidential update with φ is belief in φ. But this belief may just be belief relative to the source, i.e. within σ_i for source i. 'Genuine' belief requires global belief wrt the global state. Essentially: $B_a\varphi$ iff $\{s \in \sigma_T | s \in best_a(s(a))\} \subseteq \phi$, where $best_a\phi := \{s \in \phi | t \leq_a s \text{ for all } t \in \phi\}$. The total belief state is derived by lexicographic merge, so the content of our examples will be believed unless some higher-ranked source disagrees. What happens when a conflict arises? Consider a case of conflicting agents. Agent a claims ϕ and agent b claims $\neg\phi$. a, let's suppose, is pretty trustworthy. b is unknown; let's suppose that he looks somewhat untrustworthy. The result is that $a \succ b$ in the priority ordering for lexicographic merge. Thus the merge of σ_a and σ_b verifies ϕ.

So far: update of substates, substates unified via merge, merge priority determined by ordering. But what's the source of the ordering? Without a substantive theory of how the ordering is derived, the theory seems to have little empirical content. The claim of [13] is that the ordering is probability-based. The probabilities in question are probabilities of *reliability*. They indicate the (perceived) likelihood that information derived from the source is correct.

These probabilities arise from two factors. The first factor is experience with reliability of the source, as derived from histories; the second is the initial probabilities of reliability. These come in two types: prior beliefs about the reliability of different evidence sources, and beliefs about the reliability of the providers of testimony based on various aspects of their presentation. For an example of the first, one generally can take direct evidence to be more reliable than hearsay: if I see that it's raining outside, I am likely to discount the fact that this morning's weather report said it would be sunny. For the second, as mentioned above, judgements about the reliability of individuals are often made on the basis of stereotypical factors about their appearance and how they are categorized [4]. One might judge the kempt to be more reliable than the unkempt, the professional to be more reliable than the amateur, or someone from the same social group as you to be more reliable than someone from an outgroup. As we'll see in the next section, these kinds of judgements can be manipulated, yielding effects on the attribution of epistemic authority.

The two factors are taken to interact as follows: given an initial probability and a sequence of events of information acquisition, conditionalize on the initial probability for each new acquisition event, with respect to truth-tracking. The idea is to modify the probability that the source is reliable based on whether the new information is correct or not:

$$\frac{P_I(R \cap C)}{P_I(C)}$$

The whole notion of authoritativeness analyzed here is (in a sense) a passive one. One becomes authoritative by speaking the truth and by looking reasonably trustworthy. This is a kind of authority acquired by being a good citizen in the testimonial sense, essentially that of [5]. But is there a more active way to acquire epistemic authority by linguistic means? We think yes: by use of argumentative and other linguistic devices. Some of these will be explored in the next section.

3   Using Expressive Content for Authority Negotiation

How can one actively try to acquire epistemic authority (or deny it to others), as opposed to simply acquiring it by living a virtuous testimonial life? One way, of course, is just to assert one's authority:

(1) (You should believe me because) ...
  a. I know all about this topic.
  b. I'm your teacher.
  c. I'm your dad.

This strategy will be effective to precisely the degree that the speaker already has epistemic authority, because in the absence of epistemic authority, either the hearer won't accept what is said (1a), or the speaker's external authority is already rejected (1b, c). Consequently, a less direct strategy (or set of strategies) is needed. In the remainder of this section, we examine the use of expressive content [21] in the assumption of epistemic authority, considering several cases.

We are choosing to focus on expressive content for two reasons. Expressive content is often talked about as 'inflicted' on the hearer [16,21], which means (if correct) that the content of the expressive cannot easily be contested. This is an important feature when it comes to manipulations of epistemic authority (and in argumentation in general), as it removes the need to have epistemic authority already in order to have one's claims accepted, as with (1) above. This feature is not universally present in not-at-issue content either; [26] notes that presuppositions for example can be challenged in discourse, meaning that their content lacks the key feature of expressives we are interested in here. The second reason is the close connection of many expressives to social meanings, which are obviously relevant for epistemic authority. This point will be detailed as we proceed.

In this section, we will briefly consider the cases of particles, honorifics, and, finally, our main concern, those expressives which serve to indicate membership in various social groups.

First, particles like the Japanese *yo* (with falling intonation) work to try to 'force' the hearer to accept the content of the sentence [3,11]. Indeed, [17] presents an analysis of this particle in terms of epistemic authority. His idea is that *yo* indicates that the speaker has at least as much epistemic authority as anyone else present with respect to the content of the sentence. This implies that the particle can be used strategically to try to claim such epistemic authority for the speaker; use of the particle (if unchallenged) indicates that the speaker already has epistemic authority.

This view has some empirical effects. In the following example, the speaker requests belief via the claim of teacherhood.

(2)   watashi-wa   anata-no   sensei desu   yo
       1P.Formal-Top 2P.Formal-Gen teacher Cop.Hon PT
       'I am your teacher, don't forget.'

However, the use of strengthening *yo* implicates that the speaker doesn't have authority already, which further implies that the speaker takes his epistemic authority qua teacher to be insufficient, resulting in a failed authority grab. Compare here the observation of [25] that falling *yo* infelicitous in e.g. instructions from a commanding officer in the army, because the attempt at claiming authority represented by *yo* (in the terms of this paper) is not compatible with the presence of absolute authority.

The second case is honorifics, which, although they on a separate dimension from epistemic claims (at least according to [8,12,14,22], and others), to the extent that one's social status influences her epistemic authority the use of (anti-) honorifics should count as a strategy for assuming it, or taking it from others. Notably: 'raising' the addressee could cede some epistemic authority to them. In terms of examples, while the following are both grammatical and felicitous, there is a sad mismatch between content, honorific tone and particle: it's as if the speaker is desperately trying to assert himself. This is unlikely to yield genuine epistemic authority.

(3)   watashi-no   itteiru koto-o shinjite kudasai   yo
        1P.Formal-Gen saying thing   believe please.Pol PT
        'Believe what I'm saying, please.'

vs. the pure authority grab:

(4)   ore-no   itteru   koto-o shinjiro
        1P.Inf-Gen saying thing   believe-Imp
        'Believe what I'm saying!'

Finally, many expressives tag aspects of character which can be relevant to determinations of epistemic authority via social status; we can call these *social expressives*. This strategy is less direct than the above in that it is entirely a side effect. The main method here is to ascribe other individuals membership in groups which are or are not privileged in a social sense, and use that (lack of) privilege to implicate something about their epistemic authority. The same is true for slurs: by placing the addressee or other individual in a subordinate group, explicitly or implicitly (cf. [24]), it becomes possible to emphasize one's own epistemic authority over them. It is widely noted in the feminist philosophy literature (and elsewhere on the internet etc.) that the overt or covert primary position of males in society, and their consequent authority, can lead to differences in epistemic authority as well. For instance, the claims of men are often believed over the claims of women, all else being equal. If this is true, the use of e.g. gendered 3P pronouns in situations where other options are available (cf. [23]) could lead to the changes in who is taken to have epistemic authority, meaning that the use of gendered language can be a strategy for its assumption.

Here, we are interested in testimony: the main question in ceding epistemic authority involves how one should assign probabilities of likely reliability to individuals.

As mentioned above, [4] cites one technique, which is to make use of stereotypes about groups, for example that 'women are not logical', 'Asians are well educated', and so on; she presents some compelling examples of such cases, though examples which operate at the level of at-issue claims rather than expressive implications. However, many expressives tag aspects of character which can be relevant to determinations of epistemic authority via social status. We can call these *social expressives*; they are mainly terms which categorize individuals into categories that—at least on a stereotypical or prejudicial level—are relevant to the (non)attribution of epistemic authority. The basic method is to ascribe other individuals membership in groups which are associated with some stereotype, and then use that (lack of) privilege to implicate something about their epistemic authority.

Two examples of social expressives are slurs and gendered language. By definition, slurs are negative and subordinating (cf. [24]), so can be used to emphasize one's own epistemic authority over categorized individual, given that other relevant individuals share the prejudices the slurs express. With gendered language, the situation is more subtle, because gender is not in any sense pejorative in the way of slurs. Still, the deployment of stereotypes about gender to acquire epistemic authority. It is a truism (and a common claim in feminist philosophy as well [4]) that the overt or covert primary position of males in society, and their consequent authority, can lead to differences in epistemic authority as well. For example, it is often said that the claims of men are often believed over the claims of women, all else being equal. If this is true, the use of e.g. gendered 3P pronouns could easily lead to the changes in who is taken to have epistemic authority, meaning that the use of gendered language can be a strategy for its assumption. In order to see whether this is correct, we conducted several experiments, focusing on the use of gender stereotypes in argumentation.

4   Experiments: Gender in Argumentation

4.1   Experiment 1: English

We ran an experiment to test the relation between gendered speech and epistemic authority in argumentation. We tested two different types of argument which involve the authority of a source: the direct, or abusive, form of the *ad hominem* argument and the argument from authority (or position to know). Schematically these arguments are as follows [27]:

- Ad-hominem:
  - Source a is a person of bad character/has bad character for veracity
  - a argues that α
  - **Conclusion:** α should not be accepted
- Argument from authority (position to know):
  - Source a is in a position to know about things in a certain subject domain S containing proposition α
  - a asserts that α is true
  - **Conclusion:** α is true

In each case, the source a is part of one of the premises of the argument.

The goal of the experiment was to test whether manipulating the gender of the source induces a difference in the convincingness of the argument. We followed a protocol similar to the one used by [7] to investigate the argument from authority.

First, a preliminary experiment was run to determine three distinct sets of topics according to their gender bias. This was done as a categorization task on Amazon Mechanical Turk. Participants were presented with a topic and asked to choose which category most closely matched that topic: Men, Women or Both. 17 topics in total were tested, out of which 15 were selected, 5 for each gender category. Each topic had an agreement of 80% or above, meaning that four participants agreed the topic was associated with the relevant category. Participants could categorize multiple topics and were paid 0.05 USD for each categorized topic.

These topics were then used to produce 15 distinct arguments, in two forms: the *ad hominem* one, and the argument from authority one. Examples of each form follow (using a male biased topic):

- Authority argument
  - A and B are friends. A wants to buy a power drill and is thinking about which one to buy. A wants a high performance drill to perform heavy duty work.
  - *A:* I wonder if this one is a good choice.
  - *B:* I have a friend who says he knows a lot about power tools, and he says this model is really powerful.
- *Ad hominem*
  - A and B are friends. A wants to buy a power drill and is thinking about which one to buy. A wants a high performance drill to perform heavy duty work.
  - *A:* I heard from Jamie that this model is really powerful.
  - *B:* She doesn't know anything about it.

The factors investigated in the experiment were thus the following three:

- Arg.type: the type of argument being used (two levels: ad-hom./auth.)
- Source: the gender of the source of the information, indicated by the use of a gendered pronoun (*he, she*) or *that friend/Jamie* to use a neutral reference (three levels: maleSrc/femSrc/neutSrc).
- TopicBias: the gender bias of the topic, based on the results of the preliminary categorization task (three levels: femB/maleB/neutB).

450 US-based participants were recruited on the Amazon Mechanical Turk and paid 0.2 USD for their participation. They judged the convincingness of 5 different arguments (4 fillers + 1 target item) presented in pseudo-random order. Convincingness was rated on a 5 points Likert scale. Linear mixed effect models with maximal random effect structure were fitted to the data using the lmer package in R. The effects of condition and group were confirmed by likelihood-ratio tests.

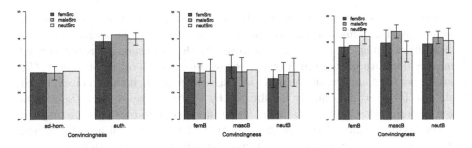

Fig. 1. Judgments of convincingness for `ArgType` vs. `Source` (left panel); `TopicBias` vs. `Source` for the *ad hominem* argument (middle panel) and authority argument (right panel).

Results. The results are shown in Fig. 1. They show that, generally, authority arguments are judged more convincing than *ad hominem* (Fig. 1, left panel, $\chi^2 = 145.38, p < 0.01$) and that the gender of the source and the gender bias of the topic have no main effect. Further analyses showed that these variables have no effect in the case of the *ad hominem* argument (Fig. 1, middle panel). However, the results of the argument from authority show that there is a significant interaction between the gender of the source and the gender bias of the topic (Fig. 1, right panel, $\chi^2 = 11.023, p = 0.026$). It was observed that, as expected, men are generally more trusted for topics biased towards men (in the `mascB` case, the difference between the masc-source and neutral-source is significant, $W = 168.5, p = 0.005$) but that women are not more trusted than men for topics biased towards women, and that there was no significant preference for neutral topics.

Discussion. To explain why authority arguments are preferred to *ad hominem* ones, we argue that when considering authority arguments the only question is how reliable the source of the argument is. The reliability of the speaker is not directly relevant. This can readily be integrated in an approach like that of [7,20] who propose a Bayesian treatment of argumentation. In that approach, the convincingness of an argument is proportional to how much the content of the argument affects the audience's prior belief in the conclusion targeted by the argument. The reliability of the source is factored in the likelihood of using an argument a to target a conclusion C. There the speaker's reliability remains constant across possible sources and does not weigh into the evaluation of the argument.

However, in the case of the *ad hominem* argument the speaker's reliability is at odds with that of the source, which might explain why those arguments are generally dispreferred since they pit the speaker's credibility against that of the source. We make the hypothesis that people are generally reluctant to overtly endorse arguments which directly attack other people's credibility, even though they might actually be unconsciously persuaded by them.

As stated above, gender biases can be integrated into the Bayesian approach of argumentation. This amounts to modifying the belief that the source is reliable by conditionalizing on its gender. However in the Bayesian approach the *ad hominem* and authority arguments are seen as dual to each other: one lowers the reliability of the source while the other increases it. As such, it should be expected that both forms would equally be affected by gender biases, contra the results of our experiment. One way to model that difference is to explicitly distinguish between the reliability of the speaker and that of the source of an information in the way an argument is evaluated, and account for the fact that they may potentially be at odds. The approach lends itself to such a modification, but further experimentation are needed to validate whether this move is an effective way to account for the data presented here.

4.2   Experiment 2: Cantonese

A second experiment similar to the one presented above was run using Cantonese material rather than English. This second experiment used a within-participants design, meaning that participants saw examples of each condition to be tested rather than just one single condition. The experiment aimed at reproducing the results of the English experiment with participants from a different socio-cultural background, and also attempted to overcome some of the flaws of the first experiment. First, the English experiment did not control for the stakes involved in the arguments. Some topics might have been interpreted as involving life or death situations (e.g. the safety of a car) while others were much more trivial (e.g. the authenticity of Japanese food). This was controlled for by only using topics which intuitively involved low stakes. Second, we ignored the case of neutral sources of information. This is because the value of such cases is difficult to interpret, as inaccessible participant biases, assumptions, and interpretation metrics may play roles in how the stimuli are processed. It is plausible that participants attributed a gender to the source matching the bias of the topic being discussed (e.g. male in the case of power tools), but there is no way to make sure of it. Third, since the *ad hominem* argument yielded no results and our analysis hypothesizes that it involves more complex reasoning, we only focused on the authority argument in this experiment. Fourth, the gender of the participant was also included in the analysis of the results.

As with the English experiment, a preliminary categorization task was run. Eleven voluntary participants, all native speakers of Cantonese, were recruited. They were shown 24 concepts paired with a property (e.g. the performance of a power drill) were shown to participants in Cantonese and they had to select the category which fitted the topic the best (Men, Women, Both). The 12 items with the highest agreement scores level were selected for the core experiment (4 in each category).

For the core experiment, we thus considered the following independent factors:

– Source: the gender of the source of the information (mascSrc/femSrc), marked by the use of gendered terms for older cousins (*biu2go1* 表哥 for male cousin, *biu2ze2* 表姐 for female cousin). Older cousins were chosen because they hold no intrinsic authority (unlike older siblings or parents who enjoy authority or younger siblings who lack it).

- `TopicBias`: the bias of the topic (`mascB/femB/neutB`), based on the categorization task.
- `GenderResp`: the self-declared gender of the respondent (`maleResp/femaleResp/otherResp`)

The experiment was run using an online questionnaire which contained 12 target items along with 24 fillers. Items and fillers were presented in a pseudo-random order with a latin-square design. 97 voluntary participants (64 `female`, 32 `male`, 1 `other`, mean age 27 years old) received a link to a questionnaire by e-mail or instant messaging. The questionnaire was hosted on the `IbexFarm` platform.

Results. Overall the results confirm the results of the first experiment (Fig. 2).

There is no main effect of `Source`: masculine sources of information were overall not judged more reliable than female sources. There is a significant interaction between `Source` and `TopicBias` ($\chi^2 = 6.8$, $p = 0.048$): female sources were overall less trusted for masculine biased topics, but male sources were not less trusted for feminine biased topics.

There is a marginal effect of `GenderResp` ($\chi^2 = 5.30$, $p = 0.07$), i.e. male respondents tend to give lower scores than female participants. There is furthermore a *significant interaction* bw `GenderResp`, `Source` and `TopicBias` ($\chi^2 = 36.74$, $p = 6.27e{-}05$): male respondents are more skeptical about male sources in the case of male oriented topics.

Discussion. The second experiment confirms the results of the first one: there is an interaction between the gender of the source of an information and the gender bias of the topic being discussed. However, there is an asymmetry between male and female sources. Male sources seem to enjoy an overall credibility, irrespective

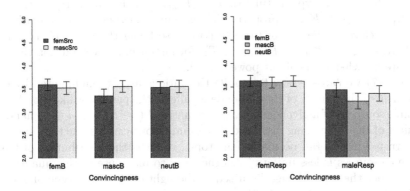

Fig. 2. Judgments of convincingness of the authority arguments: `TopicBias` vs. `Source` (left panel) and `GenderResp` vs. `TopicBias` (right panel).

of the gender bias of the topic, whereas female sources are mostly credible on female biased topics. An effect of the gender of the respondent was also observed, with male respondents being more critical in some conditions.

To show how to handle these observations, let's consider an example like (5) under the Bayesian view on argumentation mentioned above.

(5) I have a friend who says he knows a lot about power tools, and he says this model is really powerful.

Two distinct pieces of information are given about the source in (5):

– the friend is male: $i \in T_{\text{male}}$
– the friend knows about power tools: $i \in K_{\text{powertools}}$

Let's assume that agents are given a default probability of being reliable sources of information about a domain D: $P(R_{i,D})$ (possibly high in some cases, if we follow the charity assumption postulated by [20]). When observing that i is of type T_{male} we have (via Bayes' rule):

(6) $P(R_{i,D}|i \in T_{\text{male}}) = \frac{P(i \in T_{\text{male}}|R_{i,D}) \times P(R_{i,D})}{P(i \in T_{\text{male}})}$

$P(i \in T_{\text{male}}|R_{i,D})$ is the likelihood of being of male if the agent is assumed to be reliable. This can be seen as a measure of the judge's (in our case the participant's) personal biases, which might be linked to the gender of the respondent, e.g. men have a tendency to distrust other men in general (maybe because they believe they are more competent).

The same account allows to factor in both pieces of information given about the source in (5). Equation (7) shows how to integrate two pieces of information to update a prior belief on the reliability of the speaker.

(7) $P(R_{i,D}|i \in K, i \in T) = \frac{P(i \in K|R_{i,D}, i \in T) \times P(i \in T|R_{i,D}) \times P(R_{i,D})}{P(i \in K, i \in T)}$

Equation (7) expresses the posterior probability that i is reliable in domain D, knowing that i is of type T (in (5): $T = T_{\text{male}}$) and has property A (in (5): $K = K_{\text{powertools}}$). If K is a property that is typical of type T, the quantity in (7) is very close to the one in (6), the limit case being that $T \subset K$, in which case $P(R_{i,D}|i \in K, i \in T) = P(R_{i,D}| \in T)$ (for instance, the assumption that all males are knowledgeable about power tools).

Using (7) we can see how to handle the various aspects highlighted by our experiments. The gender of the respondents affects the perceived likelihood that a reliable source is of a given gender. The quantity $P(i \in K|R_{i,D}, i \in T)$ handles the effect of the topic being discussed. For example one can assume that men are more trusted in general, no matter the topic, whereas the distribution of trust for female sources is less uniform. Another way to make use of this quantity is to reconsider the topics being discussed in the light of the exact type of gender bias involved for each topic. As the discussion above made clear, there are two situations:

1. A bias corresponding to universal competence on the part of one gender but partial competence elsewhere (e.g. all women know about cooking but only some men do)
2. A bias corresponding to nonuniversal competence on the part of one gender, but lack of competence on the part of the other gender (e.g. only some men know about power tools, but no women do)

For a topic of the first type, the information about the competence on the topic should not have further effect on the credibility of the speaker since the gender information entails it. For a topic of the second type, the two pieces of information give convergent evidence that the speaker is reliable. As of now, the categorization task we used does not allow us to distinguish between the two types of bias. In future work, we will rely on a more complex categorization of each topic which will provide that information (for example by asking participants to indicate their intuitions about the proportion of men and women who are knowledgeable about the topic). Another judgment task will then be used to check whether the topics with a bias of the second type are judged differently (e.g. more convincing) than the ones of the first type.

5 Conclusion

This paper has considered the nature of epistemic authority and two methods for acquiring and modifying it. The first passive method involves being generally perceived as reliable; for an analysis, we reviewed the view of reliability of [13]— a combination of stereotype-based probability ascriptions and examination of communicative histories—and proposed it as one means of acquiring epistemic authority. The other method is more proactive: to manipulate stereotypes and other aspects of the context via the deployment of expressive content. We looked at one such instance in detail via experimental methods: the use of gendered pronouns to influence judgements about reliability, both in English and Cantonese. The results are intriguing, but still preliminary. We then proposed a Bayesian framework to account for the results.

Several directions suggest themselves for the future. The first, immediate steps involve additional experiments. We suggested above that not-at-issue content plays a different role in the manipulation of authority than at-issue content, because the efficacy of the latter already depends on the presence of epistemic authority. This difference remains to be experimentally verified, which we plan to do in the immediate future. Second, the experiments carried out so far involve subjective judgements and self-reporting tasks on the part of the experimental subjects. Avoiding potential biases, both implicit and explicit, which may be confounds for the experimental results is important and is a well-known problem in this area of research. There are several methods for addressing this problem, but the one we plan to pursue involves visual world experiments using eye-tracking; we intend to implement an experiment in this area in the near future. More generally, questions of the results on epistemic (and other) authority of the use of not-at-issue content are intriguing, especially for other sorts of expressive content such as honorification and particles; experimental approaches to these domains are also of interest, as is the examination of phenomena of other sorts such as presupposition and conversational implicature.

References

1. Baltag, A., Smets, S.: A qualitative theory of dynamic belief revision. In: Bonanno, G., van der Hoek, W., Wooldridge, M. (eds.) Logic and the Foundations of Game and Decision Theory. Texts in Logic and Games, no. 3, pp. 13–60. Amsterdam University Press, Amsterdam (2008)
2. Baltag, A., Smets, S.: Talking your way into agreement: belief merge by persuasive communication. In: Baldoni, M., Baroglio, C., Bentahar, J., Boella, G., Cossentino, M., Dastani, M., Dunin-Keplicz, B., Fortino, G., Gleizes, M.P., Leite, J., Mascardi, V., Padget, J.A., Pavón, J., Polleres, A., Fallah-Seghrouchni, A.E., Torroni, P., Verbrugge, R. (eds.) MALLOW. CEUR Workshop Proceedings, vol. 494 (2009). CEUR-WS.org
3. Davis, C.: Decisions, dynamics and the Japanese particle yo. J. Semant. **26**, 329–366 (2009)
4. Fricker, M.: Epistemic Injustice. Oxford University Press, Oxford (2007)
5. Grice, H.: Logic and conversation. In: Cole, P., Morgan, J. (eds.) Syntax and Semantics III: Speech Acts, pp. 41–58. Academic Press, New York (1975)
6. Groenendijk, J., Stokhof, M.: Dynamic predicate logic. Linguist. Philos. **14**, 39–100 (1991)
7. Hahn, U., Harris, A.J., Corner, A.: Argument content and argument source: an exploration. Informal Log. **29**(4), 337–367 (2009)
8. Harada, S.: Honorifics. In: Shibatani, M. (ed.) Japanese Generative Grammar, pp. 499–561. Academic Press, New York (1976)
9. Kennedy, C.: Vagueness and gradability: the semantics of relative and absolute gradable predicates. Linguist. Philos. **30**(1), 1–45 (2007)
10. Lewis, D.: Counterfactuals. Basil Blackwell, Oxford (1973)
11. McCready, E.: What man does. Linguis. Philos. **31**, 671–724 (2008)
12. McCready, E.: A semantics for honorifics with reference to Thai. In: Aroonmanakun, W., Boonkwan, P., Supnithi, T. (eds.) Proceedings of PACLIC 28, pp. 513–521. Chulalongkorn University, Thailand (2014)
13. McCready, E.: Reliability in Pragmatics. Oxford University Press, Oxford (2015)
14. McCready, E.: The semantics and pragmatics of honorification (2015). Manuscript, AGU
15. McCready, E.: Rational belief and evidence-based update. In: Hung, T.W., Lane, T.J. (eds.) Rationality: Constraints and Contexts, pp. 243–258. Elsevier, Amsterdam (2016)
16. Murray, S.: Varieties of update. Semant. Pragmat. **7**(2), 1–53 (2015)
17. Northrup, O.: Grounds for commitment. Ph.D. thesis, UCSC (2014)
18. Nowak, M., Sigmund, K.: The dynamics of indirect reciprocity. J. Theor. Biol. **194**, 561–574 (1998)
19. Nowak, M., Sigmund, K.: Evolution of indirect reciprocity by image scoring. Nature **393**, 573–577 (1998)
20. Oaksford, M., Hahn, U.: Why are we convinced by the ad hominem argument?: Bayesian source reliability and pragma-dialectical discussion rules. In: Zenker, F. (ed.) Bayesian Argumentation. Synthese Library, pp. 39–58. Springer, Netherlands (2013). doi:10.1007/978-94-007-5357-0_3

21. Potts, C.: The expressive dimension. Theor. Linguist. **33**, 165–198 (2007)
22. Potts, C., Kawahara, S.: Japanese honorifics as emotive definite descriptions. In: Proceedings of SALT XIV (2004)
23. Schlenker, P.: Maximize presupposition and Gricean reasoning. Nat. Lang. Semant. **20**, 391–429 (2012)
24. Stanley, J.: How Propaganda Works. Princeton University Press, Princeton (2015)
25. Kose, Y.S.: Japanese sentence-final particles: a pragmatic principle approach. Ph.D. thesis, University of Illinois at Urbana-Champaign (1997)
26. von Fintel, K.: Would you believe it? The King of France is back! In: Reimer, M., Bezuidenhout, A. (eds.) Descriptions and Beyond. Oxford (2004)
27. Walton, D.N., Reed, C., Macagno, F.: Argumentation Schemes. Cambridge University Press, Cambridge (2008)

Challenges in the Computational Implementation of Montagovian Lexical Semantics

Bruno Mery[✉]

LaBRI, Université de Bordeaux, Bordeaux, France
bruno.mery@u-bordeaux.fr

Abstract. We present and discuss a general-purpose implementation of the process of lexical semantics analysis theorised in the Montagovian Generative Lexicon ΛTY_n (hereafter MGL). The prototype software itself constitutes a proof of concept of the MGL theory. The implementation process, as well as the data structures and algorithms, also provide valuable results as to the expressive power required by MGL. While the implementation of terms and types for the purpose of meaning assembly assumed by MGL is in itself straightforward, some lexical phenomena require additional mechanisms in order to process the logical representation in order to take into account implicit common-sense world knowledge. We therefore also present a minimal architecture for knowledge representation, and how it can be applied to different phenomena. The implementation illustrates the validity of the theory, but MGL requires a stronger corpus of types and terms in order to be thoroughly tested.

Keywords: Lexical semantics · Montagovian Generative Lexicon · Knowledge representation for natural language semantics · Typed lambda calculus · Prototype software

1 Theories and Implementations of Lexical Semantics

Formal lexical semantics theories aim to integrate to the toolbox of compositional analysis of natural language developed since Montague considerations of (logical) polysemy. Based on original studies such as [5,10,27], then on the Generative Lexicon theory thoroughly developed in [29], there have been many formulations that build upon powerful type-theoretic foundations, with a generative, dynamic account of the lexicon at their heart. Such recent type-theoretic accounts of lexical meaning include Type Composition Logic (TCL) presented in [1], Dynamic Type Semantics (DTS) presented in [3], Type Theory with Records (TTR) presented in [9], Unified Type Theory (UTT) presented in [16], and the

---

Many thanks to the organisers, reviewers, speakers and participants of the LENLS 13 Workshop, in the proceedings of which the first version of this publication appeared as [20]. Thanks also to the director and adjunct directors of the LaBRI for supporting this research.

© Springer International Publishing AG 2017
S. Kurahashi et al. (Eds.): JSAI-isAI 2016, LNAI 10247, pp. 90–107, 2017.
DOI: 10.1007/978-3-319-61572-1_7

framework we helped define, the Montagovian Generative Lexicon (MGL), presented in [32]. Several partial or complete implementations of those theories have been provided for demonstration purposes. Those tend to use logical or functional programming, or theorem provers such as Coq—[8] is an example.

Concerning MGL, however, no real demonstration of the computational aspect has ever been provided. One of the stated goals of MGL was (paraphrasing slightly [32]) "to provide an integrated treatment from syntax to semantics extending existing analysers based on Montagovian semantics such as [25] with mechanisms for lexical semantics that *are easily implemented in a typed functional programming language like Haskell*". Until the present publication, however, such implementations have been purely hypothetical, except for domain-specific analogues such as evoked in [26]. Our goal in this publication is to present an actual prototype implementation of the lexical semantics of that framework. For that purpose, we have used functional and object programming in Scala, but the paradigm and language of programmation are not critical elements. We also want to analyse what MGL can do using the (quite simple) computational mechanisms involved, as well as how those should be supplemented in order to provide useful treatments of semantics and pragmatics.

We detail some of the necessary data structures and algorithms used, what we learned from this implementation on the underlying logic properties of MGL, and sketch an architecture for simple knowledge representation that is necessary for the representation of certain lexical phenomena. The demonstrably functioning prototype illustrates both the validity of type-theoretic formulations of lexical meaning (including and not limited to MGL), and the deep interaction of lexical meaning with at least some sort of knowledge representation already evoked in [6].

2   A MGL Prototype

2.1   The Montagovian Generative Lexicon

MGL is a type-theoretic account of compositional lexical semantics that uses a calculus of semantic assembly called ΛTY_n, an adaptation of the many-sorted logic TY_n (itself proposed in [28]) for the second-order λ-calculus, given in the syntax of Girard's System-F. MGL stays close to the usual Montague analysis by first performing a syntax-based analysis via proof-search, followed by the substitution of semantic main λ-terms to syntactic categories. Afterwards, lexical mechanisms are implemented in the meaning-assembly phase via a rich system of types based on ontologically different sorts and optional λ-terms that model lexical adaptation. The mechanisms, sketched in Fig. 1, are detailed in [24].

They can be roughly summarised as follows:

– First, the input utterance is super-tagged and analysed using categorial-grammar mechanisms. This is the only step of proof-search of the process, and yields a syntactic term whose components are syntactic categories. The lexicon is then used in standard Montagovian fashion to substitute semantic

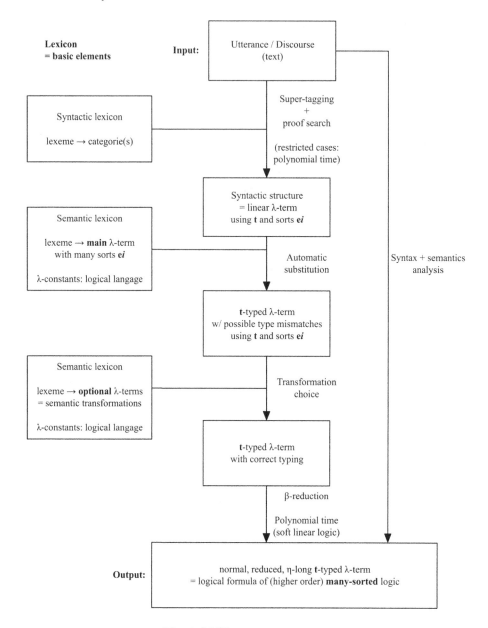

Fig. 1. MGL process summary

λ-terms to categories, yielding a main semantic term, typed with many sorts and the type **t** for propositions.

- Second, as a many-sorted logic is used, some type mismatches might occur. The direct correspondance between syntax and semantics is not guaranteed by the syntax as the arguments will be present in the correct number and position, but will not necessary be of the sort expected by the predicate. This allows the mechanisms of lexical semantics to engage and disambiguate between terms.

To that effect, the lexicon provides *optional* λ-terms that are used as *lexical transformations*. These optional terms are inserted depending on their typing, and yield a λ-term with no type mismatches.

– Finally, β-reduction yields a normal, η-long λ-term of type t (the type of propositions), i.e. a logical formula that can be used in any usual semantic theories of meaning, including as model-theoretic and game-theoretic semantics.

As the first step, syntax-based analysis, is already well-studied and implemented (we use Type-Logical Grammars and Grail for this step, as given in [25]), the object of concern is the second step: given a term reflecting the syntactic structure of an utterance and a semantic lexicon, to construct a semantic λ-term in a many-sorted logic, making use of available transformations, and yielding a suitable formula. This is the object of our prototype implementation.

2.2 Modelling Types and Terms

The data structures and algorithms responsible for implementing the *terms* and *types* of ΛTY_n are the core mechanisms of the software. They are given as two Scala sealed abstract classes, TermW and TypeW, with a flat hierarchy of case classes implementing the various possible term and type categories; this simple categorisation allows us to easily construct and detect patterns of objects.

Terms and types are constructed as binary trees (abstractions and applications of more than one argument to a given term/type can be easily curried):

– For terms:
  • Leaves are AtomicTerms (constants), TermVarIds (variables) with an identifier and type, or specific constructs for MGL, Transformations and Slots.
  • Inner nodes are TermBindings (λ-abstracted terms), or TermApplications of a predicate term to an argument term.
– For types:
  • Leaves are constant Sorts, pre-defined objects such as PropositionType for t, or second-order variable identifiers TypeVarIds.
  • Inner nodes are TypeFunctions between two types A and B (modelling $A \rightarrow B$), or TypeApplications (modelling $A \{B\}$).

A simplified UML class diagram presents this straightforward architecture in Fig. 2.

Several algorithms are provided as specialised methods and class constructors in order to work with types and terms. They are mostly simple recursive tree-walking algorithms, making the most of memoisation when possible (e.g., lists of available resources are incrementally built as terms and types are constructed in order to minimise computations). Algorithms include the type-checking of

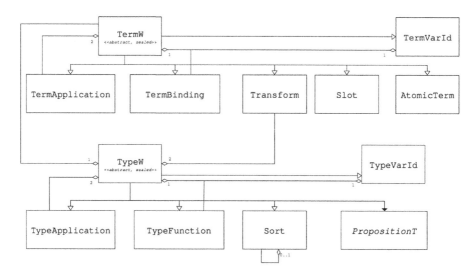

Fig. 2. Class diagram of the core package for terms and types.

applications, comparison between types, automated α-conversion of variables in order to prevent issues of scope, replacement of term and type variables, β-reduction, and the automated specialisation of types for polymorphic terms (i.e., a predicate with a type containing one or several type variables will be specialised to the correct types if applied to an argument with a compatible but specified type). In this prototype implementation, some degree of imperative programming is used in order to make some and optimisations easier.

It would be easy to convert the entire program to purely functional (or logical) programming. Most methods are linear in complexity (with exceptions in the adaptative mechanisms below); all algorithms are at most polynomial in time.

2.3   Explicit Adaptation

The core principle of MGL is to provide *transformations* as optional terms, on top of the main λ-term associated to each lexeme. A canonical example is *the book is heavy and interesting*. The usual, formal MGL analysis supposes three basic sorts:

- R for readable materials,
- φ for physical objects,
- I for informational contents;

the book can be modelled as the_book^R, *heavy* as heavy$^{\varphi \to t}$, *interesting* as interesting$^{I \to t}$. The example utterance is a case of *co-predication*, as two predicates are simultaneously asserted on two different *facets* (with different, incompatible sorts) of a same object, and MGL will resolve this by having the lexicon provide two optional terms associated with *book* in order to access these two facets: $f_{phys}^{R \to \varphi}$ and $f_{info}^{R \to I}$.

In order to apply the co-predication, a specific operator, the polymorphic conjunction *and*, is required:

$$\&^{\Pi} = \Lambda\alpha\Lambda\beta\lambda P^{\alpha\to t}\lambda Q^{\beta\to t}\Lambda\xi\lambda x^{\xi}\lambda f^{\xi\to\alpha}\lambda g^{\xi\to\beta}.(\text{and}^{t\to t\to t} \ (P \ (f \ x))(Q \ (g \ x)))$$

This yields, after suitable substitutions, application, and reduction, the term (and (heavy (f_{phys} book)) (interesting (f_{info} book))), which is normal and of type t.

In our implementation, there are several important differences with the formal account outlined above. As we distinguish between type constants and variables, there is no actual need to explicitly abstract (and Λ-bind) types. This is because the only second-order operation ever used in ΛTY_n is specialisation (i.e., the replacement – or instantiation – of type variables). Moreover, second-order (type) variables are all introduced by λ-bound first-order (term) variables.

We also distinguish between term variables that are necessary for the definition of an abstracted term (such as P, Q and x, the two predicates and the argument of the conjunction above) and *adaptation slots*, the positions where optional λ-terms (such as f and g above) can be inserted. This is because the optional terms can be provided by various different mechanisms, and might not be provided at all if the term is well-formed. There is no lexical adaptation taking place in utterances such as *heavy and black rock*; in that case, MGL provides an useful, if slightly redundant, *id* optional polymorphic term that can be inserted in order to get the identity on any type. Adding or removing adaptation slots is in fact a technical process analogue to type-raising or type-lowering in Montague semantics.

We provide optional terms (transformations) as `Transforms`, which are distinguished from other terms. Each term has a list of available transformations, constructed recursively from the leaves. The lexicon provides a list of transformations available to each lexeme, so that each atomic term has a collection of transformations. We also distinguish `Slots` for explicit adaptations; the list of slots is maintained during the construction of the terms. Our polymorphic *and* conjunction, also defined by the lexicon, then becomes:

```
lambda P^(B->t).lambda Q^(G->t).lambda x^A.((And^{(t->(t->t))}
   (P^(B->t) (f^{(A->B)} x^A))) (Q^(G->t) (g^{(A->G)} x^A)))
```

(As above, `f` and `g` are adaptation slots, distinguished from other variables and not bound, while `P` and `Q` are λ-bound predicate variables.)

During the attempted resolution of the application of the conjunction to terms for *heavy* and *interesting*, the polymorphic *and* is specialised to sorts representing φ and I, and cannot be reduced further with the application of the argument *book*. A further algorithm is provided in order to model the choice of transformations, trying to match all available transformations to the adaptation slots. As all permutations are considered, this is potentially the most costly computation taking place. The result is a *list* of possible interpretations (given as term applications with slots filled by transformations): there might be zero, one, or finitely many.

A further check on the list of terms obtained will filter those, if any, with a suitable typing, that will form the desired result(s). In the tests conduced with the input of the example, four interpretations where produced (one for every slot/transformation permutation), with only the correct one of a resolvable type (t):

```
((And^{(t->(t->t))}
    (heavy^{(P->t)} (morph_R->Phy^{(R->P)} book^{R})))
    (interesting^{(I->t)} (morph_R->I^{(R->I)} book^{R})))
```

2.4 Implicit Adaptation

Polymorphic operators such as *and*, with explicit adaptation slots, are needed for co-predications. However, most lexical adaptations can take place implicitly, simply by reacting to a type mismatch and applying any suitable transformation to resolve it. This requires to adapt the terms by adding correctly-typed and well-positioned adaptation slots. In the case of an application such as $(p^{A \rightarrow B}\ a^{C})$, there are two possibilities to resolve the type mismatch: by adapting the predicate, yielding $((f^{(A \rightarrow B) \rightarrow (C \rightarrow B)})\ p)\ a)$, or the argument, resulting in $(p\ (f^{C \rightarrow A}\ a))$. In the slightly different case of a partial application $(\lambda x^{A}.\tau\ a^{C})$, in which the argument can be adapted as before (yielding $(p\ (f^{C \rightarrow A}\ a)))$, but the typing of the predicate might not be fully determined at the moment of the adaptation.

A procedure analyses such applications with type mismatches and no explicit adaptation slots, and inserts suitable, automatically generated adaptation slots, then proceeds as with explicit adaptations. For example, a simple term application such as `(P^{(e->t)} a^{A})` with a transformation `f_{A->e}` available to the atomic term a yields the straightforward (and only felicitous) interpretation `(lambda x^A.(P^{(e->t)} (f_{A->e}^{(A->e)} x^A)) a^{A})`, that reduces to `(P^{(e->t)} (f_{A->e}^{(A->e)} a^{A})`.

Implicit adaptations are necessarily reduced to those simple cases. Trying to account automatically for co-predications would imply to try any possible permutation of types and transformations at all nodes of a term, which would be exponential in complexity; thus, the need for lexical operators with explicit adaptation slots such as the polymorphic *and*.

2.5 Lexicalisation

In addition to the core mechanisms, a `tecto` package provides support for a tectogrammatical/syntactic structure in the form of an unannotated binary tree of lexemes; this serves as a factory for the input of already analysed text, and as a more streamlined form of output for adapted terms.

A `lexicon` package enables the storage of lexical entries that associate lexemes (as strings) to terms, complete with typing, transformations and ambiguities. Lexica can be merged, in order to have combine the treatment of different phenomena, treated as standalone modules, for complex sentences. Lexica also provide automated translations from a syntactic structure (a `tecto` term) to a semantic one (a `TermW` term, initially not adapted, reduced or even type-checked). Semantic terms can be presented either by a straightforward translation to syntactic terms, or printed to a string in the usual fully-parenthesed prefix notation with apparent typing (as in the examples of this article).

2.6   Phenomena Coverage

Many lexical phenomena discussed in [18,29] can be modelled using the simple mechanisms of ΛTY_n in their prototypal implementation given above; some others require additional mechanisms. The following is a short overview of how some classical phenomena are handled in MGL and our prototype.

Lexical adaptations, including alternations, meaning transfers, grinding, qualia-exploitation and "Dot-type"-exploitation are all supported by the adaptation mechanisms, as given previously.

    Simple predications only require to have suitable transformations available, and to use the implicit adaptation mechanisms.

    Co-predications require explicit adaptation using polymorphic operators (such as the higher-order conjunction "and" above). Theoretical grounds have been laid in [18,32].

Constraints of application are required in order to perform co-predications correctly. As explained in [24], the simultaneous reference to different facets of a same entity can be infelicitous in some circumstances, such as the use of destructive transformations (grinding, packing) or metaphorical use of some words. Thus, the following co-predications are infelicitous to some degree:

    – *The salmon was lighting-fast and delicious,*

    – *? Birmingham won the championship and was split in the Brexit vote.*

In order to block such co-predications, we have proposed to place constraints on transformations in order to block their usage depending on the other transformations that have been used on the same term.

    The first version of this system given in, e.g., [18], distinguishes between *flexible* (allowing all other facets) and *rigid* (blocking all other facets) transformations, as well as relaxable constraints depending on syntactic features. The latest version, given in [19], proposes a revised calculus named $\Lambda^{-\circ}TY_n$, a system with terms of the linear intuitionistic logic as types, that (among other things) allow any arbitrary type-driven predicate to act as a constraint on the use of transformations, thus allowing the complex variability of felicity of co-predications with deverbals examined by [13] and detailed for MGL in [30].

    In this prototype implementation, all transformations are equipped with a member function that can be defined as an arbitrary constraint, the default being the boolean constant `true` (that simply models flexible transformations). A compatibility one-on-one check of all transformations can be performed using every constraint. As the constraint can effectively be any function, the precision is the same as in [19].

Ontological inclusion, called *type accommodation* in [29] and modelling the lexical relation of *hyponymy* (as evoked with different solutions in other sub-typing accounts, such as [4]), can be supported by tweaking the system of sorts. The theoretical and empirical basis for doing so are discussed in [23], in which we argue that *coercive sub-typing* is an accurate and helpful mechanism for resolving ontological inclusion, but no other lexical phenomena; the latter are implemented using word-level transformations. In order to support sub-typing, each sort can be defined with an optional `parent` sort. A careful review of the typing comparison mechanism will then be enough in order to support sub-typing.

This is not implemented yet (it requires a refactoring of the notion of equality for sorts), but does not require (much) additional processing power.

Performative lexical adaptations, such as quantification, Hilbert operators for determiners, and the alternate readings of plurals and mass nouns, are supported as far as the meaning assembly phase is concerned. However, in order to be useful, this category of lexical phenomena (as well as hypostasis and several others) require additional mechanisms in order to incorporate the knowledge gathered from the analysis of the sentence into the logical representation. The basic architecture is supported, but mechanisms of resolution remain preliminary and will be discussed next, especially in Sect. 3.3.

3 Layers of Lexica and Knowledge Representation

3.1 The Additional Layers

Theories of semantics deriving from [29] generally encompass some degree of common sense world knowledge. For example, it is considered known that a *committee* (and other such group nouns) is made of several people, and is thus a felicitous argument of predicates requiring a plural argument such as *to meet*. It is also known that *engines* are part of *cars*, and that predicates such as *powerful* or *fuel-guzzling* can apply to *cars* via their constitutive quale; all such "common-sense metaphysics" have been part of generative lexical theory from the start, as detailed by [2]. It has been argued (e.g. in [11]) that such complex knowledge does not belong in a semantic lexicon; we will ignore such claims, paraphrasing Im and Lee from [12] in defining semantics to be the meaning conveyed by an utterance to a competent speaker of the language in itself, excluding, for instance, the specific situation in which the utterance is made, but including any previous discourse. In this view the full contents of, for example, a given fairy tales, should be able to be described within semantics, while texts such as political essays will probably require additional knowledge about the position of the author and the specifics of the period of writing. From our point of view, the lexicon includes that minimal knowledge as part of the semantic terms and types involved. To design a complete tool for type-theoretic lexical semantics, we must first complete the careful definition of various lexica that can convey the necessary, elementary world-knowledge for each word. A lexicon for general use will associate to all relevant lexemes their semantics (in the form of main and optional λ-terms) as can be given in a dictionary of a language. However, there are two arguments to be made for additional lexica beyond the general language lexicon.

First are the specific lexica, that detail vocabularies relevant only to a community, such as professional jargons, local dialects, and other linguistic constructs specific to small groups of people; and/or words or word uses restrained to a specific literary universe, such as fairy tales, space opera, mythology, politic speeches, etc. Such lexica are activated on an as-needed basis, and are more specific than the general-use lexicon.

Lexical semantics also requires a lexicon used for the current enunciation. A competent speaker of any language is able to use generative mechanisms in order to introduce new lexical concepts, either by the use of a new word, the meaning of which can be inferred from context and morphology, or by creative use of an existing word.

In our view, the lexicon of the enunciation starts empty and can be augmented when the analysis of the discourse encounters words that are not present in the current active lexica. We think that such mechanisms can enable the *learning* of lexical semantic data. In addition to these lexical layers of meaning, we tend to implement different lexical phenomena using different lexica for simplicity's sake, and create a merged lexicon from every relevant one when processing text.

3.2 Individuals, Facts and Contexts

To summarise our argument in Sect. 3.1 above, in addition to mostly static lexical data, some sort of knowledge representation is needed to process even simple lexical phenomena such as collective and distributive readings for plurals. Namely, we need to keep track of the *individuals* mentioned in a given discourse, and of the *facts* asserted of those individuals. To be complete, we would also need to keep track of *agents*, in order to model dialogues or multiple points of view in which certain agents assert certain facts. Our implementation prototype supports individuals, as atomic terms of type A (for named entities: human agents, towns...), as well as some individuals as atomic terms of type $A \rightarrow t$ (for common nouns, that can be resolved to a specific individual of type A by the means of an Hilbert-based determiner) for any suitable sort A. This dual typing for individual is adopted by MGL for complete compatibility with the classical Montagovian analysis, and is well-suited for the treatment of many phenomena, using operators inspired by Hilbert for the representation of determiners (as given in [22]) in order to select a specific individual from a common noun. We think that this approach is justified; however, this is not the chosen modelisation for other type-theoretic accounts of lexical semantics, and the implementation provided is easily adapted to other types for common nouns and individuals. See [24] for a detailed discussion of our choice of types and [17] for a contrasting opinion.

We also account for *facts*, as predicates (TermBindings or atomic terms) of type $\alpha \rightarrow t$ for any arbitrarily complex type α, that are used in a term application, and apply to an individual. In the analysis of a term, individuals and types are extracted and added to the context of enunciation. The hierarchy of lexical layers given above can be implemented as a hierarchy of *contexts*, some containing initial individuals and facts relevant to each lexicon; in a such complete system, the context of the real world would, to resolve the paradox mentioned in [33], include the fact that there is no King of France (and therefore that *The king of France is bald*, while grammatical, is not felicitous because there are no qualifying referents for the entities described, and thus cannot be assigned a truth value). Such contexts are specific objects (aggregating individuals, facts and a related lexicon) in our implementation.

3.3   The Parsing-Knowledge Loop

We use a specific lexicon to list some common semantic terms for quantifiers (universal/existential, needed for some classical interpretations), counting terms (needed for plurals), logical constants (for truth-valued semantic interpretations) of grammatical constraints) and Hilbert operators (used for the smooth modelling of determiners, as detailed in e.g. [31] and more recently in [22]).

Other lexica can make use of these terms in order to construct, for instance, Link-based semantics for plurals (originally given in [15]), using lexical transformations as detailed in [21]. Some functions associated to the logical lexicon then resolve the glue operators, given a term and a context. This updated process of analysis is given in Fig. 3.

To explain what the analysis of plural readings in MGL entail, consider the following example from [21]:

- *Jimi and Dusty met* is analysed as
  $|\lambda y^e.(y = j) \vee (y = d)| > 1 \wedge meet(\lambda y^e.(y = j) \vee (y = d))$.

One elementary issue is that the predicate *met* applies to group individuals (such as *a committee*) and constructions made of more than one individuals (such as *Jimi and Dusty*) but not to singular individuals (such as *a student*). Thus, the lexical entry for the predicate is $\lambda P^{e \to t}.|P| > 1 \wedge meet(P)$ – a logical conjunction with a cardinality operator.

Those two simple elements can be defined in System-F (the calculus in which ΛTY_n, the logic of MGL, is implemented). The issue is that, in order for our system to infer correctly that Jimi and Dusty are two different individuals, and thus that the above term resolves to $meet(\lambda y^e.(y = j) \vee (y = d))$, we must use processing power beyond the simple construction and reduction of terms: a minimal system of *knowledge representation* and *logical inference*.

Within our architecture encompassing individuals and facts, and with a functional lexicon for logical connectives (including the logical *and* operator of that example), as well as quantification and counting (including the cardinality operator), this example can be treated.

However, this requires a given term to be parsed at least twice. The first time, the syntactic structure is converted into a semantic term and lexical transformations are applied.

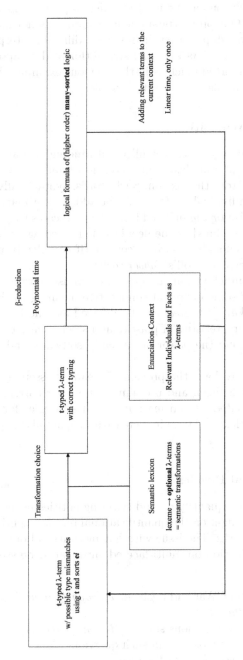

Fig. 3. Parsing-knowledge representation feedback

The second time, facts that emerge from the transformations, such as alternate distributive/collective readings for plural predicates, are added to the lexicon, and the logical lexicon can be used in order to process the operators that have been introduced. Our prototype implementation does not incorporate such feedback yet, as the first step can result in several different interpretations; this remains a work in progress. As a result, straightforward composition for plurals are tentatively supported (such as in the previous example), but ambiguous covering readings for plurals are not yet available.

3.4 World Sense Acquisition

An enunciation-context lexicon that is filled with individuals and facts inferred from the primary semantic analysis can serve, in a limited way, to account for words that are absent from the lexicon. Such words, syntactically placed in the position occupied by individuals or facts, will be added as primary entities to the lexicon, and their precise *typing* inferred from the predicates they are applied to. The typing can then be refined as the new lexeme is used again. An elementary mechanism should be enough to have a correct (if completely underspecified) representation from Lewis Carroll's *Jabberwocky*.

Of course, most competent human speakers also use morphosyntactic inference to attach at least some degree of connotative meaning to the words being proposed (e.g., *Star Wars*'s *plasteel* can be inferred as a fictive material somehow combining the characteristics of plastics and steel by any English speaker). This is completely beyond the power of our early software, and simple syntax-and-typing inferences.

A first easy step is to have the process of meaning assembly outline which lexemes are not in the lexicon, and use human input for correcting the precise types and terms associated. Promising automated strategies for learning more about new lexemes, or lexemes used in a creative way, are also explored in richly typed settings by [7].

3.5 Quantificational Puzzles

The process of counting, quantifying and selecting entities using Hilbert operators can also shed some light on the quantificational puzzles mentioned in [1] and several other related works. The issue with having universal quantification used together with co-predication on multi-faceted entities can be seen in examples such as:

– *There are five copies of* War and Peace *and a copy of an anthology of Tolstoï's complete works on the shelf.*
  What is the answer to questions such as *How many books. . . ?*
  What exactly is the type of *book* in such questions?
– *I read, then burnt, every book in the attic.*
  The entities being predicated form two different sets.

In order to resolve such quantificational puzzles satisfactorily, the methods for counting and quantifying must be adapted to each predicate, and only apply to individuals of the appropriate type. For our purpose, this implies a close monitoring of the entities introduced by lexical transformations and their context of appearance.

As the quantifiers are adapted to the transformed and original entities, maintaining expected truth values for straightforward sentences poses a challenge. This is also a work in progress.

4   Results

4.1   A Fragment of Second-Order

With this prototype software, we have proven that MGL can actually be computationally implemented. This was not really in doubt, but the way that the combination of types and terms are implemented illustrates that the time and space complexity of most of the process is limited: the algorithms used are mostly linear tree walks, with a few quadratic worst-case operations.

The most complex step is the choice of optional terms for adaptation slots, of complexity $|t| \times |s| \times n$ at worst (the product of the number of optional terms available, adaptation slots, and length of the term); the hypothesis behind MGL is that the number of available optional terms at any point remains "manageable". Thus, the step not actually implemented in this prototype (but for which many implementations exist), syntactic analysis, is the costliest of the steps detailed in Fig. 1, and the complete process of parsing is polynomial in time, as explained in [24].

MGL accounts such as [32] point out that the whole expressive power of second-order λ-calculus is not used, and that every possible operation could be implemented using first-order terms if all possible adaptations were listed at each step (which is syntactically much longer to write). Indeed, our implementation only supports the single second-order operation of type specialisation (by distinguishing type variables from other types and using pattern matching to recognise and rewrite types), which is required for having polymorphic terms. There are no features of ΛTY_n that require additional power: sub-typing can be implemented by an optional **parent** field in **Sorts**, arbitrary complex on co-predications are supported by including a check on transformations that can be any arbitrary function, quantification, counting and Hilbert operators can be included.

Formally, it has been pointed out by Ph. de Groote (pc.) that the smallest type calculus encompassing the features provided by ΛTY_n is a simply-typed calculus supporting type collections and sub-typing, such as the typing system used by OCaML (http://ocaml.org/). Moreover, the many efforts made by proponents of other related type-theoretic accounts of generative lexical semantics to provide implementations of their own theories reinforce our belief in the computational feasibility of such analyses. The chain of treatment sketched in Fig. 1 and detailed in [24], and the computational steps provided in the present paper, are very similar to, for instance, the process detailed in [14] for Dependent

Type Semantics. The many differences between the theories and representations notwithstanding, the operational steps used to analyse similar phenomena are nearly identical (despite the research process on both accounts being parallel and candid, and the presentation of the results having occurred on the same day). Such spontaneous convergence between theories is, in our view, an indication of the pertinence of both accounts.

4.2 Minimal Processing Architecture

Our prototype implementation includes the skeleton of an architecture that represents the individuals, facts and agents appearing during the semantic analysis.

This goes beyond the straightforward process of producing a logical representation for an utterance, as some of the terms of that logical representation might be analysed differently depending on the context; we argue that this process should still be part of a semantic analysis. The individuals, facts and agents are stored in objects called *contexts*, organised in a hierarchy that includes the most specific context (modelling the analysis of the current discourse), universe-specific contexts (describing whether the discourse is part of a fictional, historical or activity-specific setting), dialect- and language-specific contexts, each associated to an appropriate lexicon.

A complete analysis would minimally involve the construction of the logical representation of an utterance, the update of the enunciation context with individuals and facts introduced by that utterance, and a re-interpretation of the logical representation in the active contexts. This minimal processing architecture can be completed with no difficulties; our implementation includes relevant data structures and algorithms, but requires significant work on examples of performative lexica in order to be thoroughly tested.

4.3 Perspectives

This prototype implementation has already served its primary purpose: to illustrate that MGL can be computationally implemented, and that the examples usually given with the theory actually work. As it stands, however, this implementation is more of a proof of concept than useful software.

To be actively used by the community, more work would be required to give it an helpful interface, both for the user and for existing analysers; we also would like to convert from and to representations of the other most active type-theoretic accounts of lexical semantics. The knowledge-representation architecture remains a work in progress, and requires solid efforts in order to correspond to our ambitions. The completion of this software is not, however, an end in itself.

In fact, what MGL (and other related accounts of lexical semantics) really requires in order to be useful is a large-cover library of types and terms. The analyses, whether formal or computational, are justified by toy linguistic examples or on domain-restricted phenomena; without a significant step in the definition of a system of sorts and types, and subsequently of a large, semantically rich lexicon, type-theoretic proposals have very little value compared to, e.g., "deep" neural techniques trained using massive corpora. The prototype software presented in this publication is a nice illustration of the possibilities provided by MGL, but our main hope is that it can help to build software that can learn types and lexemes.

The first step will not be on the software side but, as suggested in [24], the establishment of a linguistically motivated kernel system of sorts.

References

1. Asher, N.: Lexical Meaning in Context: A Web of Words. Cambridge University Press, Cambridge (2011)
2. Asher, N., Pustejovsky, J.: Word meaning and commonsense metaphysics. Semantics Archive, August 2005
3. Bekki, D.: Dependent type semantics: an introduction. In: Christoff, Z., Galeazzi, P., Gierasimczuk, N., Marcoci, A., Smet, S. (eds.) Logic and Interactive RAtionality (LIRa) Yearbook 2012, vol. 1, pp. 277–300. University of Amsterdam, Amsterdam (2014)
4. Bekki, D., Asher, N.: Logical polysemy and subtyping. In: Motomura, Y., Butler, A., Bekki, D. (eds.) JSAI-isAI 2012. LNCS, vol. 7856, pp. 17–24. Springer, Heidelberg (2013). doi:10.1007/978-3-642-39931-2_2
5. Bierwisch, M.: Wördliche Bedeutung - eine pragmatische Gretchenfrag. In: Grewendorf, G. (ed.) Sprechakttheorie und Semantik, pp. 119–148. Surkamp, Frankfurt (1979)
6. Bosch, P.: The bermuda triangle: natural language semantics between linguistics, knowledge representation, and knowledge processing. In: Herzog, O., Rollinger, C.-R. (eds.) Text Understanding in LILOG. LNCS, vol. 546, pp. 241–258. Springer, Heidelberg (1991). doi:10.1007/3-540-54594-8_64. http://dl.acm.org/citation.cfm?id=647415.725191
7. Breitholtz, E.: Are widows always wicked? Learning concepts through enthymematic reasoning. In: Cooper, R., Retoré, C. (eds.) ESSLLI Proceedings of the TYTLES Workshop on TYpe Theory and LExical Semantics, Barcelona, August 2015
8. Chatzikyriakidis, S., Luo, Z.: Natural language inference in Coq. J. Log. Lang. Inf. **23**(4), 441–480 (2014). http://dx.doi.org/10.1007/s10849-014-9208-x
9. Cooper, R.: Copredication, dynamic generalized quantification and lexical innovation by coercion. In: Fourth International Workshop on Generative Approaches to the Lexicon (2007)
10. Cruse, D.A.: Lexical Semantics. Cambridge University Press, New York (1986)
11. Fodor, J.A., Lepore, E.: The emptiness of the lexicon: reflections on James Pustejovsky's the generative lexicon. Linguist. Inq. **29**(2), 269–288 (1998)
12. Im, S., Lee, C.: A developed analysis of type coercion based on type theory and conventionality. In: Cooper, R., Retoré, C. (eds.) ESSLLI Proceedings of the TYTLES Workshop on TYpe Theory and LExical Semantics, Barcelona, August 2015
13. Jacquey, E.: Ambiguïtés lexicales et traitement automatique des langues: modélisation de la polysémie logique et application aux déverbaux d'action ambigus en français. Ph.D. thesis, Université de Nancy 2 (2001)

14. Kinoshita, E., Mineshima, K., Bekki, D.: An analysis of selectional restrictions with Dependent Type Semantics. In: Proceedings of the 13th International Workshop on Logic and Engineering of Natural Language Semantics (LENLS 2013), pp. 100–113, November 2016

15. Link, G.: The logical analysis of plurals and mass terms: a lattice-theoretic approach. In: Portner, P., Partee, B.H. (eds.) Formal Semantics - The Essential Readings, pp. 127–147. Blackwell, Oxford (1983)

16. Luo, Z.: Contextual analysis of word meanings in type-theoretical semantics. In: Pogodalla, S., Prost, J.-P. (eds.) LACL 2011. LNCS, vol. 6736, pp. 159–174. Springer, Heidelberg (2011). doi:10.1007/978-3-642-22221-4_11

17. Luo, Z.: Common nouns as types. In: Béchet, D., Dikovsky, A. (eds.) LACL 2012. LNCS, vol. 7351, pp. 173–185. Springer, Heidelberg (2012). doi:10.1007/978-3-642-31262-5_12

18. Mery, B.: Modélisation de la Sémantique Lexicale dans le cadre de la Théorie des Types. Ph.D. thesis, Université de Bordeaux, July 2011

19. Mery, B.: Lexical semantics with linear types. In: NLCS 2015, The Third Workshop on Natural Language and Computer Science, Kyoto, Japan, July 2015

20. Mery, B.: Lessons from a prototype implementation of Montagovian lexical semantics. In: Proceedings of the 13th International Workshop on Logic and Engineering of Natural Language Semantics (LENLS 2013), pp. 73–85, November 2016

21. Mery, B., Moot, R., Retoré, C.: Computing the semantics of plurals and massive entities using many-sorted types. In: Murata, T., Mineshima, K., Bekki, D. (eds.) JSAI-isAI 2014. LNCS, vol. 9067, pp. 144–159. Springer, Heidelberg (2015). doi:10.1007/978-3-662-48119-6_11

22. Mery, B., Moot, R., Retoré, C.: Typed Hilbert operators for the lexical semantics of singular and plural determiner phrases. In: Epsilon 2015 - Hilbert's Epsilon and Tau in Logic, Informatics and Linguistics, Montpellier, France, June 2015

23. Mery, B., Retoré, C.: Are books events? Ontological inclusions as coercive subtyping, lexical transfers as entailment. In: LENLS 2012, jSAI 2015, Kanagawa, Japan, November 2015

24. Mery, B., Retoré, C.: Classifiers, sorts, and base types in the Montagovian generative lexicon and related type theoretical frameworks for lexical compositional semantics. In: Chatzikyriakidis, S., Luo, Z. (eds.) Modern Perspectives in Type-Theoretical Semantics. SLP, vol. 98, pp. 163–188. Springer, Cham (2017). doi:10.1007/978-3-319-50422-3_7

25. Moot, R.: Wide-coverage French syntax and semantics using Grail. In: Proceedings of Traitement Automatique des Langues Naturelles (TALN), Montreal (2010)

26. Moot, R., Prévot, L., Retoré, C.: A discursive analysis of itineraries in an historical and regional corpus of travels. In: Constraints in discourse. http://passage.inria.fr/cid2011/doku.php. Ayay-roches-rouges, France, September 2011. http://hal.archives-ouvertes.fr/hal-00607691/en/

27. Moravcsik, J.M.: How do words get their meanings? J. Philos. **78**(1), 5–24 (1982)

28. Muskens, R.: Meaning and partiality. In: Cooper, R., de Rijke, M. (eds.) Studies in Logic, Langage and Information. CSLI, Stanford (1996)

29. Pustejovsky, J.: The Generative Lexicon. MIT Press, Cambridge (1995)

30. Real-Coelho, L.M., Retoré, C.: A generative Montagovian lexicon for polysemous deverbal nouns. In: 4th World Congress and School on Universal Logic - Workshop on Logic and Linguistics, Rio de Janeiro (2013)

31. Retoré, C.: Sémantique des déterminants dans un cadre richement typé (2013). CoRR abs/1302.1422. http://arxiv.org/abs/1302.1422

32. Retoré, C.: The Montagovian generative lexicon lambda Ty_n: a type theoretical framework for natural language semantics. In: Matthes, R., Schubert, A. (eds.) 19th International Conference on Types for Proofs and Programs (TYPES 2013). Leibniz International Proceedings in Informatics (LIPIcs), vol. 26, pp. 202–229. Schloss Dagstuhl-Leibniz-Zentrum fuer Informatik, Dagstuhl, Germany (2014). http://drops.dagstuhl.de/opus/volltexte/2014/4633

33. Russell, B.: On denoting. Mind **14**(56), 479–493 (1905). http://www.jstor.org/stable/2248381

Truth Conditionals and Use Conditionals: An Expressive Modal Analysis

Lukas Rieser[(✉)]

Kyoto University, Kyoto, Japan
lukasjrieser@gmail.com

Abstract. I propose that on some non-standard interpretations of conditionals, the antecedent influences not the truth-conditional, but the use-conditional evaluation of the consequent by restricting the modal base of a necessity operator introduced by the conditional form in the expressive dimension of utterance meaning. On this view, conditionals can be grouped into **truth conditionals** and **use conditionals**, depending on which interpretation is salient. I argue that such an analysis allows to predict properties of hypothetical conditionals, biscuit conditionals, and conditional hedges within a unified account.

1 Interpretations of Conditionals

The three examples of conditionals given in (1) through (3) below each have distinct salient interpretations, which I seek to account for in a unified analysis.

(1) If John remembered to go shopping, there's beer in the fridge.

(2) If you are thirsty, there's beer in the fridge.

(3) If I'm not mistaken, there's beer in the fridge.

The salient interpretation of (1) is what I take to be the standard interpretation of conditionals, the **hypothetical conditional** interpretation, on which the truth of the consequent (=proposition of the main clause) is evaluated under the assumption that the antecedent (=proposition of the *if*-clause) holds. Variants of analyses on these lines have been prominent in the formal literature at least from Stalnaker [19]. The salient interpretation of (2) is that of a **biscuit conditional** (term due to Austin's [1] original example illustrating tis interpretation "There are biscuits on the sideboard if you want some"), on which the information the consequent provides is relevant for the addressee only in case the antecedent holds. Finally, (3) is most plausibly interpreted as a **conditional hedge**, a kind of disclaimer, on which interpretation the conditional form indicates the possibility of the consequent not being true, that is possible error on part of the speaker, in order to avoid the consequences of providing (potentially) false information.

There are differences in how the conditional antecedent and consequent relate to each other on the three interpretations, as summarized below. The analysis

© Springer International Publishing AG 2017
S. Kurahashi et al. (Eds.): JSAI-isAI 2016, LNAI 10247, pp. 108–122, 2017.
DOI: 10.1007/978-3-319-61572-1_8

aims at accounting for these differences while deriving the interpretations the three examples intuitively receive.

Conditional (In)dependence, Conditional Perfection

The hypothetical conditional in (1) conveys (truth-)conditional dependence between the antecedent and the consequent in the sense that the truth of the consequent is contingent on that of the antecedent in a specific way. On the lines of the aforementioned prominent line of analyses in the Stalnakerian spirit, this means that the truth of the consequent is (only) to be evaluated under the assumption that the antecedent holds. The intuition underlying such analyses of hypothetical conditionals in possible-world semantic analyses is that for instance (1) expresses that the actual world is such that there's beer if John went shopping, but not necessarily if this is not the case—in this way, the consequent is **truth-conditionally dependent** on the antecedent in hypothetical conditionals.

In the biscuit conditional in (2), on the other hand, there is no such dependence, as the addressee's thirst is irrelevant for the truth of whether or not there is beer in the fridge—modifying Austin's [1, 158] wording of this observation, it would be folly to think that the addressee being thirsty is enough to cause there to be beer in the fridge. That is, the hypothetical interpretation relies on a certain kind of dependence between the antecedent and the consequent that is lacking in biscuit conditionals. Franke [10, 92] defines this property of biscuit conditionals as **epistemic independence**, formally implemented as the property of a conditional that it is not sufficient to learn the truth value of the antecedent in order to find out that of the consequent, as shown below in slightly simplified form.[1]

(4) Ψ and Φ are epistemically independent (on the epistemic state of an agent)

  iff for 'if Ψ then Φ': $\Diamond\Psi \wedge \Diamond\neg\Phi \rightarrow \Diamond(\Psi \cap \neg\Phi)$

This states that antecedent and consequent are epistemically independent iff an agent who considers the consequent and the negation of the antecedent possible independently of each other also considers it possible that both hold at the same time. This is a property that differentiates (1) and (2): from the premise that the speaker of (1) considers it possible that John went shopping and also considers it possible that there is no beer in the fridge, it does not follow that the speaker considers it possible that John went shopping but there is no beer in the fridge. From the premise that speaker of (2) considers it possible that the addressee is thirsty and that there is no beer in the fridge, on the other hand, it follows that the speaker considers it possible that that the addressee

---

[1] Franke formulates the requirement not for Φ and Ψ only, but for all $X \in \{\Phi, \neg\Phi\}$ and all $Y \in \{\Psi, \neg\Psi\}$; I show the part relevant for discussion of (2).

is thirsty but there is no beer in the fridge.[2] Thus, the hypothetical conditional (1) is conditionally dependent: finding out whether or not John went shopping is sufficient to find out whether there's beer. The biscuit conditional (2), on the other hand, is **conditionally independent**: finding out whether or not the addressee is thirsty is *not* sufficient to find out whether or not there's beer.

The intuition with regard to the communicative effect of conditional independence I aim to account for in the analysis is as follows. What biscuit conditionals like (2) assert is the consequent only, while expressing that that the consequent is only relevant to the addressee in case the antecedent holds. I use Relevance essentially in a Gricean sense, *i.e.* relevant to the conversational goals of the participants.[3] Following this intuition, previous analyses in the philosophical literature have assumed that what the antecedent of biscuit conditionals conditions on is the assertion itself, *i.e.* the consequent is only asserted when the antecedent holds, *c.f.* DeRose and Grandy [6], or that biscuit conditionals indicate the existence of a "potential literal act" (such as the assertion that there is beer) in case the antecedent holds, *c.f.* Siegel [17]. Similar to these views, the analysis I propose is one where assertion is conditioned on as the contribution of the conditional enters the derivation later than the speech-act operator, but its effect is modeled building on extant analyses of conditionals within formal semantics translated to the expressive dimension of utterance meaning.

Next, the intuitions regarding the conditional hedge in (3) are parallel to intuitions on the biscuit conditional in the sense that in both cases, conditioning by the antecedent targets felicity rather than truth. The intuitions differ, however, in that the conditional hedge targets Quality (of the speaker's belief that the consequent holds), rater than Relevance (both in the sense of Gricean maxims determining utterance felicity). There is an important property of (3), however, which differentiates it from both the hypothetical conditional (1) and the biscuit conditional (2), namely **conditional perfection**. Following van Canegem-Ardijns and Van Belle [3, 350], the truth of the following sentences demonstrates that (3) exhibits this property:

(5) a. If I'm not mistaken, there's beer in the fridge.

b. If I'm mistaken, there's no beer in the fridge.

c. Only if I'm not mistaken, there's beer in the fridge.

In fact, the truth of consequent and antecedent in (3) are in a material biconditional relation $\leftrightarrow$, as the reverse conditional "If there's beer in the fridge, I

---

[2] To derive the biscuit conditional interpretation, Franke takes conditionals to convey a speaker belief that the antecedent entails the consequent, which, together with the assumption that the antecedent is possible and epistemic independence, allows for the conclusion that the speaker believes the consequent to be true. While this derives the biscuit conditional interpretation, it is incompatible with conditional hedges (see below), and assumes a different view of conditionals than this paper.

[3] The main concern of this paper are the compositional aspects of non-canonical conditionals, and there is not much space to discuss the nature of Relevance. See Sperber and Wison [18] for extensive discussion of issues around defining Relevance.

am not mistaken", while a somewhat odd thing to say, intuitively has the same truth conditions as (3), *i.e.* (3) clearly conveys that *if and only if* the speaker is not mistaken, there is beer in the fridge. The relation $\leftrightarrow$ is essentially the logical equivalent of conditional perfection, as diagnosed by the test above, but does not necessarily hold for all cases that have been labeled conditional perfection in the literature. I aim to account for conditional perfection only in this sense, and only in the case of conditional hedges, but see Van der Auwera [2] for a proposal of how conditional perfection can be derived as a Gricean scalar Q-implicature in other cases. The formal analysis I develop below is capable of predicting conditional perfection in conditional hedges, while also being applicable to biscuit and hypothetical conditionals.

2   Conditionals as Modals

I take conditionals to be modal constructions, in which the antecedent restricts the modal base of the consequent, following a possible-world semantics analysis of conditionals as modals, *c.f.* Kratzer [14] for an overview, which accounts for the Stalnakerian intuitions on hypothetical conditionals as described in the discussion of (1) above. I further assume that when there is no overt modal in the consequent, the conditional introduces a covert "**human necessity**" modal as defined by Kaufmann and Schwager [13], the conversational backgrounds being a modal base reflecting the relevant circumstances and a stereotypical ordering source to control for issues like those raised by strengthening of the antecedent.

I take the relevant circumstances to be those relevant to the connection between antecedent and consequent conveyed by the conditional, which in the case of a hypothetical conditional is truth-conditional dependence, in the case of non-standard interpretations a connection mediated by Gricean maxims (or a use-conditional dependence). I further assume the modal base to be the propositions compatible with the speaker's beliefs regarding the relevant circumstances, rather than the relevant circumstances as such, making the modal base doxastic as well as circumstantial.[4] To illustrate the standard interpretation of a (truth) conditional on this view, a paraphrase for the meaning of a hypothetical conditional is given in (6).

(6) A truth conditional "If Ψ then Φ" is true iff at all worlds (stereotypical,

compatible with the speaker's beliefs) in which Ψ holds, Φ also holds.

It is worth noting that the details of what kind of modal base and ordering source are chosen are not central to my analysis of truth- and use conditionals, as long as the modal force is necessity, and the proposition in the conditional antecedent is used to restrict the modal base, thus excluding worlds in which the antecedent does not hold. This sets the stage on which the conditional consequent is evaluated. Also, my analysis is fully compatible with a double-modal analysis

---

[4] In the paraphrases for the modal base, I will use "the speaker's beliefs" to mean "the speaker's beliefs regarding the relevant circumstances" for brevity.

of conditionals as argued for by Frank [9], on which the modal the consequent introduces a modal base and ordering source potentially different from that introduced by the conditional form.

2.1   Use Conditions

The view of use conditions I base my analysis on connects them to truth conditions in the following way. Taking assertion as an example, the meaning of an utterance is split into descriptive, or truth-conditional, and expressive, or use-conditional, dimensions. Truth conditions of a proposition are determined by valuation against worlds. When the proposition (usually one, if possibly complex) in the descriptive, truth-conditional dimension of utterance meaning is true in this sense, this is to say that something true is asserted. When the propositions (typically many) in the expressive, use-conditional meaning dimension are true, this means that the assertion is felicitous. Thus, when the source of the propositions in the expressive meaning dimensions are lexical conventional implicatures, for instance a proposition representing the negative attitude arising from *cur* vs.*dog* as discussed by Gutzmann [12], or the propositional content of a parenthetical, the utterance's use conditions are straightforwardly determined by the truth or falsity of these propositions, so that the propositions themselves have, in a way, become use-conditional.

   In addition to such use-conditional propositions derived from the lexical content, which are valuated against worlds in the same way as truth conditions are, I propose that there are types of propositions in the expressive dimension, which are not valuated against worlds, but evaluated with regard to Gricean conversational maxims (Grice [11]). Concretely, I propose that the entire propositional content in the descriptive dimension is evaluated in such a way in the expressive dimension. It is this part of the expressive dimension where conditioning on Relevance, Quality, etc., occurs in non use conditionals, and thus in the non-standard interpretations of conditionals to be accounted for here. This essentially amounts to propositions based on Gricean maxims such as "Φ is relevant...", "Φ is backed by adequate evidence", etc. being valuated against worlds in the usual way, and the utterance being felicitous if these derived propositions are true. According paraphrases for the truth- and use conditions of an assertion with the prejacent proposition Φ are shown in (7) and (8).

(7)   Truth conditions: ASSERT(Φ) is true w.r.t w iff Φ holds at w.

(8)   Use conditions: ASSERT(Φ) is felicitous w.r.t w iff Φ is relevant to the
      participants' goals, as informative as required, backed by adequate
      evidence, ... at w.

Notice that (8) makes no mention of expressive content originating in the lexical content of the utterance. This is for ease of exposition—for the same reason, I will only consider examples where no expressive content arises from the lexical material, *i.e.* examples without implicature triggers. Also, no mention is made of the truth of Φ in the use conditions, as I assume that asserting a false proposition is not necessarily infelicitous, provided that the speaker is not aware of its falsity. I will briefly return to this latter point in Sect. 4.1. In this section, I implement the view of use conditions sketched above in a model of indicative truth- and use conditionals, starting with the standard interpretation of truth conditionals.

2.2   Truth Conditionals

The descriptive, truth-conditional meaning of a hypothetical conditional "if Ψ, (then) Φ" in the formalization I propose is shown in (9), where $\|\mathcal{A}(\Phi)\|^t$ stands for the truth-conditional denotation of an utterance where a speech act $\mathcal{A}$ with the prejacent proposition Φ is performed. As for the notation representing the conditional, $\Box^H$ stands for a human necessity modal as outlined above, w for the actual world, $f(w)$ for the (doxastic/circumstantial) modal base (worlds compatible with the speaker's beliefs regarding the relevant circumstances at w), g for the (stereotypical) ordering source. The subscript $[\Psi]$ on the modal $\Box^H$ with the conditional consequent Φ in its nuclear scope indicates restriction of the modal base $f(w)$ by the conditional antecedent Ψ, yielding a restricted modal base f^+ in which non-Ψ worlds have been discarded.

(9) $\|\text{ASSERT}(\Box^H_{[\Psi]}[\Phi])\|^t = \|\text{ASSERT}(\Box^H[\Phi])\|^t$ w.r.t w, f^+, g

   (where $f^+ = \lambda w.f(w) \cup \|\Psi\|^t$)

The truth-conditional meaning of an assertion of (10) thus comes out as (11)

(10) If John remembered to go shopping, there's beer in the fridge.

(11) (10) is true iff at all worlds (stereotypical, compatible with the

   speaker's beliefs) in which John remembered to go shopping, there's

   beer in the fridge.

Next, in order to derive the use conditions of truth conditionals, I adopt the general framework of Potts [16], distinguishing between descriptive (=at-issue), truth-conditional (types marked with superscript a for "at-issue"), and expressive, use-conditional (types marked with superscript c for "conventional implicature") types on the respective levels of utterance meaning. I use the types t^a and t^c for propositions within the descriptive and expressive meaning dimensions, t^a being the type for truth-conditional content, t^c for lexical CIs and parentheticals.[5]

   In addition to these types, I introduce the use-conditional meaning type u^c (utterance) following McCready [15]. This is the type which I propose is evaluated in terms of Gricean maxims, rather than valuated against worlds in the usual manner for propositions. Also following McCready, elements of type

---

[5] As mentioned, however, the examples will not contain any of the latter, and I remain agnostic in regard to the question of whether or not content of type t^c gets evaluated for felicity in terms of Gricean maxims.

u^c arise from the type-shifting operation **utterance lifting** (UL), by which the propositional content, once asserted, is moved into the expressive domain, *i.e.* undergoes type shifting from type t^a to type u^c. I use a simplified version of UL as a function from (at-issue) propositions to (expressive) utterances in the present proposal.[6]

(12) $\text{UL}^{\mathcal{A}} = \lambda \Phi. \mathcal{A}(\Phi) : <t^a, u^c>$

   (where $\mathcal{A}(\Phi)$ is a speech act based on proposition Φ)

With this rule in place, performing a speech act based on a proposition containing no lexical expressives or parentheticals can be represented as in (13), where (a) shows the two dimensions of meaning before assertion, containing propositional (descriptive, truth-conditional) content only, and (b) shows utterance meaning[7] after assertion. The representation follows the convention $\langle \tau^a, \tau^c \rangle$ where all truth-conditional elements τ^a which are part of the utterance's meaning are shown on the left, all use-conditional elements τ^c on the right. A parallel representation for a speech act based on a conditional proposition (*i.e.* a truth-conditional) is given in (14).

(13) a. $\langle \Phi, \varnothing \rangle$

   b. $\langle \Phi, \mathcal{A}(\Phi) \rangle$

(14) a. $\langle \Box^H_{[\Psi]}[\Phi], \varnothing \rangle$

   b. $\langle \Box^H_{[\Psi]}[\Phi], \mathcal{A}(\Box^H_{[\Psi]}[\Phi]) \rangle$

In order to derive the use conditions which $\mathcal{A}(\Phi)$ and $\mathcal{A}(\Box^H_{[\Psi]}[\Phi])$ respectively contribute in the expressive dimension, $\mathcal{A}$ needs to be resolved to a specific speech act. I discuss the case of assertions of conditionals below, aiming to arrive at a formalization of use-conditions as paraphrased in (8) above, and to account for the differences between indicative use- and truth conditionals.

2.3   Use Conditionals

In the case of truth conditionals, the modal base of the covert modal introduced by the conditional form on truth-conditional level is restricted by the conditional antecedent, reflecting (truth-)conditional dependence. In the case of use-conditionals, not the truth, but the felicity of the consequent depends on the truth of the antecedent. I explain this as restriction of the modal base on the expressive rather than the descriptive level, as paraphrased in (16) for a use-conditional, parallel to the paraphrase for a truth conditional in (15), repeated from (6).

(15) A truth conditional "If Ψ then Φ" is true iff at all worlds (stereotypical,

   compatible with the speaker's beliefs) in which Ψ holds, Φ also holds.

---

[6] McCready's assumption that u is of a resource-sensitive shunting type u^s, making an additional operation of assertion-to-content necessary to reintroduce at-issue meaning, is ignored here for ease of exposition.

[7] Note that I use the label "utterance meaning" to refer to the meaning of proposition used in a speech act, rather than just for the parts of its meaning of type u^c.

(16) A use conditional "If Ψ then Φ" is felicitous iff at all worlds (stereotypical, compatible with the speaker's beliefs) in which Ψ holds, $\mathcal{A}(\Phi)$ is felicitous.

The next question to be addressed is how to evaluate the felicity of $\mathcal{A}(\Phi)$ (which in indicative conditionals is assertion), and where in the derivation of utterance meaning to introduce conditional restriction of the modal base in order for it to operate on the expressive level of meaning in use conditionals. I assume that what has to happen compositionally to get the interpretations we are after is what (17) schematically shows for a truth conditional, (18) for a use conditional, where M stands for the modal I will subsequently propose to occur in this position. The schema in (19) shows the assumption I make for what happens on the use-conditional level in the case of the speech-act $\mathcal{A}$ being an assertion, namely restriction of the modal base of a non-asserted human necessity modal on the use-conditional level.

(17)   [tree diagram] $\mathcal{A}$ / if Ψ $\Box^H$ Φ (18)   [tree diagram] if Ψ M $\mathcal{A}$ Φ (19)   [tree diagram] if Ψ $\Box^H$ $\mathcal{A}$ Φ

The crucial question with regard to whether or not (19) is on the right track can be put as follows: what does Ψ restrict when modifying a speech act $\mathcal{A}(\Phi)$? In (18), the placeholder is labeled M for modal, anticipating the analysis I propose, but there are other options—one could follow Siegel's proposal and have the conditional antecedent somehow quantify over potential speech acts, or Grandy's proposal, making the assertion of φ depend on the truth of the antecedent. I propose an approach I consider more straightforward in light of extant formal theories of conditionals, in which speech acts are neither quantified over nor suspended in this sense. Rather, the modal operator is introduced not on the descriptive, but on the expressive level. As shown in (19), in the case of $\mathcal{A}$ being an assertion, only a human necessity modal, as the one familiar from propositions of truth conditionals, is introduced (other modals could be introduced in the case of speech-acts like imperatives).

The denotation of the two dimensions of meaning of a truth conditional under the standard, hypothetical, interpretation, corresponding to the structure in (17), is shown in (20) below; the denotation of a use conditional, corresponding to the structure in (19), is shown in (21), alongside the denotation of a plain, non-conditional assertion of Φ in (22). Each denotation is represented before (a) and after (b) assertion with utterance lifting occurs.

(20) a. $\langle \Box^H_{[\Psi]}[\Phi], \varnothing \rangle$ (21) a. $\langle \Phi, \varnothing \rangle$ (22) a. $\langle \Phi, \varnothing \rangle$

b. $\langle \Box^H_{[\Psi]}[\Phi], \Box^H_{[\Psi]}[\Phi] \rangle$ b. $\langle \Phi, \Box^H_{[\Psi]}[\Phi] \rangle$ b. $\langle \Phi, \Phi \rangle$

The denotation of the use conditional in (21) is derived as follows. The prejacent Φ of type t^a is asserted and undergoes utterance lifting to u^c. At this point, the human necessity modal introduced by the conditional, with the modal base restricted by the antecedent Ψ, enters the derivation, taking Φ in its nuclear scope. The resulting representation of utterance meaning in (21b) contains the proposition Φ on the descriptive side, the modal expression $\Box^H_{[\Psi]}[\Phi]$ of type u^c on the expressive side. On the view of use-conditions sketched above, elements of type u^c are evaluated according to Gricean maxims to determine the utterance's use-value, or felicity, here against the modal base restricted by Ψ.[8]

Note that the felicity conditions of the standard, truth-conditional interpretation shown in (20b) appear the same as those of the non-standard, use-conditional interpretation as shown in (21b). There is an important difference, however: the basis of evaluation regarding Relevance, Quality, etc. is assertion of the modalized proposition in case of the truth conditional (20): as conditioning has occurred on the descriptive level, the proposition to fill in the blank [...] in statements of the form "if [...] is relevant, [...] is as informative as required,...the speaker has adequate evidence for [...]..." is the *conditioned-on consequent*, and the evaluation world for felicity is the actual world—felicity is not conditioned on in the truth conditional. In the case of the use conditional (21), on the other hand, the conditional enters the derivation later than assertion. The relevant proposition for the evaluation statements is thus the *bare consequent*, but the evaluation statements are conditioned on, adding the antecedent to the actual world before evaluating felicity. In the case of the use conditional, conditioning thus only occurs on the expressive, but not on the descriptive level, as the modal introduced by conditional form enters the derivation only after utterance lifting has occurred. In the case of the truth conditional, on the other hand, there is no conditioning on the expressive level, as the conditional has already done its work on the descriptive level.

3   Explaining Interpretations

The crucial difference for accounting for intuitions with regard to the respective interpretations of (20) and (21) lies in their respective contents in the descriptive dimension predicting what is asserted: on the truth conditional interpretation, the conditioned-on consequent is asserted, but on the use conditional interpretation, the base consequent is asserted. Differences in the expressive dimension, as discussed above, predicts *how* the respective descriptive contents are asserted.

3.1   Biscuit Conditionals and Conditional Hedges

In the case of the biscuit conditional in (2), repeated here as (23), this means that the truth conditions of the conditional are the same as those of the consequent, and that its felicity conditions, on which modal base restriction by the antecedent occurs, are as paraphrased in (24).

(23) If you are thirsty, there's beer in the fridge.

---

[8] See Sect. 4.1 for discussion of the alternative idea that propositions in elements of type u^c are valuated like those of type t^c, but with respect to speaker belief.

(24) Asserting "there is beer in the fridge" is felicitous (w.r.t to w, f^+, g) iff

at all stereotypical worlds compatible with the speaker's beliefs *where the addressee is thirsty*, it is relevant to the participants' goals, as informative as required, . . . that there is beer in the fridge.

The prediction this makes with regard to the intuitions to be accounted for is that (23) is felicitously asserted even if the addressee is not thirsty, as conditioning on the expressive level prevents the use-conditional evaluation of assertion of the consequent in worlds where the antecedent does not hold.

Next, the felicity conditions of the conditional hedge (25), repeated from (3), as predicted from the model proposed here are paraphrased in (26).

(25) If I'm not mistaken, there's beer in the fridge.

(26) Asserting "there's beer in the fridge" is felicitous iff

at all stereotypical worlds compatible with the speaker's beliefs

where the speaker is not mistaken, it is relevant to the participants'

goals . . . that there is beer in the fridge.

This is a special case of a use conditional in that, given a belief of the speaker that there is beer in the fridge, eliminating worlds in which the speaker is mistaken (*i.e.* only considering worlds in which the speaker's beliefs are true) only leaves worlds in which the consequent holds.[9] Thus, when (25) receives a truth-conditional interpretation, its truth depends solely on the existence of a speaker belief that Φ be true: given that there is such a belief, the consequent is true at all worlds where this is a true belief, *i.e.* where the antecedent holds. The observation that exchanging antecedent and consequent predicts conditional perfection in this example: "If there's beer in the fridge, I am not mistaken" on a truth-conditional interpretation says that, given there is beer in the fridge, *i.e.* modal base restriction to worlds where the antecedent Φ holds leaves only worlds in which there is beer in the fridge, which, again, assuming that the speaker has a belief that φ, means that the speaker is not mistaken. Thus, the conditional in (25) is a material biconditional $\Psi \leftrightarrow \Phi$ on the truth-conditional level.

As this case shows, the truth-conditional and the use-conditional interpretations can happily coexist even though only the use-conditional interpretation as in (26) seems to be informative: when interpreted as a use conditional, (25) asserts that there is beer in the fridge, and overrides a possible violation of Relevance on the expressive level.

3.2   Choosing an Interpretation

As for the question of how it is decided which interpretation a conditional receives, I suggest that the use-conditional interpretation is be preferred when

---

[9] On a side note, this is potentially an argument against making the modal base realistic, *i.e.* assuming a non-doxastic circumstantial modal base.

restricting the modal base is uninformative on the at-issue level due to conditional independence. It should be noted here that I assume that both a use- and a truth-conditional reading are in principle available for all conditionals, but that one will usually be more salient. For illustration, consider the following ambiguous example.

(27) If you are interested in art, we will go to the museum.

This conditional has a reading on which the information that the speaker and some third party will go to the museum is relevant if the addressee is interested in art, and a reading on which the speaker intends to take the addressee to the museum only if the addressee is interested in art. The former is a plausible use-conditional, the latter a plausible truth-conditional reading. Different contexts can be constructed to bring out the salient interpretation. Once we have settled on a use-conditional interpretation, we need to decide which part of felicity, or which Gricean maxim, is targeted by modal base restriction with the antecedent. In the case of a use-conditional interpretation of (27), for instance, Relevance is a salient option, if the consequent is intended as an invitation to join.

Examples for ambiguity arising from the possibility of different parts of felicity being targeted include the contrast between discourse-structuring and problem-solving conditionals Csipak [5] proposes. Examples she provides for a contrast between the two types with regard to past tense are given below.

(28) If you were hungry yesterday, there was pizza in the fridge.

(29) #If I was being frank yesterday, you looked tired.

Csipak proposes that the difference lies in the nature of (28) as a problem solving conditional as opposed to the discourse-structuring conditions (29), claiming that the felicity of the latter depends on the speaker making an honest statement, such as "you looked tired". However, uttering "you looked tired" *now* does not count as an instance of having made honest statement *in the past*. (28), on the other hand, is fine (if not very helpful as a communicative act) as that there was pizza was potentially relevant to the addressee if the addressee was hungry.

The main observation of interest here is that the relation between the antecedent and the consequent, which can presumably be recovered by world knowledge, determines which part of felicity is targeted by the conditional antecedent. It is not immediately clear which part of felicity appropriateness hedges like "If I am being frank...", or "If I may be frank..." are targeting. McCready [15] suggests that they target the same part of felicity as parenthetical disclaimers, which is not easy to grasp with Gricean maxims, but could potentially be subsumed under Manner. Without a definitive answer to this issue, it can be noted that the properties of use conditionals differ depending on which part of felicity is targeted, as constrained by the relation between antecedent and consequent, but a more thorough survey of which aspects of felicity use conditionals can target has to be left for further research.

4   Conclusion and Outlook

Summing up, I have proposed that the properties of conditionals with non-standard salient interpretations such as biscuit conditionals and conditional hedges can be accounted for in a unified analysis with hypothetical conditionals when assuming that a covert modal, the modal base of which is restricted by the conditional antecedent, can not only be introduced by the conditional form on the descriptive (truth-conditional), but also on the expressive (use-conditional) level. The latter interpretation is made salient by conditional independence, that is when modal base restriction on the descriptive level does not result in truth conditions that differ from that of the bare consequent, thus rendering restriction uninformative.

In the remainder of this section, I briefly discuss two alternative approaches within the current proposal, followed by possibilities for integration with other analyses and expansion of data coverage.

4.1   An Assertion Modal

There is an additional possibility for the introduction of a modal which can be modified by the conditional antecedent on the use-conditional level, namely the assumption that assertion itself introduces a modal like human necessity, with a doxastic modal base. A possible argument for this goes as follows. Assuming that elements of type u^c, introduced by utterance-lifting, are evaluated by Gricean maxims to determine felicity of the utterance, how does truth factor in? It would potentially be welcome to have a way of including the descriptive content of an utterance in form of utterance-lifted elements of type u^c in propositional valuation to determine their use-values parallel to elements of type t^c, rather than limiting elements of type u^c to maxim-related evaluation. The intuition this is accounts for is that a proposition is felicitously asserted if the speaker *believes* it is true, whether or not this is actually the case. This can easily be reflected in utterance meaning by assuming that utterance lifting introduces a covert modal with a doxastic modal base and a stereotypical (or possibly empty) ordering source, similar to the human necessity operator introduced by the conditional form. An according representation of a plain assertion before and after utterance meaning is shown in (30), modified from (22).

(30) a. $\langle \Phi, \varnothing \rangle$

     b. $\langle \Phi, \Box^H[\Phi] \rangle$

On this analysis of assertion, the paraphrase for use-conditions of assertions given in (8) can be revised as in (31), for assuming that felicitous utterance requires the speaker to believe Φ,[10] so that Φ is required to hold at all worlds compatible with the speaker's beliefs for felicity.

---

[10] This is possibly a simplification not in line with the relevant Gricean maxim of Quality, which bans assertion of propositions believed to be false, rather than requiring asserted propositions to be believed to be true.

(31) Use conditions (modified): ASSERT(Φ) is felicitous w.r.t. w, f, g iff Φ is

   true at all stereotypical worlds compatible with the speaker's beliefs,

   relevant to the participants' goals, as informative as required, ...

This modified view of use conditions has another possibly welcome effect on the analysis presented so far: there is no need to assume that the conditional form introduces a human necessity modal on the non-propositional part (assuming that elements of type u^c are no propositions) use-conditional level, just as it does on the truth-conditional, propositional level. Rather, the conditional antecedent, now not restricting a modal base on propositional level, restricts the modal base of the conditional introduced by the speech act assertion.

4.2   Symmetric Use- and Truth Conditionals

There is a potentially more natural view of use conditionals (as well as truth conditionals) on which their meanings do not differ from truth conditionals, only their interpretations do. Above, I proposed that because of what amounts to uninformativity of restriction of the conditional modal base on the truth conditional, the conditional is interpreted on the use-conditional level in the case of biscuit conditionals and conditional hedges. There seems no harm, however, in assuming that the standard truth-conditional interpretation goes for cases of conditional independence as well, it just happens to have the same truth-value as the consequent. On such a view, truth- and use conditionals to not only share the same expressive meaning, but also the same descriptive meaning.

There is a potential issue with this view, however. While modal base restriction on the descriptive level is inert in the case of biscuit conditionals and conditional hedges (as the resulting proposition is truth-conditionally equivalent to the consequent), modal base restriction on the expressive level is not necessarily inert in the case of hypothetical conditionals: by example of (1), for asserting that there's beer at all worlds where John went shopping to be felicitous, it is for instance arguably not sufficient that this information is relevant in worlds where John went shopping. While discussion of conditional hedges has shown that in the non-canonical case, the use conditional and truth conditional interpretations can coexist, this may not be the case on the standard, hypothetical interpretation. If this (substantial) issue can be resolved, a symmetric approach to use- and truth conditionals would be a direction to consider for further research as a variant of the present proposal.

4.3   Connections Other Analyses, Broadening Data Coverage

An obvious starting point for integration of the present proposal with other analyses of related phenomena, is McCready's [15] analysis of such parenthetical hedges and disclaimers within the same framework. The proposal also naturally integrates with analyses of conditionals within a framework of possible-world semantics of modals, and can be readily applied to double-modal analyses of conditionals. I have built on Kaufmann and Schwager's [13] analysis of conditionals, so that expansion of the scope to the conditional imperatives they propose a modal analysis for is straightforwardly possible. The examples they give for use- and truth-conditional (which they label "relevance-" and "factual conditional") imperatives are shown in (32).

(32) a. If you are so smart, then do it yourself.

b. If I may give you some advice, don't go.

Combining my proposal with the double-modal analysis of conditional imperatives Kaufmann and Schwager propose predicts that in the case of the truth conditional imperative (32a), the modal base of a covert human necessity modal introduced by conditional form is restricted on the descriptive level, and takes the prejacent modalized by the imperative form in its nuclear scope. In the case of the use conditional imperative (32b), on the other hand, the prejacent modalized by the imperative form is asserted, *i.e.* an imperative speech act is performed, with the caveat that its felicity is only evaluated in worlds where the speaker may give the addressee some advice due to restriction of the modal base of the conditional modal on the expressive level.

Analyses of other non-standard conditionals in a similar possible-world semantic framework such as Condoravdi and Lauer's [4] analysis of anankastic conditionals are also possible targets for expansion, especially considering that there are variants of anankastic conditionals which share properties with biscuit conditionals, as discussed in Francez [8] under the label "chimerical conditionals". An example of a chimerical conditional is given in (33a) below.

(33) If you enter the museum from the south,...

a. ...there are no guards.

b. ...there are no guards where you enter.

c. ...there are no guards there.

Francez observes that (33a) has interpretations corresponding to (33b) and (33c), which I take to correspond to hypothetical and biscuit conditional interpretations, respectively. If this is on the right track, chimerical conditionals can be accounted for on the present analysis as well: the salient interpretation for (33b) is that of a hypothetical truth conditional with conditioning in the descriptive dimension—the speaker expresses that the consequent is only true at worlds where the antecedent holds; the salient interpretation for (33c) it that of a biscuit use conditional with conditioning in the expressive dimension—the consequent (there are no guards [in the south]) is relevant only if the addressee plans to enter in the south.

Finally, in order to test the limits of which kinds of conditionals can be accounted for on the present analysis, expansion to the non-standard conditionals with properties apparently differing from better-studied case like the ones discussed in this paper, as well as sentences "expressing conditional thoughts", but without conditional form, as for instance discussed in Elder and Jaszczolt [7], is an interesting perspective for further research.

References

1. Austin, J.: Ifs and cans. In: Philosophical Papers. Oxford University Press (1979)
2. Van der Auwera, J.: Pragmatics in the last quarter century: the case of conditional perfection. J. Pragmat. **27**(3), 261–274 (1997)
3. van Canegem-Ardijns, I., Van Belle, W.: Conditionals and types of conditional perfection. J. Pragmat. **40**(2), 349–376 (2008)
4. Condoravdi, C., Lauer, S.: Anankastic conditionals are just conditionals. Semant. Pragmat. **9**(8), 1–69 (2016)
5. Csipak, E.: Free factive subjunctives in German – ich hätte da eine Analyse. Ph.D. thesis, Georg-August-Universität Göttingen (2015)
6. DeRose, K., Grandy, R.E.: Conditional assertions and "biscuit" conditionals. Noûs **33**(3), 405–420 (1999)
7. Elder, C.H., Jaszczolt, K.M.: Towards a pragmatic category of conditionals. J. Pragmat. **98**, 36–53 (2016)
8. Francez, I.: Chimerical conditionals. Semant. Pragmat. **8**(2), 1–35 (2015)
9. Frank, A.: Context dependence in modal constructions. Ph.D. thesis, Universität Stuttgart (1996)
10. Franke, M.: The pragmatics of biscuit conditionals. In: Proceedings of the 16th Amsterdam Colloquium, pp. 91–98 (2007)
11. Grice, H.P.: Logic and conversation. In: Syntax and Semantics. Speech Acts, vol. 3, pp. 41–58. Academic Press, New York (1975)
12. Gutzmann, D.: Use-Conditional Meaning: Studies in Multidimensional Semantics. Oxford Unviersity Press, Oxford (2015)
13. Kaufmann, S., Schwager, M.: A unified analysis of conditional imperatives. In: Proceedings of Semantics and Linguistic Theory (SALT), vol. 19, pp. 239–256 (2009)
14. Kratzer, A.: Modals and Conditionals. Oxford University Press, Oxford (2012)
15. McCready, E.: Reliability in Pragmatics. Oxford University Press, Oxford (2015)
16. Potts, C.: The Logic of Conversational Implicatures. Oxford University Press, Oxford (2005)
17. Siegel, M.E.A.: Biscuit conditionals: quantification over potential literal acts. Linguist. Philos. **29**(2), 167–203 (2006)
18. Sperber, D., Wison, D.: Relevance: Communication and Cognition. Blackwell, Oxford (1996)
19. Stalnaker, R.C.: A theory of conditionals. In: Rescher, N. (ed.) Studies in Logical Theory, pp. 41–55. Blackwell (1968)

On the Interpretation of Dependent Plural Anaphora in a Dependently-Typed Setting

Ribeka Tanaka[1(✉)], Koji Mineshima[1,2], and Daisuke Bekki[1,2]

[1] Ochanomizu University, Bunkyō, Japan
[2] CREST, Japan Science and Technology Agency, Kawaguchi, Japan
{tanaka.ribeka,mineshima.koji,bekki}@is.ocha.ac.jp

Abstract. Anaphora resolution is sensitive to dependency relations between objects. One example, which is well known in the plural anaphora literature, is the dependent interpretation of the pronoun *it* in the mini-discourse *Every boy received a present. They each opened it.* The standard account of the dependent interpretation records dependency relations using sets of assignment functions (van den Berg [4,5], Nouwen [17], Brasoveanu [7]). This approach, however, requires substantial changes to the central notion of context and gives special treatment to dependent interpretations. In this paper we provide an alternative account from the perspective of dependent type theory (Martin-Löf [16]). We account for dependency relations in terms of dependent function types (Π-types), which are independently motivated objects within dependent type theory. We will adopt Dependent Type Semantics (Bekki [1], Bekki and Mineshima [2]) as a semantic framework and illustrate how dependent function types encode dependency relations and naturally provide a resource for dependent interpretations.

1 Introduction

Interpretation of pronouns can be sensitive to linguistically introduced dependency relations between objects. Consider the following examples discussed in the literature (Kamp and Reyle [12], van den Berg [4,5], Krifka [14], Nouwen [17], Brasoveanu [7]).[1]

(1) a. If every[1] boy received a[2] present, they[1] opened it[2].

  b. Every[1] boy received a[2] present. They[1] opened it[2].

In (1a), given a reading where *every boy* receives wide scope over *a present* (henceforth, the $\forall$–$\exists$ reading), the whole sentence can mean that if every boy received a present, each boy opened the present he received. Similarly, the second sentence in (1b) can be understood to mean that each boy opened the present he received. In both cases, the $\forall$–$\exists$ reading induces a dependency relation between boys and presents. This quantificational dependency plays a crucial role in the

---

[1] An anaphor is subscripted by an index, while its antecedent is superscripted by the same index.

© Springer International Publishing AG 2017
S. Kurahashi et al. (Eds.): JSAI-isAI 2016, LNAI 10247, pp. 123–137, 2017.
DOI: 10.1007/978-3-319-61572-1_9

interpretation of the singular pronoun *it* in the consequent of (1a) and the second sentence of (1b).

More generally, the reference to a dependency relation is possible when a semantic link between the restrictor of the universal quantifier and the subject of a subsequent sentence can be established.

(2)   a. If every boy receives a^1 present, some boy will open it$_1$.

b. If every boy receives a^1 present, every young boy will open it$_1$.

c. If every boy receives a^1 present, John will open it$_1$.

In (2a), the subjects of the antecedent and the consequent share the same noun phrase *boy*. The consequent can be understood to mean that some boy will open the present he received (Hintikka and Carlson [10], Ranta [18]). In (2b), *young boy* is a subset of *boy*. Again, a similar interpretation is allowed (cf. van den Berg [5]). There is no explicit link in the case of (2c), but if we have the background information that John is a boy, i.e., the information that links *John* to *boy*, the sentence can mean that John will open the present he received. The same observation applies to the following examples.

(3)   a. Every boy will receive a^1 present. Some boy will open it$_1$.

b. Every boy will receive a^1 present. Every young boy will open it$_1$.

c. Every boy will receive a^1 present. John will open it$_1$.

However, if it is difficult to establish a link between the two NPs, the dependent interpretation of the pronoun in question is impossible. The following examples demonstrate this contrast.

(4)   a. Every man will receive a^1 present. Some wife will open it$_1$.

b. Every man will receive a^1 present. *Some woman will open it$_1$.

In (4a), since it is relatively easy to find a relation between *man* and *wife*, the second sentence can be understood to mean that some man's wife will open the present he received. This contrasts with (4b), where a dependent reading is not possible unless a strong relation between *man* and *woman* is provided by the context.

A similar observation can be made about the so-called *quantificational subordination* phenomenon, which was originally discussed by Karttunen [13].

(5)   a. Harvey courts a^1 girl at every convention. She$_1$ is very pretty.

b. Harvey courts a^1 girl at every convention. She$_1$ always comes to the banquet with him. The$_1$ girl is usually also very pretty.

Although this example is more complicated than those we have considered so far, a similar structure seems to be involved. (5a) can only mean that there is one specific girl such that Harvey courts her at every convention and she is very pretty. If we discard this reading and force ourselves to keep the $\forall$–$\exists$ reading, there is no way to establish an anaphoric link between *a girl* and the singular pronoun *she*, and hence the discourse becomes infelicitous. In (5b), however, *she* can refer to a girl at each convention, since the subsequent discourse contains quantificational adverbs such as *always* and *usually*, which provide links to *every convention*.

This observation suggests that if the dependency relation between objects is used later on to interpret a pronoun, it must be tracked through discourse as an anaphoric resource. Since dependency relations are crucially involved in plural anaphora phenomena in general, constructing a formal mechanism to account for dependencies is one of the central issues in the dynamic semantics literature. The standard approach is to model dependencies as sets of assignments (van den Berg [4,5], Nouwen [17], Brasoveanu [7]). Another approach is to model it using an extended notion of assignment functions called *parametrized sum individuals* (Krifka [14]). However, since integrating functional relations directly into the underlying semantics is not straightforward, both approaches require substantial changes to the central notion of context to account for dependent interpretations.

In this paper, we propose an alternative account. We account for dependency relations in terms of *dependent function types* (Π-types) in dependent type theory (Martin-Löf [16]). In contrast to the mechanisms introduced in previous model-theoretic approaches, Π-types are independently motivated objects that are already provided in dependent type theory. We will adopt Dependent Type Semantics (Bekki [1], Bekki and Mineshima [2]; henceforth DTS) as a semantic framework and illustrate how Π-types encode the dependency relations in question and are readily provided as anaphoric resources in discourse. In the following section, we will first provide an overview of DTS. In Sect. 3, we describe our approach to handling the reference to dependency relations and show how it can be applied to the examples discussed above. In Sect. 4, we compare our technique with existing approaches.

2   Dependent Type Semantics

2.1   Dependent Types and Natural Language Sentences

DTS is a proof-theoretic natural language semantics based on dependent type theory. Dependent type theory (Martin-Löf [16]) is a formal system that extends simple type theory with the notion of *types depending on terms*. This rich type structure provides a foundation for handling context dependence in natural language. One of the distinctive features of DTS, compared with other frameworks based on dependent type theory, is that it is augmented with under-specified terms called @-terms. DTS uses @-terms to provide a unified analysis of entailment, anaphora, and presupposition from an inferential and computational perspective. DTS also gives a compositional account of inferences involving anaphora; see Bekki [1] and Bekki and Mineshima [2] for details on compositional semantics within the framework of DTS.

Dependent type theory uses two type constructors Π (dependent function type) and Σ (dependent product type) to construct dependent types. The type constructor Π is a generalized form of the functional type. A term of type (Πx :

$A)B(x)$ is a function f which takes any element a of A and returns a term $f(a)$ of type $B(a)$ dependent on the choice of the argument a. In other words, a dependent function is a function whose codomain is dependent on the given argument. The type constructor Σ is a generalized form of the product type. A term of type $(\Sigma x : A)B(x)$ is a pair (m, n) which consist of a term m of type A and a term n of type $B(m)$, where the type of the second element n depends on the choice of the first element m.

Dependent type theory is based on the Curry-Howard correspondence, where a type can be regarded as a proposition and a term of the type can be regarded as a proof of the proposition. Accordingly, a Π-type corresponds to a universal quantifier; a proof term of the universal sentence is a function. If x does not occur free in B, i.e., there is no dependencies involved, $(\Pi x : A)B$ corresponds to implication. A Σ-type corresponds to an existential quantifier; a proof term of the existential sentence is a pair. If x does not occur free in B, $(\Sigma x : A)B$ corresponds to conjunction. See, e.g., Martin-Löf [16] for more details, including inference rules for Σ and Π constructors. Figure 1 shows the notation of Σ-type and Π-type adopted in DTS.

	Π-type	Σ-type
Martin-Löf (1984)	$(\Pi x : A)B(x)$	$(\Sigma x : A)B(x)$
DTS	$(x : A) \rightarrow B(x)$	$\begin{bmatrix} x : A \\ B(x) \end{bmatrix}$

Fig. 1. Notation of Π-type and Σ-type.

Since Π-types correspond to propositions with the universal quantifier, the sentence *every boy entered* can be represented as follows.

$$(6) \quad \left(u : \begin{bmatrix} x : \textbf{entity} \\ \textbf{boy}(x) \end{bmatrix} \right) \rightarrow \textbf{enter}(\pi_1 u)$$

Here, **entity** is a basic type for all entities. The restrictor *boy* is analyzed as a Σ-type. A term u having this Σ-type would be a pair (e, b), where e is a term of type **entity** and b is a proof term of the proposition $\textbf{boy}(e)$. Σ-types are associated with projection functions π_1 and π_2. These functions allow one to access the first and second elements of the pair, respectively: for any pair (m, n), $\pi_1(m, n) = m$ and $\pi_2(m, n) = n$. Thus, the term $\pi_1 u$ picks up from u the term e of type **entity**. Therefore, (6) corresponds to the proposition that for every entity that is a boy, that entity entered.

A sentence with an existential quantifier such as *a boy entered* is represented in terms of Σ-types. Again, $\pi_1 u$ corresponds to an entity which is a boy, and thus, (7) corresponds to the proposition that there exists an entity which is a boy and which entered.

(7)
$$\left[u : \begin{bmatrix} x : \textbf{entity} \\ \textbf{boy}(x) \end{bmatrix} \atop \textbf{enter}(\pi_1 u) \right]$$

One advantage of using Σ-types is that they can capture an *externally dynamic* property of existential quantifier and conjunction (Groenendijk and Stokhof [9]). For instance, a discourse such as (8a) is problematic in the sense that its syntactically-corresponding formula in predicate logic, (8b), fails to represent an anaphoric link between *a boy* and *he*.

(8)   a. A boy entered. He whistled.
      b. $\exists x (\textbf{boy}(x) \wedge \textbf{enter}(x)) \wedge \textbf{whistle}(x)$

The Σ-type, by contrast, can straightforwardly provide the semantic representation of this discourse as follows.

(9)
$$\left[v : \left[u : \begin{bmatrix} x : \textbf{entity} \\ \textbf{boy}(x) \end{bmatrix} \atop \textbf{enter}(\pi_1 u) \right] \atop \textbf{whistle}(\pi_1 \pi_1 v) \right]$$

Although a term u is no longer accessible from the argument position of **whistle**, one can still pick up the term via a newly introduced term v, since v is a pair and each of its parts is accessible by applying (a sequence of) the projection function. In this way, the Σ-type can pass a variable binding relation to a subsequent discourse.

2.2   DTS and Anaphora Resolution

The remaining question is how one can obtain the term $\pi_1 \pi_1 v$ in (9) for the representation of the pronoun *he*. In DTS, anaphoric expressions are represented in terms of underspecified terms called @-terms. Anaphora resolution in DTS is therefore defined as a process that replaces the @-term with the specific term that is constructed via type checking and proof construction (Bekki and Satoh [3], Bekki and Mineshima [2]). For instance, the pronoun *he* is assigned the semantic representation in (10), where the type annotation of the @-term represents the requirement that *he* refers to some entity being male.

(10) $\pi_1 \left(@_i \begin{bmatrix} x : \textbf{entity} \\ \textbf{male}(x) \end{bmatrix} \right)$

Dynamic conjunction between sentences is defined in terms of Σ-type. Thus, the semantic representation of the whole discourse in (8a) is given as follows.

(11)
$$\left[v : \left[u : \begin{bmatrix} x : \textbf{entity} \\ \textbf{boy}(x) \end{bmatrix} \atop \textbf{enter}(\pi_1 u) \right] \atop \textbf{whistle}\left(\pi_1 \left(@_i \begin{bmatrix} x : \textbf{entity} \\ \textbf{male}(x) \end{bmatrix} \right) \right) \right]$$

This underspecified representation is required to be well formed, that is, to have type **type**. This condition (called the *felicity condition* of a sentence) invokes type checking and leads to the proof construction associated with the @-term. In the current example, one needs to find a proof term that satisfies the following inference.

$$(12) \quad \Gamma, \; v : \begin{bmatrix} u : \begin{bmatrix} x : \mathbf{entity} \\ \mathbf{boy}(x) \end{bmatrix} \\ \mathbf{enter}(\pi_1 u) \end{bmatrix} \vdash \boxed{?} : \begin{bmatrix} x : \mathbf{entity} \\ \mathbf{male}(x) \end{bmatrix}$$

Here, Γ is a global context that represents background knowledge, and v is a term accessible from the position of the @-term, which corresponds to the information provided up to this point of the mini-discourse. From these premises Γ and v, one needs to construct a proof term of the consequent that fills in the position marked by $\boxed{?}$. Now, suppose that the global context contains the proof term in (13), which corresponds to the knowledge that every boy is male.

$$(13) \quad k_b : \left(u : \begin{bmatrix} y : \mathbf{entity} \\ \mathbf{boy}(y) \end{bmatrix} \right) \to \mathbf{male}(\pi_1 u)$$

By using this knowledge k_b together with v, one can eventually construct a proof term of the required type that can replace the underspecified term in (11), yielding the fully-specified representation in (9).

Note that this anaphora resolution procedure in DTS can account for the following *externally static* property of universal quantifiers.

(14)　　a.　Every[1] boy received a present. *He$_1$ looks happy.

　　　　b.　Every boy received a[1] present. *It$_1$ was a toy car.

In these cases, the first sentences are universal sentences, so proof terms provided to the subsequent discourse are functions. Thus, neither an entity being a boy embedded in the domain of the function (i.e., an entity in the restrictor), nor an entity being a present embedded in the codomain of the function (i.e., an entity in the nuclear scope) can be picked up by the operation that is available in the case of existential sentences represented as Σ-types.

In this way, Σ-types and Π-types, together with the anaphora resolution process in DTS, provide a proof-theoretic account of the dynamic properties of the existential and universal quantifiers.

3　Dependency Relations and Dependent Interpretation

As we have seen so far, Σ-types are externally dynamic in that they introduce pairs of objects as discourse referents which can be picked up by projection functions; by contrast, Π-types are externally static in that they do not introduce individual discourse referents. Because of this difference, one might think that Π-types do not contribute to establishing any discourse referents. Ranta [18], however, describes exactly such a case in the following example.[2]

---

[2] This example is attributed to Lauri Karttunen in Hintikka and Carlson [10].

(15) If you give every child a present, some child will open it.

In (15), the antecedent clause is analyzed as a Π-type, so it introduces a *function* as a discourse referent. The functional discourse referent introduced by this Π-type can be used to give the interpretation of the pronoun *it* in the consequent clause. Although Ranta's brief discussion is confined to the example in (15), we will show below that the idea that proof terms of Π-types serve as functional discourse referents can apply to examples (2)–(4) discussed in Sect. 1, as well as to the case of plural anaphora involving the pronoun *they* and to quantificational subordination. In Sect. 3.1, we will first focus on the basic case of the dependent interpretation of a pronoun and show how the idea of functional discourse referents couched within the framework of DTS can capture the quantificational dependency.[3] In Sect. 3.2, we will generalize this idea to other cases including plural anaphora involving *they*.

3.1 Basic Example

Let us consider the simplest example in (3a), which is repeated below.

(3a) Every boy will receive a[1] present. Some boy will open it$_1$.

Since a universal quantifier corresponds to a Π-type, the first sentence can be represented as follows.

$$(16) \quad \left(u : \begin{bmatrix} x : \textbf{entity} \\ \textbf{boy}(x) \end{bmatrix} \right) \rightarrow \left[v : \begin{bmatrix} y : \textbf{entity} \\ \textbf{present}(y) \end{bmatrix} \right]$$
$$\textbf{receive}(\pi_1 u, \pi_1 v)$$

The terms $\pi_1 u$ and $\pi_1 v$ pick up the entity being a boy and the entity being a present, respectively. The type as a whole represents the proposition that, for every boy, there exists a present such that the boy received it. This representation corresponds to the distributive reading in question. Thus, a term of this type is a function that receives a pair consisting of an entity and a proof of that entity being a boy, and then returns a tuple that consists of an entity, a proof of that entity being a present, and a proof of the boy and the present being in the receiving relation. This means that the representation of the first sentence introduces a function that corresponds to the dependency relation between boys and presents.

The second sentence is represented by the Σ-type, where the pronoun *it* can be defined as an underspecified term of type **entity**. Thus, by combining the semantic representation of the two sentences in terms of dynamic conjunction, (3a) is represented as the following Σ-type.

---

[3] There are important differences between Ranta's [18] framework and that of DTS. First, while Ranta did not adopt the framework of compositional semantics, DTS provides a compositional derivation of the semantic representations involving anaphora. Another difference between Ranta's and our analysis is that Ranta interprets common nouns as *types*, while DTS treats them as *predicates*. More discussion on these points can be found in Bekki and Mineshima [2].

$$
(17) \quad
\begin{bmatrix}
f : \left(u : \begin{bmatrix} x : \textbf{entity} \\ \textbf{boy}(x) \end{bmatrix} \right) \rightarrow \begin{bmatrix} v : \begin{bmatrix} y : \textbf{entity} \\ \textbf{present}(y) \\ \textbf{receive}(\pi_1 u, \pi_1 v) \end{bmatrix} \end{bmatrix} \\
\begin{bmatrix} z : \begin{bmatrix} x : \textbf{entity} \\ \textbf{boy}(x) \end{bmatrix} \\ \textbf{open}(\pi_1 z, @_1 \textbf{entity}) \end{bmatrix}
\end{bmatrix}
$$

In this way, the proof term f of the first sentence that corresponds to a dependency relation between boys and presents serves as an anaphoric resource. In the current case, anaphora resolution of the pronoun *it* yields the following inference.

$$
(18) \quad \Gamma, \, f : \left(u : \begin{bmatrix} x : \textbf{entity} \\ \textbf{boy}(x) \end{bmatrix} \right) \rightarrow \begin{bmatrix} v : \begin{bmatrix} y : \textbf{entity} \\ \textbf{present}(y) \\ \textbf{receive}(\pi_1 u, \pi_1 v) \end{bmatrix} \end{bmatrix}, \, z : \begin{bmatrix} x : \textbf{entity} \\ \textbf{boy}(x) \end{bmatrix} \vdash \boxed{?} : \textbf{entity}
$$

There are two proof terms accessible from the position of the @-term: the term f, which is a proof term of the first sentence, and z, which is a term corresponding to the subject of the second sentence. The proof construction goes as follows: first, by applying z to the function f, one obtains the proof term fz that is a pair corresponding to the present received by the boy, $\pi_1 z$; second, by taking the first projection of the first projection of fz, one obtains a term $\pi_1 \pi_1 (fz)$ of type **entity**. Therefore, by replacing the @-term with the obtained term $\pi_1 \pi_1 (fz)$, the second argument of **open** will be filled with an entity which depends on the term z, namely, an entity which depends on the subject of the second sentence. In this way, we can account for the dependent interpretation of the pronoun *it* in (3a).[4]

3.2   More Examples

In Sect. 1, we have observed that an anaphoric link can be established even when the subject of a subsequent discourse does not exactly match the restrictor of the universal quantifier of an earlier sentence. These examples, (3b) and (3c), are repeated below.

(3)   b.   Every boy will receive a[1] present. Every young boy will open it$_1$.

c.   Every boy will receive a[1] present. John will open it$_1$.

Both the first and the second sentences in (3b) can be represented in terms of Π-types. Thus, the whole sentence receives the following semantic representation.

---

[4] Some readers may think that proof terms have something in common with discourse referents in Discourse Representation Theory (Kamp and Reyle [12], Kamp et al. [11]) in that both objects are introduced by sentences and referred to afterward to resolve anaphora. There are at least two crucial differences. Firstly, as Ranta [18] discussed, while discourse referents are limited to individuals without any inner structure, proof terms can have any type. Secondly, together with the anaphora resolution mechanism provided in DTS, proof terms can contribute to logical inference, which yields a new proof term serving as an antecedent.

$$(19) \quad \begin{bmatrix} f : \left(u : \begin{bmatrix} x : \textbf{entity} \\ \textbf{boy}(x) \end{bmatrix} \right) \rightarrow \begin{bmatrix} v : \begin{bmatrix} y : \textbf{entity} \\ \textbf{present}(y) \end{bmatrix} \\ \textbf{receive}(\pi_1 u, \pi_1 v) \end{bmatrix} \\ \left(t : \begin{bmatrix} z : \begin{bmatrix} x : \textbf{entity} \\ \textbf{boy}(x) \end{bmatrix} \\ \textbf{young}(\pi_1 z) \end{bmatrix} \right) \rightarrow \textbf{open}(\pi_1 \pi_1 t, @_1 \textbf{entity}) \end{bmatrix}$$

The premises of the inference associated with the resolution of $@_1$ are terms f and t. Since the term $\pi_1 t$, shown in (20), can be derived from the given t and can be applied to f, one eventually obtains a term $\pi_1 \pi_1 f(\pi_1 t)$, which corresponds to the present dependent on each young boy, $\pi_1 \pi_1 t$.

$$(20) \quad \pi_1 t : \begin{bmatrix} x : \textbf{entity} \\ \textbf{boy}(x) \end{bmatrix}$$

Similarly, (3c) is represented as follows.

$$(21) \quad \begin{bmatrix} f : \left(u : \begin{bmatrix} x : \textbf{entity} \\ \textbf{boy}(x) \end{bmatrix} \right) \rightarrow \begin{bmatrix} v : \begin{bmatrix} y : \textbf{entity} \\ \textbf{present}(y) \end{bmatrix} \\ \textbf{receive}(\pi_1 u, \pi_1 v) \end{bmatrix} \\ \textbf{open}(\textbf{john}, @_1 \textbf{entity}) \end{bmatrix}$$

To find a semantic link between *John* and *boy*, one needs the background knowledge that John is a boy. If the global context Γ supplies the knowledge $k_j : \textbf{boy}(\textbf{john})$, one can construct the following term.

$$(22) \quad (\textbf{john}, k_j) : \begin{bmatrix} x : \textbf{entity} \\ \textbf{boy}(x) \end{bmatrix}$$

Again, this term can serve as an argument to the function f.

If the relation between the restrictor of the universal quantifier and the subject of the subsequent discourse is not clear, then the procedure simply fails to find a proof. For instance, in the case of (4b), repeated here as (23a), there exists neither an explicit link nor an implicit link between men and women.

(23)   a. Every man will receive a[1] present. *Some woman will open it[1].

$$\text{b.} \quad \begin{bmatrix} \begin{bmatrix} f : \left(u : \begin{bmatrix} x : \textbf{entity} \\ \textbf{man}(x) \end{bmatrix} \right) \rightarrow \begin{bmatrix} v : \begin{bmatrix} y : \textbf{entity} \\ \textbf{present}(y) \end{bmatrix} \\ \textbf{receive}(\pi_1 u, \pi_1 v) \end{bmatrix} \end{bmatrix} \\ \begin{bmatrix} z : \begin{bmatrix} x : \textbf{entity} \\ \textbf{woman}(x) \end{bmatrix} \\ \textbf{open}(\pi_1(z), @_1 \textbf{entity}) \end{bmatrix} \end{bmatrix}$$

In this case, one needs to apply an argument to the function f to construct a proof of the present received by some man. Thus, unless some relation which bridges men and women is available in the global context, there is no way to obtain the required proof term from z and f.

The conditional sentences in (2a–c) can be treated in parallel to the examples in (3a–c). The sentences are reproduced below.

(2)   a. If every boy receives a[1] present, some boy will open it$_1$.

b. If every boy receives a[1] present, every young boy will open it$_1$.

c. If every boy receives a[1] present, John will open it$_1$.

For instance in the case of (2a), we can provide the following semantic representation.

$$(24) \quad \left(f : \left(u : \begin{bmatrix} x : \textbf{entity} \\ \textbf{boy}(x) \end{bmatrix} \right) \rightarrow \begin{bmatrix} y : \textbf{entity} \\ \textbf{present}(y) \\ \textbf{receive}(\pi_1 u, \pi_1 v) \end{bmatrix} \right) \rightarrow \begin{bmatrix} z : \begin{bmatrix} x : \textbf{entity} \\ \textbf{boy}(x) \end{bmatrix} \\ \textbf{open}(\pi_1 z, @_1 \textbf{entity}) \end{bmatrix}$$

The whole conditional is analyzed as a Π-type of the form $(f : A) \rightarrow B$. Here the antecedent clause *every boy receives a present* is represented as a Π-type and thus introduces a function f in the antecedent. This proof term f is accessible from the consequent. As a result, the proof term f, together with the proof term z introduced in the consequent, can be used for the resolution of $@_1$. This enables the dependent interpretation of the pronoun *it*. We can see that the resolution of $@_1$ involves essentially the same inference as that for (17). Similarly, (2b) and (2c) can be analyzed along the same lines as (3b) and (3c), whose semantic representation are given in (19) and (21), respectively.

The case of quantificational subordination repeated here can be accounted for in a similar way.

(5)   a. Harvey courts a[1] girl at every convention. She$_1$ is very pretty.

b. Harvey courts a[1] girl at every convention. She$_1$ always comes to the banquet with him. The$_1$ girl is usually also very pretty.

The $\forall$–$\exists$ reading of the first sentence of (5a,b) is analyzed in the following simplified representation.

$$(25) \quad \left(u : \begin{bmatrix} x : \textbf{entity} \\ \textbf{convention}(x) \end{bmatrix} \right) \rightarrow \begin{bmatrix} v : \begin{bmatrix} y : \textbf{entity} \\ \textbf{girl}(y) \end{bmatrix} \\ \textbf{court-at}(\textbf{harvey}, \pi_1 v, \pi_1 u) \end{bmatrix}$$

This Π-type introduces a function, so the anaphoric link between *a girl* and *she* is blocked in (5a). In the case of (5b), *always* in the second sentence introduces another Π-type, whose restrictor provides an adequate argument for the function introduced by the first sentence. The entire derivation is similar to the case of (19) above. Thus, *she* can be interpreted as a girl at each convention.

Let us now turn back to our first example (1) involving plural anaphora. Example (26) is similar, but with adjectival quantifiers (Krifka [14]).

(1)   Every[1] boy received a[2] present. They$_1$ opened it$_2$.

(26)   Three[1] students each wrote an[2] article. They$_1$ each sent it$_2$ to L&P.

Although providing a comprehensive analysis of plural anaphora including an analysis of the so-called collective reading is not the main concern of this paper, we will briefly sketch how to account for the dependency relation involved in plural anaphora. In our analysis, two factors are essential to account for the reference to the dependency relation. Firstly, the initial sentence must have the $\forall$–$\exists$ reading which induces a dependency relation between objects in terms of a dependent function. Secondly, a singular pronoun in the subsequent discourse can be interpreted anaphorically if it supplies an adequate argument to the dependent function introduced by the initial sentence. These two points are critical for our analysis of plural anaphora in (1) and (26).

As for the first point, we follow an analysis of generalized quantifiers and adjectival quantifiers in DTS (Tanaka et al. [20], Tanaka [19]) that provides semantic representation of those quantificational expressions by using a dependent function. According to this analysis, generalized quantifiers such as *most*[5] and adjectival quantifiers such as *three* are uniformly represented as involving existential quantification over dependent functions whose domain is restricted by the cardinality condition.[6] Thus, this dependent function can be used for anaphora resolution as in the cases we have seen so far.

The essential role of the plural pronoun *they* is thus to supply terms that are adequate for the arguments of the dependent function. The semantic representation of *they* is also given in terms of the @-term. In contrast to the singular pronoun *it*, the type annotation of the @-term associated with *they* requires a predicate and a proof term of the cardinality condition. This is because the domain of the dependent function provided by the quantificational expression is restricted by the predicate and the cardinality condition. Therefore, the term replacing the @-term can supply an adequate argument to the function, which enables the dependent interpretation of the singular pronoun which comes after.

4   Previous Approaches

In this section, we provide a brief overview of some of the existing solutions in dynamic semantics to handle reference to a dependency relation.

In classical Discourse Representation Theory (Kamp and Reyle [12], henceforth DRT), reference to a dependency relation is handled by using a copy mechanism. First, the first sentence in (1), *every boy received a present*, yields the following discourse representation structure (DRS).

---

[5] In the case of *every*, we can provide its semantic representation in two ways: one possibility is to treat it simply as a Π-type as we have seen above; another possibility is to represent it in the same way as other generalized quantifiers such as *most*. Since these two formulas are mutually deducible, the account of generalized quantifiers presented here can be applied to the case of *every* as well.

[6] As Π-types correspond to the $\forall$–$\exists$ reading (or distributive reading), the semantic representation of *three* provided by Tanaka [19] should correspond to the semantic representation of *three...each*. To obtain the semantic representation of *three...each* in a compositional way, we can integrate the existing analysis of plural objects into our framework (see Link [15] for the standard approach; for the treatment of plural objects in a dependently-typed setting, see Boldini [6] and Chatzikyriakidis and Luo [8]). A full discussion of this phenomenon is beyond the scope of this paper.

(27)

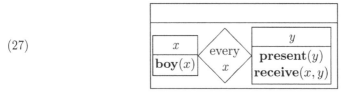

The construction of this DRS triggers the operation called *abstraction*, which constructs a new plural discourse referent X' consisting of an object that satisfies the condition of x. The pronoun *they* refers to this X' and yields the DRS in Fig. 2a, where universal quantification over X' takes place. In this DRS, however, there is no discourse referent which can be associated with singular y in **open**(x, y). In such a case, there is an option to apply a copy operation, which copies the conditions of x constituting X' to the restrictor part of the duplex condition. The corresponding DRS is given in Fig. 2b. In this way, the singular variable y in **open**(x, y) can refer to each present associated with each boy.

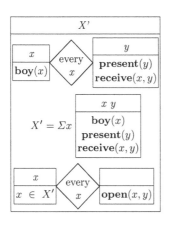
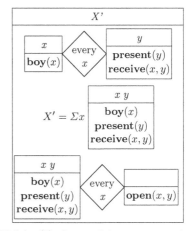

a. DRS for (1) before applying copy operation. b. DRS for (1) after applying copy operation.

Fig. 2. DRS associated with (1).

Krifka [14] criticizes using a representation-based copying operation as ad hoc, and proposes an analysis based on an enriched assignment function called *parametrized sum individuals*. Parametrized sum individuals are sets of pairs of an individual and a variable assignment associated with that individual. A possible instance of parametrized individuals for *every boy received a present* may have the following representation.

$$\langle x, \{\langle b_1, \{\langle y, p_1\rangle\}\rangle, \langle b_2, \{\langle y, p_2\rangle\}\rangle, \langle b_3, \{\langle y, p_3\rangle\}\rangle, \ldots\}\rangle$$

The individuals can be either singular or plural. Since individuals are followed by assignments associated with them, this structure captures dependency relations between objects. In the case of the distributive interpretation, each parametrized individual is independently evaluated against predicates. Thus, singular pronouns can be interpreted along each parametrized individual, which produces an effect of interpretation sensitive to the dependency relation.

The standard way to encode dependency relations is to adopt *information states for plurals*, as proposed by van den Berg [4,5] in Dynamic Plural Logic. In this approach, formulas are interpreted relative to information states, which are sets of assignments, instead of to assignments. A possible information state for *every boy received a present* may have the following representation.

$$\{\{\langle x, b_1\rangle, \langle y, p_1\rangle\}, \{\langle x, b_2\rangle, \langle y, p_2\rangle\}, \{\langle x, b_3\rangle, \langle y, p_3\rangle\}, \ldots\}$$

When distribution over x is involved, predicates are evaluated against each assignment of information states. The assignment of new values takes place independently of each assignment function; thus, the variables introduced may be dependent on x. This is the source of dependency.

Our intuition about the $\forall$–$\exists$ reading of *every boy received a present* is that it introduces a quantificational dependency, that is, a function f such that x is a boy receiving a present $f(x)$. However, there is no natural place in standard dynamic semantics theory to store such a function for subsequent anaphora. Therefore, each of the three approaches mentioned above need to capture dependency relations in an indirect way, which requires integrating a special mechanism or structure into the underlying framework.

There are also several empirical issues to consider. First, the copy mechanism in DRT is triggered by the resolution of the plural pronoun *they*. However, we have seen that there are cases such as (3a–c), where a plural pronoun does not appear but still reference to a dependency relation takes place. A stipulation or operation is needed in DRT to handle more general cases including these examples. Second, there exists no proof theory for either of the frameworks proposed by Krifka or van den Berg. Van den Berg's [5] analysis can account for cases such as (3b), where a subset relation allows reference to a dependency relation. In general, however, a semantic link between the restrictor of the universal quantifier and the subject of the subsequent discourse is not limited to the subset relation, as we can observe in example (4a). Rather, the dependent interpretation involves a more general kind of inference, of which a semantic link in terms of subset relations is a special instance.

An advantage of the proposed DTS analysis is that Π-types are independently motivated objects already provided in dependent type theory, and thus, we do not need to extend our framework to account for dependency relations. By following the standard dynamic conjunction operation and anaphora resolution procedure, DTS can naturally provide a function as a discourse referent, which straightforwardly leads to the dependent interpretation of singular pronouns. In addition, because the anaphora resolution process in DTS involves a proof search, it provides a more general and uniform account of semantic links between the restrictor of the universal quantifier and the subject of the subsequent discourse.

5 Conclusion

In this article, we have argued for an account of dependency relations between objects as dependent functions in dependent type theory. This contrasts with approaches in the dynamic semantics tradition, where a function does not serve as a discourse referent, and the enriched notion of assignment functions plays an essential role in handling dependencies. We have seen that the proposed account is capable of explaining the dependent interpretation of pronouns by integrating with the anaphora resolution mechanism of DTS. This new account may also offer a basis for the proof-theoretic analysis of plural anaphora.

Acknowledgments. This paper is a revised and expanded version of a paper presented at the Logic and Engineering of Natural Language Semantics 13 (LENLS13). We would like to thank Alastair Butler, Yusuke Kubota, Robert Levine, the audience of the workshop 'New landscapes in theoretical computational linguistics,' two anonymous reviewers, and the audience of LENLS13 for their valuable comments, suggestions, and discussions. We are also grateful to Robin Cooper for helpful discussions on an earlier version of this paper. The first author acknowledges the financial support from JSPS (Grant-in-Aid for JSPS Research Fellow; 15J11772).

References

1. Bekki, D.: Representing anaphora with dependent types. In: Asher, N., Soloviev, S. (eds.) LACL 2014. LNCS, vol. 8535, pp. 14–29. Springer, Heidelberg (2014). doi:10.1007/978-3-662-43742-1_2
2. Bekki, D., Mineshima, K.: Context-passing and underspecification in dependent type semantics. In: Chatzikyriakidis, S., Luo, Z. (eds.) Modern Perspectives in Type Theoretical Semantics. Studies in Linguistics and Philosophy, vol. 98, pp. 11–41. Springer, Heidelberg (2017). doi:10.1007/978-3-319-50422-3_2
3. Bekki, D., Satoh, M.: Calculating projections via type checking. In: Cooper, R., Retoré, C. (eds.) ESSLLI Proceedings of TYTLES Workshop on Type Theory and Lexical Semantics, ESSLLI, Barcelona (2015)
4. van den Berg, M.: Dynamic generalized quantifiers. In: van der Does, J., van Eijck, J. (eds.) Quantifiers, Logic, and Language, pp. 63–94. CSLI publications, California (1996)
5. van den Berg, M.: Some aspects of the internal structure of discourse: the dynamics of nominal anaphora. Ph.D. thesis, University of Amsterdam (1996)
6. Boldini, P.: The reference of mass terms from a type theoretical point of view. In: Proceedings of Forth International Workshop on Computational Semantics (2001)
7. Brasoveanu, A.: Donkey pluralities: plural information states versus non-atomic individuals. Linguist. Philos. **31**(2), 129–209 (2008)
8. Chatzikyriakidis, S., Luo, Z.: An account of natural language coordination in type theory with coercive subtyping. In: Duchier, D., Parmentier, Y. (eds.) CSLP 2012. LNCS, vol. 8114, pp. 31–51. Springer, Heidelberg (2013). doi:10.1007/978-3-642-41578-4_3
9. Groenendijk, J., Stokhof, M.: Dynamic predicate logic. Linguist. Philos. **14**(1), 39–100 (1991)

10. Hintikka, J., Carlson, L.: Conditionals, generic quantifiers, and other applications of subgames. In: Saarinen, E. (ed.) Game-Theoretical Semantics. Synthese Language Library, vol. 3, pp. 179–214. Springer, Heidelberg (1979). doi:10.1007/978-1-4020-4104-4_7

11. Kamp, H., van Genabith, J., Reyle, U.: Discourse representation theory. In: Gabbay, D.M., Guenthner, F. (eds.) Handbook of Philosophical Logic. Handbook of Philosophical Logic, vol. 15, pp. 125–394. Springer, Heidelberg (2011). doi:10.1007/978-94-007-0485-5_3

12. Kamp, H., Reyle, U.: From Discourse to Logic: Introduction to Modeltheoretic Semantics of Natural Language, Formal Logic and Discourse Representation Theory. Studies in Linguistics and Philosophy, vol. 42. Springer, Heidelberg (1993)

13. Karttunen, L.: Discourse referents. In: McCawley, J.D. (ed.) Syntax and Semantics, vol. 7, pp. 363–386. Academic Press, Cambridge (1976)

14. Krifka, M.: Parametrized sum individuals for plural anaphora. Linguist. Philos. **19**(6), 555–598 (1996)

15. Link, G.: The Logical Analysis of Plurals and Mass Terms: A Lattice-Theoretical Approach. de Gruyter, Berlin (1983)

16. Martin-Löf, P.: Intuitionistic type theory: notes by Giovanni Sambin of a series of lectures given in Padua, Bibliopolis (1984)

17. Nouwen, R.: Plural pronominal anaphora in context: dynamic aspects of quantification. Ph.D. thesis, Utrecht Institute for Linguistics OTS (2003)

18. Ranta, A.: Type-Theoretical Grammar. Oxford University Press, Oxford (1994)

19. Tanaka, R.: A proof-theoretic approach to generalized quantifiers in dependent type semantics. In: de Haan, R. (ed.) Proceedings of the ESSLLI2014 Student Session, pp. 140–151 (2014)

20. Tanaka, R., Nakano, Y., Bekki, D.: Constructive generalized quantifiers revisited. In: Nakano, Y., Satoh, K., Bekki, D. (eds.) JSAI-isAI 2013. LNCS, vol. 8417, pp. 115–124. Springer, Heidelberg (2014). doi:10.1007/978-3-319-10061-6_8

HAT-MASH 2016

Healthy Aging Tech Mashup Service, Data and People

Ken Fukuda

AI Research Center, National Institute of Advanced Industrial Science
and Technology, Tsukuba, Japan
ken.fukuda@aist.go.jp

1 The Workshop

The 2nd International Workshop HAT-MASH 2016 (Healthy Aging Tech mashup service, data and people) was successfully held on the 14th and 15th of November, 2016 in Kanagawa, Japan as part of JSAI-isAI 2016. It was the second international workshop following the first one in 2015 that brings people from healthy aging and elderly care technology, information technology and service engineering all together.

The main objective of this workshop is to provide a forum to discuss important research questions and practical challenges in healthy aging and elderly care support to promote transdisciplinary approaches. The workshop welcomes researchers, academicians as well as industrial professionals of different but relevant fields from all over the world to present their research results and development activities. The workshop will provide opportunities for the participants to exchange new ideas and experiences, to establish research or business networks and to find global partners for future collaboration.

This year, we featured two keynote session with three keynotes and three oral sessions with twelve submissions.

2 Papers

In "Toward Sentiment Analysis in Elderly Care Facility", Ken Fukuda, Satoshi Nishimura, Huizhi Liang and Takuichi Nishimura report the impact of introducing an ICT system into elderly care facilities. Hand-over notes are extremely important to share information about the care-receiver's ADL status and provide high-quality services. However, taking notes is a time-consuming task. Moreover, handwritten hand-over notes make it difficult to search the required information. To solve this issue, a handover support system for elderly care facilities was installed into a facility and evaluated. The authors propose to utilize this system as a platform to share and analyze verbal information in the facility. The paper reports results of the system installation and text-mining analysis. Furthermore, the authors explore to apply sentiment analysis to hand-over messages to sense the atmosphere of their working environment and defined sentiment categories for elderly care.

3   Acknowledgment

As the organizing committee chair, I would like to thank the committee members, Dr. Taku Nishimura, Dr. Hiroyasu Miwa, Dr. Satoshi Nishimura, Dr. Kentaro Watanabe who were also members of the program committee. The organizers would like to thank Mr. Barada and Ms. Yamada Mr. Koyama for their support. The organizers would like to thank to Prof. Mihoko Ohtake, Chiba University, Japan, Prof. Pertti Saariluoma University of Jyväskylä, Finland and Dr. Jaana Leikas, VTT, Finland. Finally, the organizers would like to thank JSAI for financial support.

Toward Sentiment Analysis in Elderly Care Facility

Ken Fukuda[(⊠)], Satoshi Nishimura, Huizhi Liang,
and Takuichi Nishimura

AI Research Center,
National Institute of Advanced Industrial Science and Technology, Tokyo, Japan
ken.fukuda@aist.go.jp

Abstract. Hand-over notes are extremely important to share information about irregular incidents at elderly care facilities and provide high-quality services. However, taking notes is a time-consuming task. Moreover, handwritten hand-over notes make it difficult to pass on experience and related know-how to other workers. To curate that field community intelligence, a handover support system for elderly care facilities was installed into a facility and evaluated. The system is now in actual operation at the care facility. The authors aim to use handover support system as a communication tool that sense the feelings of the care workers and supports them to maintain motivation and cultivate self-directedness. To realize this aim, this paper reports the results of hand-over data analysis and comparison with traditional paper-based hand-over notes. Furthermore, the authors explored the possibility of applying sentiment analysis technology to hand-over messages to sense the atmosphere of their working environment.

Keywords: Information share · Elderly care · Text mining · Sentiment analysis

1 Introduction

The rate of population aging and the nursing care insurance payout of Japan have reached 25.0% and 9,200 billion yen in FY 2013, respectively [1, 2]. Baby boomers will become latter-stage elderly (75 years old and older) in 2025, so that social burdens are anticipated to increase year-by-year. However, the profitability of nursing care facility businesses is as low as 5% or less because the unit price of services is determined by the Long-Term Care Insurance Act. There is no margin for extra expense paid for the duties of employees. The burden on employees is so heavy that most nursing care fields are suffering from the chronic shortage of human resources (inadequacy of human resources is 59.3%; the turnover rate is 16.5% in FY 2014 [3]). Securing reliable nursing care services is an important social subject. Therefore, it is extremely important to conduct an approach to mitigate burdens on employees and to improve their fulfillment and ability to respond promptly at the work site as well as to provide high-quality services.

This article reports results of quantitative analysis of non-routine task records (information related to findings about care-receiver's daily state) for ten months

© Springer International Publishing AG 2017
S. Kurahashi et al. (Eds.): JSAI-isAI 2016, LNAI 10247, pp. 143–154, 2017.
DOI: 10.1007/978-3-319-61572-1_10

accumulated by introducing an information-sharing support system, DANCE (Dynamic Action and kNowledge assistant for Collaborative sErvice fields) [4], which was developed by a design practice known as 'Participatory Design', into nursing a care facility. Then this article presents an example of business process improvement effects by feed-back of analysis results to on-site employees.

The rest of the paper is organized as the following. Section 2 presents an overview of information sharing in nursing care facilities. Section 3 explains the installation of the proposed system to a care field. Then, after discussing the results of analysis of handover data in Sect. 4, Sect. 5 explains cases in which the feedback of the analysis result to the working site has specifically led to business process improvement. Section 6 explores the possibility of applying sentiment analysis technology to elderly care documents. Section 7 concludes this article and presents discussion of future subjects.

2   Information Sharing in Nursing Care Facilities

Shared information about a care-receiver at nursing care facilities is classified into "care-receiver's personal information" and "information for task performance" [4]. The "care-receiver's personal information," such as information specifying an individual (name and date of birth), and health information (weight, temperature, blood pressure, etc.), is information to be logged necessarily as official information, such as electrical medical records in the nursing care duties. The "information for task performance" includes proper care procedures, special notices, requests from the family, etc. for each care-receiver, and is private information that employees should know for performing duties more properly on the scene. The information for task performance can be classified further into "information for task performance for each care-receiver," which is information updated less frequently to some extent and which can be written down on a regular format such as the meal or toilet procedure of a care-receiver, and "handover information," which describes responses to atypical duties that are difficult to define in advance, such as daily conditions and changes of the mental state of a care-receiver and requests and communications from a care-receiver or the care-receiver's family. Usually, this information is handed over to the next shift in a briefing or notebook.

This article specifically examines such "handover information," and quantitative and qualitative changes in sharing of "information for task performance" by introduction of an information-sharing support system developed by the authors and colleagues.

3   On-Site Implementation of the System

3.1   Nursing Home with Introduced System

The DANCE system was introduced at Wakoen Long-Term Care Health Facility of Keiju Healthcare System, Tosenkai, a Soocial medical care corporation in Nanao-shi, Ishikawa, Japan. Wakoen consist of four business divisions (and offices) of three inpatient houses (one house for dementia patients) and one visiting rehabilitation block.

Nursing care facilities have total capacity of 190 patients comprising 150 inpatients and 40 outpatients for rehabilitation, with over 120 employees [4].

3.2 System Introduction Process

The possibility that the time required per day for both logging duties and confirmation duties of handover information could be shortened significantly was demonstrated by evaluation of the prototype system [4]. On the occasion of full introduction of the system to the care facility, an introduction action committee comprising employees was organized, and collection of requests for system improvement and education to employees was conducted. All employees were provided a period of about three months when the system was freely available for test operation. Then, the conventional paper-based handover procedure was dis-allowed thoroughly, by switching to duty using this proposed system. No overlap period was provided in switching from the conventional paper-based handover to the system-based handover, which was a key for smooth transfer.

3.3 DANCE System

The DANCE system is a messaging app that transmits hand-over information as a message and shares it. DANCE comprises a DANCE server that administers data, a DANCE application for mobile terminals for inputting and browsing data (Fig. 1), and a web application for inputting and browsing data using a web browser. It is assumed that an employee conducts data logging and search quickly at a spare time during field work using a small personal digital assistant. It performs editing work such as elaboration and fair copy of recorded information using a large tablet at the office. Every employee requires a different login account to use the system. Table 1 presents some handover items. A "Like!" function is installed as a part of an approach to let an on-site

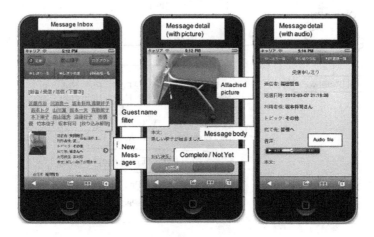

Fig. 1. DANCE app screenshot [4]

Table 1. Hand-over information attributes

Attribute name	Attribute value
Handover state	Sent, draft, deleted
Transmission date	Date, time
To	To all members, to an occupational branch, etc.
Care-receiver	Applicable cere-receiver name
Message body	Description of contents of handover
Photograph	Attached file
Audio	Attached audio file
Importance	High, middle, low, N/A
Task state	Done/Not yet
Read	List of employees who read the message
Comment	Note by corresponding employee, etc.
Like!	Whether "Like!" button is pressed or not

employee express subjective assessments of the information. Specifically, a browsing employee can express gratitude easily for sharing of a useful finding by clicking the "Like!" button.

3.4 Hardware System Configuration

Figure 2 shows the system configuration of the introduced system. A DANCE server is installed in the office of the nursing care facilities, which is a Mac OS X Server running on Mac mini with an SSD. Each nurse station has four iPod touch and one iPad mini

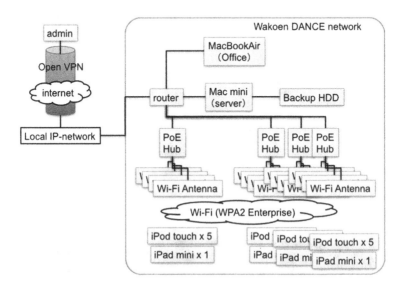

Fig. 2. DANCE system diagram.

with the DANCE application installed. Four switching hubs equipped with Power over Ethernet (PoE) supply function are installed in each nurse station. Then four PoE powered Wi-Fi access points are installed in each business area. This configuration enables access from a terminal over all the buildings. PoE powered devices improve the degree of freedom and maintenance.

This system is connected by OpenVPN to a dedicated terminal installed in the office building of a company in charge of the maintenance service operation of this system. Consequently, this system cannot be accessed from an exterior source other than this terminal.

Handover information includes important personal information that is fundamentally necessary for performance of smooth nursing care duties. It is therefore extremely important to adopt proper measures for security and a business continuing plan (BCP). Security is sustained at a proper level under the assumption of crises of various levels, such as physical attack, mischief, theft, loss, and attack over the network to a server or each mobile terminal, incorporation with the medical information and department of the nursing care facilities. Regarding support for BCP, backup equipment is reserved for the networking gear and the server machine. Mirroring of the disks of the server machine is always conducted to hard disks installed at a physically isolated place.

4   Quantitative Analysis of Sharing of Handover Information

4.1   Target Data

The employees' user accounts and handover information are grouped according to business divisions. The business divisions consist of four groups: a group that integrates two blocks, aside from the dementia house, among three inpatient houses (hereinafter, Group A); the dementia house (Group B); the visiting rehabilitation block (Group C); and the Office group. The two blocks in Group A are in charge of different operations but share information because they are located on the ground and second floor of the same building (interoperability is required). Therefore, the handover system is determined to operate as a single group based on the examination result of the introduction action committee. Groups A and B provide services with overnight stays, while Group C provides day care services without an overnight stay.

The target data of this analysis is handover data accumulated during the ten months from Feb. 1, 2014 to Nov. 30 (for 303 days) after the start of full inauguration. The results are reported in comparison to handover data recorded on handover notebooks for three months immediately before the inauguration (Nov. 2013–Jan. 2014).

The total of transmitted handover items from all groups was 5,566. The daily average was 18.4. In detail, Group A: 2,998 (monthly average, 299.8; standard deviation, 34.9; and daily average, 9.9); Group B: 916 (monthly average, 91.6; standard deviation, 12.1; and daily average, 3.0); Group C: 1,264 (monthly average, 126.4; standard deviation, 11.6; and daily average, 4.2); and Office: 388 (monthly average, 38.8; standard deviation, 11.6; and daily average, 1.3). Group A deals with more cases because job divisions have been merged as described previously, whereas the Office Group accommodates fewer cases because it deals with handovers not for sharing of findings related to routine or non-routine work but for liaison. A cumulative number of 152 employees transmitted a handover during the period.

Figure 3 depicts changes in the number of monthly handovers. Decline in the number of handovers by employees unfamiliar with the operation of mobile terminals such as a smart phones or tablets was a concern on the occasion of full inauguration. However, no phenomenon was observed in which the number of handover cases fell immediately after system introduction as compared with three months before intro-duction. The number of handovers of Group A dropped considerably in January immediately before introduction and then turned upward gently. However, no great change was found in the number of reports on the sickbed utilization rate, accidents, or near miss cases in January or the previous December. Consequently, this was con-cluded to be affected by a special situation in which January was mostly occupied by IT system renewal in the whole medical corporation including nursing care facilities.

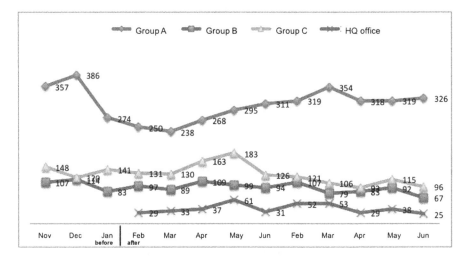

Fig. 3. The number of hand-over messages.

A questionnaire survey conducted in advance of the system introduction revealed that sharing a single notebook placed in the nurse station in the paper-based handover yielded time and spatial deviation between the times of occurrence and logging of a handover issue [5]. Although it was expected that this deviation would be dissolved by setting up multiple sets of mobile terminals and that the number of handovers would increase, no upward tendency was observed. This point will be discussed in qualitative analysis of the following section.

Aside from the number of handovers, the effect of system introduction on the handover length was also analyzed. Both paper-based and system-based handovers had 25 or more and fewer than 50 characters in most handovers for all groups except the Office Group. Also, 95.7% of system-based and 94.9% of paper-based handovers had fewer than 150 characters (Fig. 4).

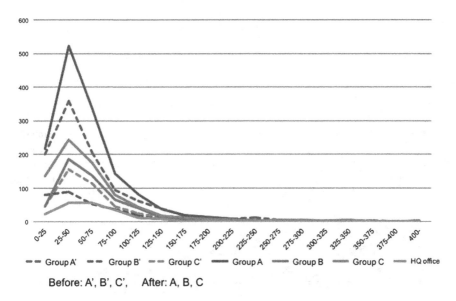

Before: A', B', C',   After: A, B, C

Fig. 4. Message length distribution of hand-over messages.

These results suggest that employees managed the transfer from hand-written hand-over duties to the system based handover duties satisfactorily. Active utilization of Photo was observed as a new trend (Table 2).

Table 2. Number of photographs and attachment rate.

	With photograph	Without photograph	Photograph attachment rate
Group A	581	2417	19.4%
Group B	59	857	6.4%
Group C	72	1192	5.7%
Office	60	328	15.45
Total	772	4794	13.9%

5   Analysis Results Led to Operational Improvement

This chapter reports one of several cases in which the analysis results of handover data improved business processes [6, 7].

Actually, 3903 handovers concerned a specific care-receiver (including those designating multiple care-receivers), accounting for 70.1% of all the 5,566 handovers. A small number of care-receivers had particularly many handovers, as shown in the example of Fig. 5.

An accompanying large number of handovers might imply that special attention is necessary for providing sufficient nursing care service.

We presented the data analysis results discussed in Sect. 4 to on-site employees three months after inauguration of the system and asked them to have a group

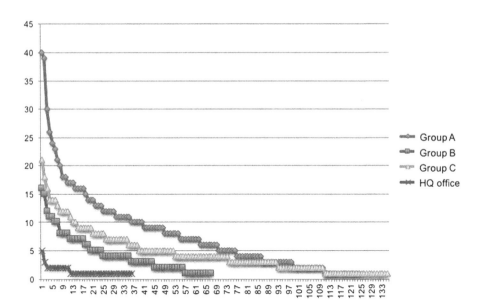

Fig. 5. # of messages per care-receiver (Y-axe: # of messages, X-axe: care-receiver).

discussion. As a result, they requested that they like to review the contents of handovers of care-receivers with especially many handovers. Specifically, handover texts of care-receivers of most handovers in each week and the top ten care-receivers in each group were reviewed. The co-occurrence network of KHcoder [8] was used to visualize the contents of handover texts.

Figure 6 presents the co-occurrence network of cases that lead to business improvement. Employees who verified the co-occurrence network focused on the co-occurrence of the terms of "carried-in foods," "husband," "silence," and "eat" as shown in the upper right of Fig. 6, whereas terms related to the ordinary nursing care coincided because different terms derived from ordinary nursing care duties, and the care-receiver's family "husband" appeared. Follow-up to the requests of a care-receiver's family is so important for providing better service in inpatient facilities as well as care of the care-receiver, as discussed in the preceding section.

The body of handovers in which the key terms appeared were re-examined in a group comprising the related staff members, including those who prepared the handovers. That re-examination revealed that the care-receiver needed feeding support by an employee because of weak ability to swallow and the risk of incorrect swallowing. However, the care-receiver's husband had carried in foods and fed the care-receiver while no nursing care employee was present. Discussion of measures by related on-site employees concluded, respecting the husband's thoughts, to ask the care-receiver and her husband to come to the dining hall, where an employee is always present when feeding food, and the employee would adjust the angle of the backrest of her chair. This business improvement certainly raised the satisfaction of this care-receiver's family greatly.

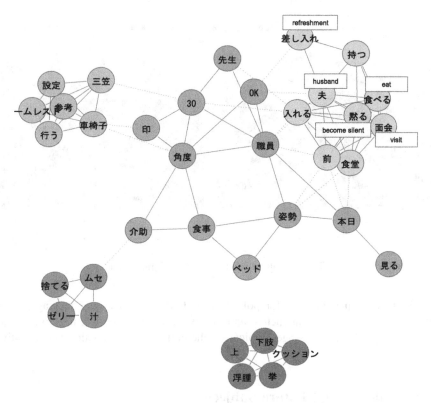

Fig. 6. Example of insightful co-occurrence network.

6   Sentiments in Handover Messages

Through the analysis of the huge archive of hand over messages, the authors are convinced that handover support system is quite effective as a communication tool that sense the feelings of the care workers. Analyzing sentiment of messages has the potential to support them to maintain motivation and cultivate self-directedness. To realize this aim, the authors explored the possibility of applying sentiment analysis technology to hand-over messages to sense the atmosphere of their working environment.

NTCIR6 [10] (Japanese news wire text) was used as training data and 27,783 sentences were analyzed. The result was 5,660 negative messages, 7,262 neutral messages and 14,861 positive messages. Although the authors have not done parameter optimization, the result was very different from what was observed in the actual hand-over messages.

Based on the sentiment analysis result, the authors reached to the conclusion that a corpus for sentiment analysis in the elderly care domain is required. Figure 7 shows the classification of hand-over messages that was discovered in this process.

		2016/08	2016/07	2016/06	2016/05	2016/04	2016/03	2016/02	Total
	Positive	3	3	3	2	3	1	2	17
内訳	改善報告/improvement repot	3	1	2	2	3	1	1	13
	気配り/caring others	0	2	1	0	0	0	0	3
	対応感謝/appreciation	0	0	0	0	0	0	1	1
	Negative	17	21	31	25	20	9	20	143
	指摘事項/pointout mistake	9	10	25	20	12	7	18	101
	表現(感情移入)/strongly emotional	6	3	5	3	3	7	0	27
	表現(強い表現)/domineering	2	8	4	3	3	0	0	20
内訳	表現(距離を置く)/keeping distance	2	0	0	0	1	0	0	3
(重複あり)	表現(言葉足らず)/lacking thoughts	1	1	2	2	2	0	0	8
	neutral	191	300	320	313	264	307	332	2027
	Total	211	324	354	340	287	317	354	2187

Fig. 7. Sentient hand-annotation 1st run result

The annotation guide line for polarity in elderly care hand-over messages is as follows: positive means constructive and kind well-meant messages, negative means injurious, unproductive messages, typically unhelpful suggestions with strong negative emotion.

7 Conclusions and Future Subjects

The authors and colleagues have conducted on-site implementation of the DANCE information sharing support system, which was developed with an on-site participatory design approach in collaboration with a nursing care facility. This article reports the results of the effect of the system introduction and the potential to conduct autonomous process improvement by the analysis of data acquired by the system.

The slight amount of change in data volume after the system introduction from the previous paper-based handover has proved that the system has remained in on-site use without confusion. The system developed with on-site employees' voluntary participation and cooperation from evaluation of real business processes to designing of a prototype system has also been evaluated at a workshop by on-site employees undertaking data analysis after the system introduction, where an important shortcoming that had not been noticed before was discovered by application of text mining, and business improvement was implemented. This result implies that "the cycle of collection and application of information" in nursing care duties [9] was completed for the first time in on-site implementation at Wakoen.

However, the problem of spatiotemporal deviation between the occurrence time and logging time of a handover issue has not been fully resolved yet from restrictions of the number of distributed mobile terminals. The same applies for the confirmation duties of information. Future subjects for this issue include measurement of the effects of distributing terminals to all employees working in a certain block and capturing the details of a terminal location using a position determination technology such as iBeacon.

Another thing to be noted is the positive effect of adopting information system in employee education. "What to write as a handover" is a typical issue for novice care givers. Better usability of browsing, searching, filtering messages resulted in more browing of handovers of other employees. And this was evaluated as not only useful by itself but also had positive effects on education of on-site employees and on encouraging motivation at these facilities.

A preliminary study on sentiment analysis in elderly care hand-over messages suggested a sentiment classification of handover messages. This information is insightful to direct further research in employee education and in developing technologies that contribute to better working environment for care-givers.

It will be desirable in the future to develop an analysis feature by which employees tackle the analysis and application of system data rather subjectively. At the same time, it is necessary to propagate knowledge obtained in this study to other nursing care facilities and service industries. Consequently, our future subjects include development of KPI that quantitatively evaluates various effects of the system introduction evaluated qualitatively in this report of our study.

Acknowledgment. This study was partly conducted under the support of "Fundamental Research For Human Centered Service Engineering", the development and demonstration testing of next-generation, highly reliable, energy-saving core IT technologies (Service Engineering R & D field) a commissioned project for FY 2011 by the Ministry of Economy, Trade and Industry (METI) of Japan, the "Project to Promote the Development and Introduction of Robotic Devices for Nursing Care" by METI, and Grants-in-Aid for Scientific Research (Subject number: 24500676). This study was partly supported by Japanese METI's Robotic Care Equipment Development and Introduction Project,NEDO's Artificial Intelligence Research Project. The authors extend their gratitude to Long-Term Care Health Facility Wakoen of Social Medical Care Corporation Tosenkai for cooperation in development and evaluation of the system.

References

1. White Paper in 2015, Cabinet Office, Government of Japan (in Japanese). http://www8.cao.go.jp/kourei/whitepaper/w-2015/html/zenbun/index.html
2. Ministry of Health, Labor and Welfare (in Japanese) http://www.mhlw.go.jp/topics/kaigo/osirase/jigyo/13/
3. The Care Work Foundation report 2014 (in Japanese). http://www.kaigo-center.or.jp/report/h26_chousa_01.html
4. Fukuhara, T., Nakajima, M., Miwa, H., Hamasaki, M., Nishimura, T.: Design and implementation of a handover support system based on information recommendation in nursing homes. Trans. Jpn. Soc. Artif. Intell. **28**(6), 468–479 (2013)
5. Nakajima, M., Fukuhara, T., Nishimura, T., Akamatu, M.: Handing-over at a nursing-care facility using mobile device. J. Mob. Interact. **3**(2), 47–55 (2013)

6. Fukuda, K., Hamasaki, M., Fukuhara, T., Fujii, R., Horita, M., Nishimura, T.: Preliminary result of hand-over text mining in care facilities. In: IEICE-NLC2014-20, vol. IEICE-114, no. 211, pp. 11–16 (2014)

7. Fukuda, K., Watanabe, K., Fukuhara, T., Hamasaki, M., Fujii, R., Horita, M., Nishimura, T.: Text-mining of hand-over notes for care-workers in real operation. In: Meiselwitz, G. (ed.) SCSM 2015. LNCS, vol. 9182, pp. 30–38. Springer, Cham (2015). doi:10.1007/978-3-319-20367-6_4

8. Higuchi, K.: Sociological theory and methods. Jpn. Assoc. Math. Sociol. **19**(1), 101–115 (2004)

9. Nakajima, M., Fukuhara, T., Miwa, H., Nishimura, T.: Development of a supporting system for handing-over at nursing-care facilities. J. Mob. Interact. **2**, 39–48 (2012)

10. Kishida, K., Chen, K., Lee, S., Kuriyama, K., Kando, N., Chen, H.-H.: Overview of CLIR task at the sixth NTCIR workshop. In: Proceedings of NTCIR-6 Workshop Meeting, Tokyo, Japan, 15–18 May 2007

AI-Biz 2016

Artificial Intelligence of and for Business

Takao Terano[1], Hiroshi Takahashi[2], and Setsuya Kurahashi[3]

[1] Tokyo Institute of Technology, Okayama, Japan
[2] Keio University, Minato, Japan
[3] University of Tsukuba, Tsukuba, Japan

1 The Workshop

In AI-Biz2016 held on November 14, three excellent invited lectures including the plenary talk of the symposium and nine cutting-edge research papers were presented with total of 20 participants. The workshop theme focused on various recent issues in business activities and application technologies of Artificial Intelligence to them.

The first invited lecture was "Big Data, AI, Smart Government" by Dr. Chang-Won Ahn who is a principal researcher of ETRI (Electronics and Telecommunications Research Institute, Korea). In his presentation, the Government 3.0 initiative based on Big data was proposed, and discussions on future governance system were held. The second invited lecture was "An Expert System for Assessing the Likelihood of Child Labor in Supplier Locations based on Bayesian Networks and Text Mining" by Professor Alfred Taudes, Vienna University (Austria). Child labor discovery technology using text mining technology was introduced in this lecture. In addition, the plenary talk of JSAI-isAI 2016 was given on "Disruptive Technologies and the Future of Society" by Prof. Fernando Koch (Korea University/The University of Melbourne).

The AI-Biz workshop was the first one hosted by the SIG-BI (Business Informatics) of JSAI and we believe the workshop was successful, because of very wide fields of business and AI technology including passenger behavior simulation, mobile app market analysis, systemic risk analysis, evacuation behavior simulation and so on.

2 Papers

Twelve papers were submitted for the workshop, and nine papers were selected to be presented in the workshop. After the workshop, they were reviewed by PC members again and five papers were finally selected. Followings are their synopses. Keiichi Ueda and Setsuya Kurahashi illustrate and demonstrate how individuals opt for SST upon decision making by replicating the ABM model of self-service adoption at the airport. Data from an airline's system were utilized to explore the external and internal factors promoting SST. By examining boarding data, they find that heuristic factors explain whether to opt for SST more than travel conditions do.

Meng-Ru Lin and Goutam Chakraborty analyse purchasing data of free game Apps to determine crucial factors of influencing in-App purchase. From available literatures and the data set, 27 potential factors have been defined. Further, they employ LASSO to select the important factors. Results indicated that LASSO can effectively recognize 6 crucial factors.

Jing Su, Mohsen Jafari Songhori, Takamasa Kikuchi, Masahiro Toriyama, and Takao Terano develop an agent-based model to study post-acquisition integration strategies for M&A according to the behavioral theory of the firm. The model conceptualizes firms conducting search over associated NK performance landscapes. Using this model, the simulation experiments indicate that strategies of personnel allocation, high level manager's feedback and the frequency of exchanging information could have impact on company's performance after M&A.

Shuang Chang, Wei Yang, and Hiroshi Deguchi deploy a bottom-up simulation approach to assess the LTC service distribution by simulating stratified individuals' care-seeking behaviors. They estimated the LTC needs across different social groups, and then simulated their selection behaviors of corresponding LTC service providers.

Morito Hashimoto and Setsuya Kurahashi propose new systemic risk index that reduce such chain reaction of failures at a minimum cost by building a model of interbank fund transaction networks. This model's structure has its basis in the Erdos-Renyi network with the network characteristics considered. They confirm that financial assistance given for the purpose of stopping chain reaction failures could result in increasing a chain reaction, and that the use of systemic risk index as references to determine financial institutions that provide financial assistance.

3 Acknowledgment

As the organizing committee chair, I would like to thank the steering committee members, The members are leading researchers in various fields:

Hiroshi Deguchi (Tokyo Institute of Technology), Manabu Ichikawa (National Institute of Public Health), Hajime Kita (Kyoto University), Keiki Takadama (The University of Electro-Communications), Shingo Takahashi (Waseda University), Takashi Yamada (Yamaguchi University). The organizers would like to thank JSAI for financial support.

Finally, we wish to express our gratitude to all those who submitted papers, PC members, reviewers, discussant and attentive audience.

The Passenger Decision Making Mechanism of Self-service Kiosk at the Airport

Keiichi Ueda[(⊠)] and Setsuya Kurahashi

Graduate School of Systems Management, University of Tsukuba,
Otsuka 3-29-1, Bunkyo, Tokyo 112-0012, Japan
ke.wader@gmail.com

Abstract. We examine how individuals decide to use self-service technology. The decisions made by individuals between options of service are to be located in various contexts, including that of their traits. We focus on the check-in process for air travelers at the airport and map the actual existing world onto the experimental space to represent the decision making process in an agent-based model (ABM). Real-world data, taken from an airline's system, is used to verify and validate the model. A cognitive model is implemented in ABM, which utilizes a fuzzy inference system to model each agent's choice. Passenger behavior is carefully designed based on the knowledge of experienced front-line airport customer-service experts and is also reviewed and clarified by on-site observations. We also discuss how to validate the effectiveness of ABM in the end.

Keywords: Agent-based modeling (ABM) · Simulation · Fuzzy · Self-service technology · Airport · Airline · Innovation

1 Introduction

1.1 Background

Developed countries, such as the G7, are facing a future in which they will be required to deal with their aging societies. With better health care and fewer children, industries are securing their workforce in new ways. In these countries, the service industry's share of economic activity and employment is increasing; the so-called "service economy" continues to develop.

Someone or something is needed to offer better service and interact with consumers. Self-service technology (SST) is a promising alternative for the future of the service workforce. This study focus on self-service kiosks at the airport, as these are a familiar alternative travelers can take to check-in.

1.2 Purpose of This Study

We pursue how individuals opt for SST. This study examines observable external facts and invisible, internal facts, which include the history and traits of the

© Springer International Publishing AG 2017
S. Kurahashi et al. (Eds.): JSAI-isAI 2016, LNAI 10247, pp. 159–175, 2017.
DOI: 10.1007/978-3-319-61572-1_11

individual to understand how consumers opt for SST. In order to understand the dynamic mechanism of decision making, we implement a new conceptual model using agent-based modeling (ABM), which illustrates the behavior of the adoption of SST, specifically of the use of a self-service kiosk at the airport.

2 Related Work and Findings

SST has been examined in various perspectives. We review innovation studies, as SST adoption is an individual decision to take a new way. The service-marketing field is given an overview to understand the development of SST studies. Then we review agent-based modeling, as a tool to explicate the dynamics of the phenomenon of SST adoption.

2.1 Innovation Diffusion

Innovation is defined as the introduction of something new: a new idea, method, or device. However, innovation is often also viewed as the application of better solutions to meet new requirements, unarticulated needs, or existing market needs. This is accomplished through more effective products, processes, services, technologies, or business models that are readily available in markets, governments and society. Rogers [12] designated variables to define the speed of diffusion. More relative advantage, higher compatibility, less complexity, higher trialability, and greater observability speed up the diffusion of innovation. The role of the change agent in promoting innovation is an important variable increasing the speed of the diffusion [12].

Since SST adoption is an individual decision to accept innovation, variables enhancing the diffusion speed indicate what we should look at for this study.

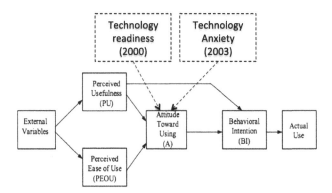

Fig. 1. Technology acceptance model (Concepts of consumer readiness and technology anxiety are added by the authors.)

2.2 Service-Marketing Framework

Convenience has been examined and discussed from two main perspectives: (1) wait time and its management and (2) what consumers find convenient [1].

There are studies that have found factors that influence the usage of SST through various means, both surveys and interviews. Meuter et al. [10] concluded that service convenience through SST brought consumer satisfaction when it was "better than the alternatives" and they appreciated "time saving" the most [10]. They also concluded that SST usage depends on customer readiness for SST [2]. Davis [5] proposed a technology acceptance model [5] (Fig. 1). He concluded that perceived usefulness and ease of use create attitudes toward SST. Liljander et al. [9] reviewed SST adoption in the perspective of consumer readiness. Another study concluded that technical anxiety explains the influence of SST adoption better than the demographics of users [11].

Dabholker and Bagozzi [4] proposed an extended attitudinal model of technology-based self-service (TBSS), which clarifies the moderating variables affecting attitude toward and intention to use SST. This model shows that consumer traits and situational factors can slow down SST usage or prompt it (Fig. 2).

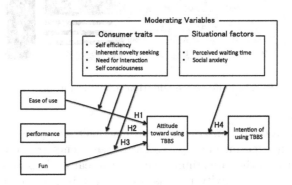

Fig. 2. An attitudinal model of TBSS (technology-based self-service)

2.3 ABM

ABM is based on technical instruments that enables each agent to behave autonomously. Agent-based simulation is developed through placing players in experimental space and approximating the experimental space to the real world. A social multi-agent system shows phenomena in complex social systems [8]. Kawai built an abstract model to explain the diffusion of the services using ABM [7].

These studies indicate important facts and concepts for the diffusion of innovation. However, they merely illustrate the concept, but do not reproduce the mechanism of decision making at the moment when one out of several options is chosen. As Kawai's model does not use observed data from the real world, it remains to show the concept but it fails to represent the actual phenomenon of diffusion and what makes consumers select a new alternative.

By mapping the real world in the experimental space using airline data, Ueda and Kurahashi (2014) created ABM that demonstrates how air travelers choose self-service kiosks at the airport [13] (Fig. 3). Their model illustrates the mechanism of SST adoption at the moment of choosing one of two options.

This model uses a fuzzy inference methodology for each created agents in the experimental space. Experienced airline staff defined simple rules (Table 1). The implement model calculates the self-service preference index (*SPI*) at the moment of decision-making. Each agent refers to its own *SPI* score to descide which direction to take (Fig. 4).

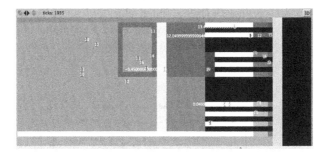

Fig. 3. Self-service adoption model in airport

SPI quantification is constructed from two main components. One copies the real world in experimental space. Passenger agents (represented by turtles) are created with the same timing with which real-world passengers arrived, according to passenger activity records. Each agent is given a variable with a random value, which represents that agent's hesitation to accept novelties. The other obtains membership scores in the experimental space. Agents move toward the conventional check-in area as their first choice. When an agent reaches the decision-making area, it counts the number of turtles already queuing in order to estimate the waiting time for this check-in area. It calculates the difference between expected queuing time for the conventional check-in area and for the self-service kiosk; thus it perceives whether its default option has a shorter waiting time (membership score W). It perceives the existence of self-service kiosk, recognizing that kiosks are there and they are for check-in (membership score V). In the application of the rules shown in Table 1 results are calculated using the max-mini inference method and the simplified centroid method for defuzzification combines these results.

The input value for calculating W is defined by equation (Eq. 1). EQT is the predicted difference in waiting time at the conventional check-in, the wait time

Table 1. Fuzzy rules

Rule 1	IF W is short and V is low, THEN SPI is negative.
Rule 2	IF W is long and V is high, THEN SPI is positive

Score W: Waiting time for conventional check-in.
Score V: Visibility of self-service kiosk.

for using the self-service kiosk. $NCCQ$ is the number of passengers waiting in the conventional check-in queue. $CCPs$ is the number of conventional check-in positions. $NSSQ$ is the number of passengers waiting in self-service queue. SSU is the number of self-service units. Finally, $p1$ and $p2$ are the weighting parameters for each member of the equation. If the preference is the same between two options, they have the same value; however, few passengers prefer the self-service kiosk.

V reflects how the passenger perceives self-service kiosks. V is low where there are no passengers using self-service kiosk. As more passengers use self-service kiosks, the value becomes higher. Once the number of passengers who are using self-service kiosk exceeds the number of self-service kiosks, V is reduced, because if many passengers occupy the self-service area, visibility of the self-service kiosks significantly deteriorates.

$$EQT = \left(\frac{NCCP}{CCPs}\right) \times p1 - \left(\frac{NSSQ}{SSUs}\right) \times p2 \qquad (1)$$

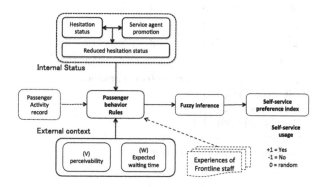

Fig. 4. Self-service preference index

Airport staff interaction leading passengers to SST use is viewed positively [3,6], this model locates customer-service agents in the check-in lobby of the experimental space.

Passenger agents must go through the designated area in order for customer-service agents to interact with them. Once the passenger-agent makes contact with the customer-service agent, the anxiety over using SST is reduced.

The model creates a passenger agents at the same time as passenger actually arrive according to airline system records, and it locates productive properties such as check-in position, self-service kiosks and customer-service staff in the same amounts, as shown in Tables 2 and 3.

Verification and validation were carried out carefully with one dataset for training out of six datasets. After fitting the parameters using the training data, we conducted experiments using the other datasets. In each experiment, the number of check-in counters and staff is mapped as same as they were in the actual situation. In addition, various parameters are set to map the real world, such as *baggage holder rate* (0.7), *frequent self-service user rate* (0.05), *non-self-service user rate* (0.2), and the *processing time* for the different service options (interpersonal service, self-service, and baggage check-in).

In these experiments, we observed a *self-service usage rate*: the quotient of passengers using self-service divided by all passengers. The result of simulation showed a less than 3% RMSE (Root Mean Squared Error) in *self-service usage rate* versus the real data. This is persuasive for the modeling actual passenger handling for managers at the airport.

Table 2. Experimental dataset

	Passenger choice		Product property			
Dataset	IPSC	SSC	SSU	Ckin	Bag	STF
406	85	46	4	3	3	2
408	100	60	4	2	3	2
409	68	39	4	2	2	3
410	67	54	4	2	2	3
411	63	62	4	2	2	3
412	67	25	4	3	2	0

IPSC: interpersonal Service (conventional).
SSC: self-service.
Ckin: check-in.
Bag: baggage check-in.
STF: custemer service agent.
Dataset 412: training dataset.

2.4   Subjects of Related Works

Innovation studies describe how people introduce new way. The literature in the service-marketing field specifies and explores factors that have the effect of promoting the use of technology-based self-service (TBSS). Such studies are based on statistical methods using pastl data. The analysis is static, not dynamic.

Dabholkar and Bagozzi [4] concluded that situational factors and consumer traits have a direct effect on promoting a positive attitude towards TBSS and the intention to use TBSS. However, if one situational factor change, the results would also alter as Dabholkar noted. Works in the field have not determined the mechanism by which predictable results can be reliably reproduced.

Our proposed ABM, a model of self-service adoption at the airport, supports the concept of technology readiness and technology anxiety. It demonstrates the dynamic mechanism of SST adoption in the moment of passenger decision of how to check-in to a flight. However, the proposed ABM does not examine the traits of individual which are moderating variables that establish attitudes toward using SST [4]. We must introduce the concept of moderating variables into our ABM in the context of choosing one service.

3   Refining the SST Adoption Model

We replicated the implemented ABM, a model of self-service adoption at the airport with the addition of a concept claimed by Dabholkar and Bagozzi [4]. This new concept is described in Sect. 3.1, and how it comes to be implemented in ABM is explained in Sect. 3.2.

3.1   Concept Expansion

Passengers are influenced by several factors when they choose a check-in option, such as their previous flight experiences, the queue length, how self-service kiosks and their surroundings appear, and their travel conditions (volume of baggage, number of passengers in their party, etc.).

It has also been observed that the guidance and support of customer-service staff promotes the use of self-service kiosks. We organized a decision-making conceptual model for self-service kiosk usage (Fig. 5).

Demographic data, travel conditions, and the historical record of departing passengers were collected from an airline. We explore and examine the efficient factors that influence SST usage by aggregate analysis, which we discuss in Sect. 4.2.

3.2   ABM Implementation Model

Departing passengers must check in for their flight. It is important for them not to have their time constrained by others.

Usefulness is defined as the expectation of reduced waiting time by comparison with conventional check-in queuing time and ease of use is defined as the passenger being able to see the kiosk and recognize that it is functioning.

ABM can allow the agent to move and queue for either service option and count how many agents are located in each queuing line. Waiting time and perceiving whether the kiosk is functioning both influence whether the air traveler chooses SST. The situation is very different at every moment, because the timing of the passenger's arrival creates queuing lines and it is unclear who will choose which option. We map real-world data onto the experimental space, including passenger traits that have not been introduced in previous work.

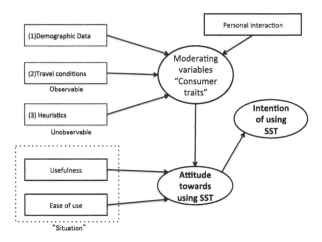

Fig. 5. SST adoption concept model

4   Experimental Results and Discussion

The details of datasets for the experiments with the replicated ABM are described in Sect. 4.1. The traits of the passengers are mentioned in Sect. 4.2. We discuss the results followed by the explanation of the detail of the experiments in Sect. 4.3.

4.1   Datasets

Passenger data was obtained from an airline's system. This data consist of demographic data, travel conditions, flight records, and chosen check-in options (Fig. 6).

We examined *DatasetB* carefully and passengers were categorized into three types. 35.2% of passengers are weak SST users: they seldom use SST; 14.8% of them have a strong preference for SST; and 50% of them are neutral.

4.2   Statistic Analysis

We randomly selected 400 samples from *DatasetB*, which contains equal numbers of SST users and non-users.

We used multiple regression analysis to observe variables explaining the use of Self-service-kiosk. The result found that P value of "Bag" and "Trvltoge" are not significant and "ssuRecency", "flightFreqClass" and "*Density_wizin_15min*" are significant (Fig. 7). It indicates that travel conditions, such as volume of baggage or traveling in a group, do not influence opting for SST. Recent use of self-service is the biggest impact for choosing self-service and frequency of flights is second. We also observe that congestion of departure lobby matter. "*Density_wizin_15min*" is the variable to express the degree of the congestion of

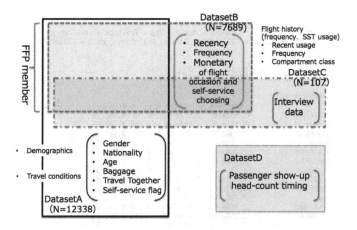

Fig. 6. Dataset for experiments

departure lobby. It is equal to the number of passengers within 15 min when the passenger checks in. The record clearly indicates that once a passenger's flight frequency reaches the premier customer status in the Frequent Flyer Program, they seldom use SST any more. This is natural, because such passengers have the privilege of accessing first class check-in without needing to wait.

Logistic regression analysis of sample data shows the rate of this judgement's being correct to be 70%.

4.3 ABM Experiment

We verify the refined ABM model of self-service adoption at the airport by observing the model's behavior and validating it with the training data from *DatasetD* (described in Tables 2, 3 and Fig. 8 as dataset412).

The validation process is conducted by calibrating the parameters. We find the closest value for *self-service usage rate* to the real world by adjusting the *speedmax* (a parameter which is the upper speed limit of moving agents) in 0.01 increments and running a simulation. The difference between the simulation results and the real world is the smallest when the *speedmax* value is 0.21. The same process is conducted with the parameter *p1* which represents interpersonal service preference. A parameter value 5.0 brings the result that is closest to the real world.

After setting the parameter values, we execute an experimental 50 runs each for five test datasets with different circumstances to observe the *self-service usage rate*. The tested datasets vary in the timing of passenger arrival; they are completely different. The experiments adjust the number of service staff, check-in positions, and self-service kiosks. The experimental results in the replicated model shows that the RMSE of simulation vs real data is less than 4% (Fig. 8).

```
Call:
lm(formula = SSUflag ~ Gender + Age10 + Bag + Trvltoge + ssuRecency +
    flightFreqClass + C21 + C31 + Density_Wizin15min, data = DSBsample)

Residuals:
    Min       1Q   Median       3Q      Max
-1.04560 -0.40138 -0.00027  0.41119  0.96950

Coefficients:
                     Estimate Std. Error t value Pr(>|t|)
(Intercept)        -1.550e-01  2.617e-01  -0.592 0.554139
Gender              6.591e-02  7.697e-02   0.856 0.392361
Age1010s            3.390e-01  3.307e-01   1.025 0.305965
Age1020s            4.419e-01  2.442e-01   1.810 0.071137 .
Age1030s            2.708e-01  2.408e-01   1.124 0.261556
Age1040s            2.798e-01  2.423e-01   1.155 0.248775
Age1050s            1.346e-01  2.421e-01   0.556 0.578555
Age1060s            2.061e-01  2.423e-01   0.851 0.395330
Age1070s            1.392e-01  2.811e-01   0.495 0.620636
Bag                 7.706e-05  2.540e-03   0.030 0.975814
Trvltoge            2.510e-01  1.060e-01   2.367 0.018404 *
ssuRecency          1.154e-01  1.536e-02   7.516 4.00e-13 ***
flightFreqClass    -9.246e-02  2.036e-02  -4.541 7.51e-06 ***
C21                -1.764e-01  8.786e-02  -2.008 0.045348 *
C31                 4.256e-02  3.057e-02   1.392 0.164617
Density_Wizin15min  1.637e-03  4.261e-04   3.843 0.000142 ***
---
Signif. codes:  0 '***' 0.001 '**' 0.01 '*' 0.05 '.' 0.1 ' ' 1

Residual standard error: 0.4502 on 384 degrees of freedom
Multiple R-squared:  0.2217,   Adjusted R-squared:  0.1913
F-statistic: 7.292 on 15 and 384 DF,  p-value: 3.355e-14
```

Fig. 7. Regression analysis

Table 3. Experimental results

Dataset	Product property			SIM ave.	realdata	RMSE
	Ckin	Bag	STF			
406	3	3	2	0.373	0.351	0.022
408	2	3	2	0.417	0.375	0.042
409	2	2	3	0.350	0.364	0.015
410	2	2	3	0.419	0.446	0.027
411	2	2	3	0.409	0.496	0.087
412	3	2	0	0.285	0.272	0.013

average of RMSE for test result: 0.0385.
SIM ave.: average results of simlation.

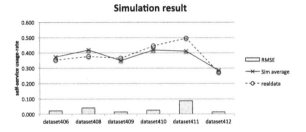

Fig. 8. Experimental results

4.4   Discussion

The previous ABM creates passenger agents and gives a random score of the variable "*hesitation*" to each of them, which contains the concept of individual heuristic and traits including "Technology Anxiety". Once a passenger agent comes into contact with the service agent, the value of this variable is reduced.

The replicated model stochastically adds individual traits to each passenger agent. When refined ABM creates passenger agents, each agent has a self-service preference, reflecting the proportions of each category of passenger as described in Sect. 4.1. In other words, the replicated model implements a new heuristic variable that contains each individual's historical experience, including SST usage and number of flights.

Though the RMSE of the replicate model experiment (RMSE < 0.039) is relatively larger than that of the original model (RMSE < 0.03), it is practically accurate enough for on-site managers. As giving stochastic traits was a major change from the previous model, it may expand the variance of the results. We were able to improve how we categorize passenger traits for future research.

Passengers select the most feasible option from their immediate perception of their surroundings. It is obvious that queuing time is a key to determining the attitudes toward self-service kiosks, because passengers value their time. In the parameter-validation process of ABM, calibration results indicate what could be done to promote the use of self-service kiosks at the airport.

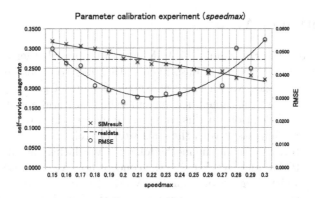

Fig. 9. Parameter fitting:speedmax

Our experiments show that each parameter works differently. One agent parameter, *speedmax*, has a linear relation to the *self-service usage rate* (Fig. 9). The other parameter, the weighting parameter of interpersonal service preference (*p1*) has a non-linear relations to the *self-service usage rate* (Fig. 10).

The graphs show that the results of calibration of *speedmax* have less variance than *p1*. It appears that even though an individual's mind-set could change, this does not control the outcome of their behavior. However, this means that if

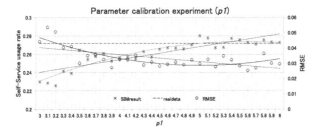

Fig. 10. Parameter fitting:p1

we can control the speed that individuals move, we may promote SST use more effectively than trying to change individual traits. If we let passengers have more time to recognize and compare their options, they might choose SST more often. We would have a greater ability to reduce individual and overall wait time by changing the factors in the environment, such as passenger flows.

ABM is a powerful tool for reproducing the dynamic situations created by the interaction of decision makers. The replicate model quantitatively supports the conceptulization of Dabholkar and Bagozzi [4], with an average RMSE less than 0.04 versus the real-world results. Through the replication process using ABM, we were able to learn how selected parameters affect outcomes by observing behaviors and employing sensitive analysis.

5   Validation of ABM

It is possible to verify the validity of ABM by evaluating whether it reproduces the feature of the stylized fact. Passengers rather like to choose interpersonal service than self-service, which is a well known feature of passenger behavior. Our experiment shows the same phenomena as the real world and its results are close to the actual situation. In this chapter, we discuss the advantage of ABM quantitatively in comparison with the statistical model.

In this chapter, we discuss the methodology of validating the effectiveness of the proposed ABM model. We describe the concept how we can observe that the ABM enhance logistic regression analysis. ABM experiments are carried out to reproduce the same condition as the data used for the logistic regression analysis, and the result of the ABM and the logistic regression analysis is compared to validate the effectiveness of ABM.

5.1   Preliminary Discussion

ABM is a useful instrument to represent the real world. It can detect the surrounding situational factors of each agent, which varies time by time and includes the interactions of agents in the experimental space. ABM has an advantage to describe more about how an individual perceives the surroundings and to represent the interaction between individuals and/or situational factors in every second. We discuss how to validate that ABM is capable of enhancing the result of the logistic regression model by following explanations.

As mentioned Sect. 4.2, it is suggested by the multiple regression model, that individual traits and situational factors in the decision making phase are significant for choosing the self-service kiosk. Because the extracted data used in the logistic regression analysis don't consider the time series, we have an issue that the ABM can't reproduce the approximated situation, which passenger experienced, by simply inputting the same data into the experimental space. In the following section, we explain how to deal with this issue.

5.2   The Experiment Outline

We compare the results of logistic regression analysis with the results of ABM experiments and verify the effectiveness of ABM. The extracted data from the ABM experiments are used, which has the same condition as the data used for logistic regression analysis.

As mentioned in the previous section, there is an issue that we needed to create approximated context, passenger traits in similar situations. We utilize the advantage of the ABM to deal with this issue. Since AMB can repeat experiments and obtain the results in various patterns, we conduct numerous simulations and extract the data which has the same combinations as the logistic regression analysis result has. Then we observe whether the ABM explicates the decision making process of the individual more effectively by comparing two results between the logistic regression analysis and the ABM.

A specific experimental procedure is described in four steps (Fig. 11).

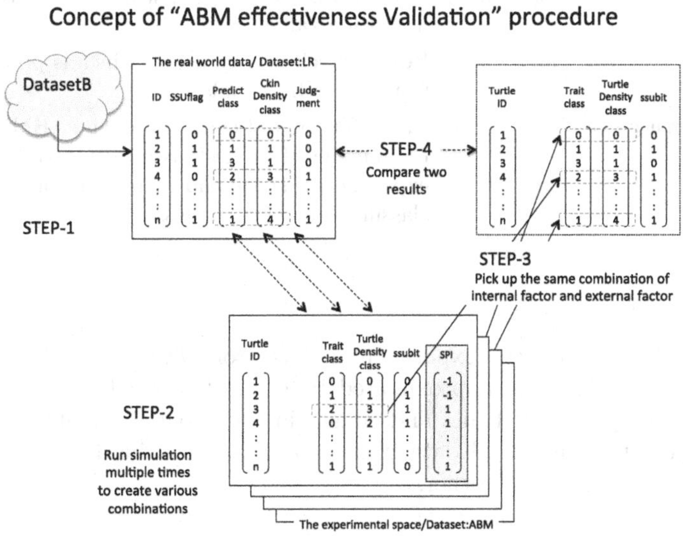

Fig. 11. ABM validation procedure

STEP-1: We randomly extract the passenger data to form a new dataset (*Datasets:LR*). The constituent elements of *Datasets:LR* include ID that identifies the passenger, and the actual result use/non-use of self-service kiosk (*"SSU-flag"*), Check-in time, and *"CkinDensity"* that means a congestion degree of departure lobby. It also includes the predicted values whether the passenger use self-service kiosk, which are calculated from logistic regression analysis and they are classified into two classes as *"Judgement"*: using or not-using self-service. *"CkinDensity"* is the number of passenger who finished check-in within 15 min timeframe. The actual number of passenger is classified into several classes (*"CkinDensity class"*).

STEP-2: We conduct simulations by ABM multiple times to collect experimental results and form the other dataset. The dataset (*"dataset:ABM1"*) has the variable *"hesitation"*, *"ssubit"*, and *"SPI"* and classified variable of *"turtle-density"*. Table 4 describes what each variable stands for.

Table 4. Experimental dataset

Variable name	Value	Explanation
SSUflag	[0, 1]	Actual result of self-service usage
Predict class	[0, 1, 2...n]	Classified passenger trait which comes from predict value
Judgement	[0, 1]	Classified value; using or not-using SST
CkinDensity class	[0, 1, 2...n]	Classified congestion, the proxy function of lobby congestion
Trait class	[0, 1, 2...n]	Classified passenger trait which comes from hesitation value
ssubit	[0, 1]	Result of self-service usage, which is produced from the ABM simulation
TurtleDensity class	[0, 1, 2...n]	Proxy function of congestion. The ABM counts the number of turtles(passenger) in the experimental space and those headcounts are classified into several classes

STEP-3: We pick up the experimental data from *Dataset: ABM1* to form *Dataset:ABM2*, which are approximated to *Dataset: LR*, which has the same combination of passenger traits and congestion of departure lobby. Therefore, we have the same amount of data from two data sets. The data of those two datasets are similar to each other.

STEP-4: We aggregate each combination of passenger traits and congestion degree, calculate the true/false judgment of self-service usage prediction for each combination of two datasets, and compare the *"Correct-predict rate"* between them.

The *"Correct-predict rate"* of two spaces are to be examined whether the ABM works to increase prediction accuracy. *"Correct-predict rate"* is obtained as follows (Fig. 11 and Eq. 2).

$$Correct\_Predict\_rate = \frac{Passenger\_with\_correct\_predict}{All \quad passenger} \quad (2)$$

$$Predict \begin{cases} Correct & (SSUflag = Judgment) \\ Incorrect & (SSUflag \neq Judgment) \end{cases}$$

$$Predict \begin{cases} Correct & (SSUflag = ssubit) \\ Incorrect & (SSUflag \neq ssubit) \end{cases}$$

5.3 Evaluation and Expectation

We compare *"Correct-predict rate"* which is the results of the four step procedure in Sect. 5.2. Guessing that difference between them comes from the behavior of an agent which moves autonomously in the experimental space. We will examine how the result of the interaction works for improving the accuracy of the prediction. Our expectation is to find the outcome of the experiment brings higher prediction accuracy.

6 Conclusion

6.1 Summary

There have been much suggestive related work and many indications for SST adoption. We implemented the essence of related work into our ABM. In the service-marketing field, conceptual models supported by quantitative surveys indicate how attitudes formed and they lead to action. However, the results of statistic explain but do not always demonstrate what an author means can be reproduced and how their mechanism actually functions.

This study illustrates and demonstrates how individuals opt for SST upon decision making by replicating the ABM model of self-service adoption at the airport. Data from an airline's system were utilized to explore the external and internal factors promoting SST. By examining boarding data, we find that heuristic factors explain whether to opt for SST more than travel conditions do. Recent self-service kiosk experience is the strongest factor for using a self-service kiosk in the dataset; higher flight frequency comes second. Even though herd behaviors were observed at the service site, these are not significant statistically.

Since the simulation results through replicating the ABM remain close to the actual data, this proves the expanded conceptual model with passenger traits reproduces the decision-making mechanism to a certain degree.

By examining parameters using sensitivity analysis, this model indicates deeper insights. This study shows that ABM is capable of analyzing each component respectively, focusing on the process and simulating different situations and conditions of self-service adoption at the airport.

6.2 Subjects for Future Study

In this study, three categories of passenger were presented and implemented in the replicate model. Individual traits may expand the variance of experimental results, as mentioned in Sect. 4.4. Passengers could be divided into groups with proper proportions with more probability. Since structuring the dynamics of internal change of individuals is challenging, we need deeper aggregate analysis of individual traits before introducing the processed data into ABM.

There are many methodologies to analyze and explicate phenomena. ABM integrates other methodologies to construct a model framework. It is capable to pursue the behavior rule of an individual. And defined autonomous behaving rule can be used to explore the macro phenomenon.

In Sect. 5 of this study, we discussed the validating procedure of ABM to clarify its effectiveness. We should shortly work on this through the four-step-procedure, hoping the ABM simulation result shows that its explanatory capability has more accuracy than the logistic regression analysis does. Even though the real world is hard to be mapped completely, we need to continue to pursue the way to extract the essence of circumstances where we see the important phenomenon occur. We hope that the insignificant data for the statistical model may somehow be significant after those are processed by ABM.

References

1. Barry, L.L., Seiders, K., Seiders, D.: Understanding service convenience. J. Mark. **66**(7), 1–17 (2002)
2. Bitner, M.J., Ostrom, A.L., Meuter, M.L.: Implementing successful self-service technology. Acad. Manag. Perspect. **29**, 2431–2437 (2013)
3. Castillo-Manzano, J.I., López-Valpuesta, L.: Check-in services and passenger behaviour: self service technologies in airport systems. Comput. Hum. Behav. **16**(4), 96–108 (2002)
4. Dabholkar, P.A., Bagozzi, R.P.: An attitudinal model of technology-based self-service: moderating effects of consumer. J. Acad. Mark. Sci. **30**(3), 184–201 (2002)
5. Davis, F.D.: Perceived usefulness, perceived ease of use and user acceptance of information technology. MIS Q. **13**(3), 318–339 (1989)
6. Gelderman, C.J., Ghijsen, P.W.T., van Diemen, R.: Choosing self-service technologies or interpersonal services the impact of situational factors and technology-related attitudes. J. Retail. Consum. Serv. **18**, 414–421 (2011)
7. Kawai, K.: Reconsidering innovation diffusion model on modeling diffusion of services. Jpn. Soc. Inf. Manag. **30**(1), 26–35 (2009). (in Japanese)
8. Kurahashi, S.: Model estimation and inverse simulation. J. Control Meas. Syst. Intergration **52**(7), 588–594 (2013). (in Japanese)
9. Liljander, V., Gillbert, F., Gummerus, J., van Riel, A.: Technology readiness and the evaluation and adoption of self-service technologies. J. Retail. Consum. Serv. **13**, 177–191 (2006)
10. Meuter, M.L., Ostrom, A.L., Roundtree, R.I., Bitner, M.J.: Self-service technologies: understanding customer satisfaction with technology-based service encounter. J. Mark. **64**(7), 50–64 (2000)

11. Meuter, M.L., Ostrom, A.M., Bitner, M.J., Roundtree, R.I.: The influence of technology anxiety on consumer use and experiences with self-service technologies. J. Bus. Res. **56**, 899–906 (2003)
12. Rogers, E.: Diffusion of Innovations, 3rd edn. Free Press, New York (1983)
13. Ueda, K., Kurahashi, S.: How passenger decides a check-in option in an airport: self-service technology adoption model in passenger process. In: ESSA (2014)

A Study of Crucial Factors for In-App Purchase of Game Software

Meng-Ru Lin and Goutam Chakraborty$^{(\boxtimes)}$

Department of Software and Information Science, Iwate Prefectural University,
Takizawa, Iwate-Ken, Japan
page55667788@gmail.com, goutam@iwate-pu.ac.jp

Abstract. Google Play and App Store registered 17.2 billion downloads of game software worldwide in the first quarter of 2016, according to a report published by Sensor Tower, a platform that supports apps for iOS and Android. Related researchers too predicted tremendous growth in gaming applications. Not only the game App developers need to know how to design products that match gamer's needs, and will continue to use it, but also allure gamers to decide in-app purchase (IAP) which is the final goal. In particular, IAP is the major revenue model. Hence, this study attempts to define the potential factors influencing IAP for gamer. We collect data for many possible features from which, using Least Absolute Shrinkage and Selection Operator (LASSO) feature selection method, we identify important factors that affect gamer IAP behavior. The extracted factors can help game developers to improve their design for increasing revenue.

Keywords: Game app · Feature selection · Least Absolute Shrinkage and Selection Operator (LASSO) · In-app purchase (IAP)

1 Introduction

According to a fresh forecast from App Annie, game sales will hit 41.5 billion USD in 2016 and reach 74.6 billion USD by 2020 [1]. This Statistics Portal pointed out that the global mobile game revenue will reach 40.6 million U.S. dollars, up from 30.1 million in 2015 [39]. From the above, we can conclude that mobile game has become an important revenue earning application on a digital platform.

The App Monetization Strategies includes in-app advertising, in-app purchase, freemium, paywalls, paid apps, sponsorship [38] etc. An in-app purchase (IAP) is when the game is bought from within the application, typically a mobile app running on a smartphone or other mobile devices. Software vendors can sell all manners of things from within apps. In games, for example, users can buy characters, upgrade abilities and spend real money on in-game currencies [41].

Recently, related issues regarding revenue models of game app has also attracted the interest of researchers. For examples, Park and Kim [13] discussed

© Springer International Publishing AG 2017
S. Kurahashi et al. (Eds.): JSAI-isAI 2016, LNAI 10247, pp. 176–187, 2017.
DOI: 10.1007/978-3-319-61572-1_12

the key successful factors of App. Koekkoek [14] discussed how successful apps make money from their user base. Roma and Ragaglia [29] empirically examine how the revenue model, adopted for a given app, affects the app revenue performance as measured by the app daily revenue rank. Gao et al. [23] discovered the continued use intension for mobile payments. Lin and Wang [15] aims to investigate the key factors underlying consumers' decision to buy Apps for their smartphones.

In recent years, many researchers hypothesized various factors that may affect IAP [3,6,7,9,16–18,22] etc. However, no previous studies have investigated which are the important factors, among all those proposed, that actually affect gamer IAP behavior [6]. In this highly dynamic and competitive environment [29], app developers need to know which factors are crucial to influence gamer's IAP [20]. This study examines a large number of potential factors as listed in Table 1, later in Sect. 4. The factors are from different works, and have overlapping meaning. It is easy to conclude that a few of them have strong correlations, and therefore it is possible to select some as important and discard others. There are various algorithms for feature selection, which we briefly reviewed in Sect. 2. In this work, we used Least Absolute Shrinkage and Selection Operator (LASSO) feature selection method to identify crucial factors. This is mainly because it is very efficient. The motivation of this work is to identify most important factors, so as to help game developers to design products that not only matches gamer's needs but also lead to higher in-app purchase (IAP).

2 Literature Review

2.1 Related Works

Gamasutra [11] and Sensor Tower [37] regularly report the sales and revenue earned by game softwares. Worldwide game software downloads and sales grew unabated. Hsu and Lin pointed the IAP have proven to be an effective monetization strategy for freemium apps [6]. All survey point that IAP has become the mainstream of revenue model for game apps. Understanding how to target people who will actually spend money on a title, and which features of a game software attract the user for IAP, is vital to success. [2] shows that 18 to 24 is the age group that spends the longest amount of time on mobile apps. [12] indicated that the majority of big spenders in Southeast Asia are teenagers while in China hardly teenagers spends money on purchasing games. There the biggest group of spenders are people around 30 years old. [27] also pointed that in smart phone gamers, the average age of gamers is 31 years old. Those between 21 to 25 years constitute 20%, 16 to 35 years old player population accounts for nearly 80% of the proportion. Therefore, we focus on 18 to 40 years old players which covers the whole age range of mobile gamers. The subjects of our experiment were chosen from different age groups as shown in Table 2.

2.2 Features Affecting On-Line Game Purchase

We surveyed a large number of works where the factors leading to IAP are proposed or hypothesized. Different factors and corresponding works are listed in Table 1. As they are unrelated works, many of the proposed factors have overlapping meaning and are correlated. Naturally, there is a scope for feature selection, to find important factors. To the best of our knowledge, there is no previous work done for selecting which factors are important.

As we are not sure what affects a gamer to make an in-application purchase decision, we started with all possible factors proposed in previous works. We started with 27 factors, detail of which are described in Table 1. Our target classification is user's decision, whether the user will do In-Application Purchase or not. We need to find the smallest set of features that would give the highest classification accuracy. As mentioned earlier, some features are correlated. Discarding one feature does not mean that that feature is unimportant for the classification task. It only means redundancy, as a similar correlated feature is included in the selected set of features. We aim to select the smallest set so that it would be easier for the designer to focus attention only to those features.

2.3 Feature Selection

In many practical applications with real data, due to the presence of noisy, irrelevant, or correlated features, feature selection is one the most important step before classification and data mining [26,30,33,35,42,43,45,46]. The basic aim is to remove redundant or irrelevant features (attributes) and thereby reduce the computational cost of training the classifier, improving classification accuracy. In addition, it would facilitate data visualization and data understanding, and improve classification accuracy (generalization) for unseen samples [42].

Feature selection method is classified into two approaches: Filter method and Wrapper method. In filter method, an individual feature is evaluated using statistical methods like Chi squared test, information gain or correlation coefficient score. Features are selected according to their scores. In wrapper method a model is used, and a subset of feature is evaluated using the model. The model could be anything, like a regression model, K-nearest neighbor, or a neural network. Searching for the optimum subset of features, could be heuristic, stochastic or forward-backward to add and remove features.

We used Least Absolute Shrinkage and Selection Operator (LASSO) feature selection method. Among existing feature selection algorithms, LASSO (Least Absolute Shrinkage and Selection Operator) is the most popular one [19,28] because of its efficiency, robustness and high accuracy performance [30]. As evaluation is by logistic regression, it is very fast. LASSO was introduced in [32] as a means of eliminating less informative variables in least squares multiple linear regression. It is a method of automatic variable selection which can select variable by shrinking the coefficient values and setting some equal to zero [10], for features to be eliminated. It has been widely used in many fields [36,40]. There are non-linear versions of Lasso [45]. Depending on the data, it will give as good or better result compared to linear version. For efficiency, in this work we used linear Lasso. We draw trace plot of coefficients for different features, varying the regularization parameter. From this plot, we manually fixed the value of regularization parameter, such that the number of non-zero coefficients are low (here 6) and MSE is also low.

For testing the classification result using the selected features, we used Support Vector Machine (SVM) as classifier. Basically, SVM is a linear classifier. Usually, the data is projected to a higher dimension, using a non-linear function like polynomial or radial-basis-function, so that in the transferred higher dimension the data is linearly separable. The non-linear transfer functions are called kernel functions. When the number of features and the total available data are large, we do not need (and it is computationally heavy) to transform the data to higher dimension using a non-linear kernel. In our experiment, the number of selected features is low, and the total data is not large. We used SVM with rbf kernel. It is known that non-linear SVM will work better or at least as good as linear SVM (direct data). If the data is linearly separable, the result using linear or non-linear SVM will be the same. For finding optimum kernel hyper-parameters we used grid-search method.

2.4 Data Structure

The data set consists of opinions from subjects, the details of which is in Sect. 4.2. For every feature, the subjects were to score them by a number from 1 to 5. Therefore, a sample consists of a 27 dimension feature vector, where all elements are numerical with values from 1 to 5. We had 361 valid responses, i.e., we had 361 samples in total.

3   Methodology

The employed approach involves 8 steps, as follows: define factors of game apps, design questionnaire, pre-test questionnaire, collect data, pre-process data, implement and run LASSO feature selection, build SVM classifier, evaluate results and make conclusions. The details of the procedure steps are explained as follows.

Step 1: Define factors of game Apps
Based on related works, we assembled all proposed potential factors of game Apps for doing In-App Purchase. We surveyed published works and defined them. Next, according to these defined factors, we go to the next step to design questionnaire for collecting data.

Step 2: Design questionnaire
We developed a set of questionnaire to estimate gamer's feeling about the level of importance for factors which will probably influence her/his in-App purchase behaviors. Briefly speaking, this questionnaire contains three parts.

- Part I: Basic information of the respondent (subject). The subject puts a score from 1 to 5 for each factor.
- Part II: The question items of different defined factors to estimate the importance levels for doing in-App purchase.

– Part III: Whether in-App purchase is done or not? In addition, we also collect information about the mode of payment as well as amount spent, though we do not use that information in our present work.

Step 3: Preliminary test
The original questionnaire is issued for preliminary testing (pretest). In this step, according to the feedbacks of respondents, we modify the questionnaire items. Then, we finalize and issue the questionnaire.

Step 4: Data collection
After pretesting, the modified questionnaire will be issued to gamers who have experiences of playing game Apps. The subjects complete the set of questionnaire over a month, as they play and purchase (or not purchase) a game software.

Step 5: Data pre-processing
The collected data is integrated [24] into a data set. For feature selection as well as classification, 5-fold cross validation is used. The part of the train data used for feature selection is also used for training the classifier, and the rest of the data used for testing.

Step 6: Implement LASSO feature selection
The LASSO (Least Absolute Shrinkage and Selection Operator) is a regression method that involves penalizing the absolute size of the regression coefficients. Let p is the number of factors, and N is the number of samples, y_i be the outcome of x_i. The objective is to solve

$$min_{\beta_0, \beta} \left[\frac{1}{N} \sum_{i=1}^{N} (y_i - \beta_0 - x_i^T \beta)^2 \right], \; subject \; to \; \sum_{j=1}^{p} |\beta_j| \leq \lambda \qquad (1)$$

Here, λ is a free parameter that determines the amount of shrinkage.

Step 7: Train Support Vector Machine (SVM) classifier
To see the effectiveness of feature selection using different methods, we use the whole attribute set (without implementing feature selection) and reduced attribute set (implementing feature selection) to build classifier. Feature selection by LASSO, χ^2, and back-propagation network (BPN) are compared. A SVM classifier is trained for checking classification performance.

Step 8: Draw conclusions
Through analysis of the results of step 7, we will identify important factors of influencing in-App purchases for game Apps.

4   Experiments and Results

4.1   Defined Factors

The 27 potential factors influencing purchase of game software are shown in Table 1. Based on that, we design questionnaire, collect responses to finally filter important factors based on user responses.

Table 1. Potential factors influencing purchase of game apps.

No	Notation	Factors	Supports
1	VM	Value-for-money	[5]
2	SV	Social value	[5]
3	AR	App rating	[5,34]
4	S	Satisfaction	[3,5,8]
5	U	Unexpectedness	[21]
6	Con	Confirmation	[21]
7	Com	Compatibility	[7,24]
8	AI	Affective involvement	[22]
9	PU	Perceived usefulness	[6,8,44]
1	PE	Perceived enjoyment	[3,5,8,44]
1	PEU	Perceived ease of use	[8,31]
1	GG	Graphics	[13]
1	GA	Animation	[13]
1	GS	Sound	[13]
1	GSC	Scenario	[13]
1	GC	Character	[13]
1	GI	Innovative	[13]
1	V	Visibility	[4,9]
1	VOL	Voluntaries	[4,9]
2	RD	Result demon	[4,9]
2	T	Trial-ability	[4,7,9]
2	IM	Image	[4,9]
2	Mm	Mass media	[7]
2	IC	Interpersonal	[7]
2	Cc	Cognitive co	[44]
2	PR	Perceived ri	[17,25]
2	UC	Use context	[44]

4.2 Collected Data

A total of 410 responses of questionnaires were collected, 271 over the Internet and 139 in the paper-and-pencil version. After removing invalid responses, 361 valid responses are kept for further analysis.

Table 2 shows the basic information of subjects and collected samples (such as gender, age, and income per month of respondents). We gathered information about their background regarding operating system, game usage time per day, playing experience, game types, and payment methods. Finally, we could know only 27% respondents who did in-App purchase and the amount of in-App purchase is more than 5 USD (35%), and then 1 USD each time (27%).

Table 2. Statistics of collected data

Variable	Distribution
Gender	Male:56%, Female:44%
Age	<18 years old (3%) 18~30 years old (46%) 31~40 years old (17%) >40 years old,(34%)
Income per month	<5K NTD (24%) 5K~10K NTD (10%) 10K~20K NTD (13%) 20K~50K NTD (39%) >50K NTD (14%)
Operating system	iOS (29%) Android (66%) Windows Phone (5%)
Game usage time per day	<3 hrs (70%) 4~6 hrs (23%) 7~9 hrs and above (7%)
Playing experience	<1 year (30%) 1~3 years (33%) 3 years and above (37%)
Game types	Sports/Simulation/Driving (28%) RPG/MMORPG/Strategy (25%) Action/Adventure/Fighting (18%) Children/Educational (7%) Above (22%)
In-App purchase	Ever (32%) Never (68%)
Payment methods	Credit card (38%) Far Eas Tone Telecommunications (20%) Google play gift card (3%), PayPal (2%) Google wallet, (2%) ATM (7%) Point card (28%)
The amount of in App purchase	<1 USD (27%) 1~1.99 USD (16%) 2~5 USD (22%) >5 USD (35%)

In addition, data set [24] have 131 valid data. To compare with it, we make use of random repeat sampling on collected data to get the same quantity.

4.3   Results of Feature Selection

In this study, 5-fold cross-validation experiment is done. Those factors whose coefficient values are not zero are picked up as important factors. Results of LASSO feature selection method is shown in Table 3. Based on occurrence frequency, we can build the important feature set. The feature subset selected by LASSO is PEnjoy2, SV2, AI, GA, GSC, GI.

Next, we train SVM classifier to evaluate the effectiveness of LASSO and compare with original full feature set (without feature selection).

Table 3. Summary of selected factors (LASSO feature selection)

Fold	Factors					
	Fold1	Fold2	Fold3	Fold4	Fold5	Occurrence frequency
PEnjoy2	0.725	0.461	0.562	0.562	0.725	5
SV2	0.564	0.476	0.508	0.508	0.564	5
AI	0.224	0.159	0.183	0.182	0.224	5
GA	0.268	0.219	0.236	0.236	0.268	5
GSC	0.058	0.044	0.050	0.050	0.058	5
GI	0.329	0.322	0.327	0.328	0.329	5
IC1	0.038	0	0	0	0.038	2
PU1	0	0	0	0	0	0
PU2	0	0	0	0	0	0
PE1	0	0	0	0	0	0
PE2	0	0	0	0	0	0
C1	0	0	0	0	0	0
C2	0	0	0	0	0	0
V1	0	0	0	0	0	0
V2	0	0	0	0	0	0
RD1	0	0	0	0	0	0
RD2	0	0	0	0	0	0
VOL1	0	0	0	0	0	0
VOL2	0	0	0	0	0	0
T1	0	0	0	0	0	0
T2	0	0	0	0	0	0
I1	0	0	0	0	0	0
I2	0	0	0	0	0	0
PR1	0	0	0	0	0	0
PR2	0	0	0	0	0	0
PR3	0	0	0	0	0	0

Table 4, shows the results of SVM. In this table, we can see LASSO selected features show better performances compared to when all features are included in classification. Table 5 provides comparison of LASSO, BPN, and chi-square methods. From this table, we can find LASSO feature selection outperforms BPN and χ^2. Based on the results, we can claim that we found 6 important factors, as listed in Table 6. Due to strong correlation, it is possible that some other important factors are not included. By including that, and discarding one of the selected factor, it is possible to achieve similar classification results. But, our aim of getting minimum subset of features is achieved.

Table 4. Evaluation results of LASSO feature selection

Index	Factor set	
	Original set	LASSO set
	Mean (StDev)	Mean (StDev)
OA (%)	75.38 (8.85)	76.15 (6.88)
F1 (%)	77.22 (9.72)	78.19 (8.26)
Time (s)	0.50 (0.04)	0.24 (0.05)

Table 5. Comparison of LASSO and [24] feature selection methods

Index	Factor set		
	LASSO set 6 factors	[24] BPN set 4 factors	[24] χ^2 set 16 factors
	Mean (StDev)	Mean (StDev)	Mean (StDev)
OA (%)	76.15 (6.88)	64.62 (7.40)	61.54 (6.08)
F1 (%)	78.19 (8.26)	67.62 (6.41)	62.32 (10.43)
Time (s)	0.24 (0.05)	4.29 (2.01)	4.26 (3.45)

Table 6. The extracted important factors

No	Notation	Factors	Definitions
1	SV	Social value	The degree to which an app is perceived as the enhancement of a person's self-concept provided by the product
2	PE	Perceived enjoyment	The extent to which the activity of using the App is perceived to be enjoyable in its own right, apart from any performance consequences that may be anticipated
3	AI	Affective involvement	The expected that consumers who connect interactivity to mobile apps may believe that using mobile apps is appealing and interesting
4	GA	Animation	Movement of characters or background
5	GSC	Scenario	Creativity of the scenario
6	GI	Innovativeness	Newness of the game to the market

In order to further evaluate the effectiveness of LASSO, we also compare the performance of LASSO with the feature selection method of [24].

5   Conclusion

The purpose of this study is to determine crucial factors of influencing in-App purchase.

From available literatures and [24] data set, 27 potential factors have been defined. Further, we employ LASSO to select the important factors. Results indicated that LASSO can effectively recognize 6 crucial factors. They are social value (SV), perceived enjoyment (PE), affective involvement (AI), animation (GA), scenario (GSC), and innovativeness (GI).

Therefore, game App developers should pay their attention to these crucial factors which can increase their revenue. As future work, we plan to use other more powerful feature selection method to identify important factors of doing in-App purchase. Additionally, more data from more subjects need to be collected for more reliable results.

References

1. App Annie (2016). http://blog.appannie.com/app-annie-releases-inaugural-mobile-app-forecast/
2. Business of Apps (2015). http://www.businessofapps.com/app-usage-statistics-2015/
3. Hsiao, C.-H., Chang, J.-J., Tang, K.-Y.: Exploring the influential factors in continuance usage of mobile social apps: satisfaction, habit, and customer value perspectives. Telemat. Inform. **33**, 342–355 (2016)
4. Hsu, C.-L., Lu, H.-P., Hsu, H.-H.: Adoption of the mobile Internet: an empirical study of multimedia message service (MMS). Omega **35**, 715–726 (2007)
5. Hsu, C.-L., Lin, J.C.-C.: What drives purchase intention for paid mobile apps?-an expectation confirmation model with perceived value. Electron. Commer. Res. Appl. **14**, 46–57 (2014)
6. Hsu, C.-L., Lin, J.C.-C.: Effect of perceived value and social influences on mobile app stickiness and in-app purchase intention. Technol. Forecast. Soc. Chang. **108**, 42–53 (2016)
7. Rogers, E.M.: Diffusion of innovations, 4th edn. Simon and Schuster, New York City (2010)
8. Park, E., Baek, S., Ohm, J., Joon Chang, H.: Determinants of player acceptance of mobile social network games: an application of extended technology acceptance model. Telemat. Inform. **31**, 3–15 (2014)
9. Moore, G.C., Benbasat, I.: Development of an instrument to measure the perceptions of adopting an information technology innovation. Inf. Syst. Res. **2**, 192–222 (1991)
10. D'Angelo, G.M., Rao, D., Gu, C.C.: Combining least absolute shrinkage and selection operator (LASSO) and principal-components analysis for detection of gene-gene interactions in genome-wide association studies. In: BMC Proceedings, vol. 3 (2009)
11. Gamasutra (2015). http://www.gamasutra.com/view/news/256997/The_top_10_games_take_25_of_global_mobile_app_revenue.php
12. Games Industry (2015). http://www.gamesindustry.biz/articles/2015-07-10-ios-game-revenues-show-top-20-dominate-newzoo
13. Park, H.J., Kim, S.-H.: A bayesian network approach to examining key success factors of mobile games. J. Bus. Res. **66**, 1353–1359 (2013)
14. Koekkoek, H.: How the most successful apps monetize their user base. Distimo Publication (2013). http://www.bbvaopen4u.com/sites/default/files/document/document/how_the_most_successful_apps_monetize_their_user_base.pdf

15. Lin, H.-H., Wang, Y.-S.: An examination of the determinants of customer loyalty in mobile commerce contexts. Inf. Manag. **43**, 271–282 (2006)
16. Kim, H.-W., Kankanhalli, A., Lee, H.-L.: Investigating decision factors in mobile application purchase: a mixed-methods approach. Inf. Manag. (2016, in press). Corrected proof
17. Al-Jabri, I.M., Sohail, M.S.: Mobile banking adoption: application of diffusion of innovation theory. J. Electron. Commer. Res. **13**, 379–391 (2012)
18. Sweeney, J.C., Soutar, G.N.: Consumer perceived value: the development of a multiple item scale. J. Retail. **77**, 203–220 (2001)
19. Duan, J., Soussen, C., Brie, D., Idier, J., Wan, M., Wang, Y.-P.: Generalized lasso with under-determined regularization matrices. Signal Process. **127**, 239–246 (2016)
20. Lee, J., Lee, J., Lee, H., Lee, J.: An exploratory study of factors influencing repurchase behaviors toward game items: a field study. Comput. Hum. Behav. **43**, 13–23 (2015)
21. Hsu, J.S.-C., Lin, T.-C., Fu, T.-W., Hung, Y.-W.: The effect of unexpected features on app users' continuance intention. Electron. Commer. Res. Appl. **14**, 418–430 (2015)
22. Kang, J.-Y.M., Mun, J.M., Johnson, K.K.P.: In-store mobile usage: downloading and usage intention toward mobile location-based retail apps. Comput. Hum. Behav. **46**, 210–217 (2015)
23. Gao, L., Waechter, K.A., Bai, X.: Understanding consumer's continuance intention towards mobile purchase: a theoretical framework and empirical study- a case of China. Comput. Hum. Behav. **53**, 249–262 (2015)
24. Chen, L.-S., Lin, M.-R.: Key factors in-app purchase for game applications. In: International Conference of Emerging Trends in Engineering and Technology, Kobe, pp. 91–95 (2015)
25. Kleijnen, M., Ruyter, K.D., Wetzels, M.: Consumer adoption of wireless services: discovering the rules, while playing the game. J. Interact. Mark. **18**, 51–61 (2004)
26. Mollaee, M., Moattar, M.H.: A novel feature extraction approach based on ensemble feature selection and modified discriminant independent component analysis for microarray data classification. Biocybern. Biomed. Eng. (2016, in press). Uncorrected proof
27. Mic (2015). http://mic.iii.org.tw/aisp/pressroom/press01_pop.asp?sno=400&type1=2
28. Connor, P., Hollensen, P., Krigolson, O., Trappenberg, T.: A biological mechanism for bayesian feature selection: weight decay and raising the LASSO. Neural Netw. **67**, 121–130 (2015)
29. Roma, P., Ragaglia, D.: Revenue models, in-app purchase, and the app performance: evidence from apple's app store and google play. Electron. Commer. Res. Appl. **17**, 173–190 (2016)
30. Zhou, Q., Song, S., Huang, G., Wu, C.: Efficient lasso training from a geometrical perspective. Neurocomputing **168**, 234–239 (2015)
31. Davis, R., Lang, B.: Modeling game usage, purchase behavior and ease of use. Entertain. Comput. **3**, 27–36 (2012)
32. Tibshirani, R.: Regression shrinkage and selection via the lasso. J. Roy. Stat. Soc. **58**, 267–288 (1996)
33. Sayed, S.A.-F., Nabil, E., Badr, A.: A binary clonal flower pollination algorithm for feature selection. Pattern Recogn. Lett. **77**, 21–27 (2016)
34. Basuroy, S., Chatterjee, S., Ravid, S.A.: How critical are critical reviews? The box office effects of film critics. J. Mark. **67**, 103–117 (2003)

35. Gunasundari, S., Janakiraman, S., Meenambal, S.: Velocity bounded boolean particle swarm optimization for improved feature selection in liver and kidney disease diagnosis. Expert Syst. Appl. **56**, 28–47 (2016)
36. Kwon, S., Lee, S., Kim, Y.: Moderately clipped LASSO. Comput. Stat. Data Anal. **92**, 53–67 (2015)
37. Sensor Tower (2016). https://sensortower.com/blog/top-ios-games-december-2015
38. Techco (2015) http://tech.co/6-app-monetization-models-make-money-2015-08
39. The Statistics Portal (2016). http://www.statista.com/statistics/536433/mobile-games-revenue-worldwide/
40. Bardsley, W.E., Vetrova, V., Liu, S.: Toward creating simpler hydrological models: a lasso subset selection approach. Environ. Model. Softw. **72**, 33–43 (2015)
41. Whatls (2015). http://whatis.techtarget.com/definition/in-app-purchase-IAP
42. Zhang, X., Mei, C., Chen, D., Li, J.: Feature selection in mixed data: a method using a novel fuzzy rough set-based information entropy. Pattern Recognit. **56**, 1–15 (2016)
43. Cong, Y., Wang, S., Fan, B., Yang, Y., Yu, H.: UDSF: unsupervised deep sparse feature selection. Neurocomputing **196**, 150–158 (2016)
44. Liu, Y., Li, H.: Exploring the impact of use context on mobile hedonic services adoption: an empirical study on mobile gaming in China. Comput. Hum. Behav. **27**, 890–898 (2011)
45. Yamada, M., et al.: High-dimensional feature selection by feature-wise non-linear lasso. Neural Comput. **26**(1), 185–207 (2014)
46. Zu, Z., et al.: Gradient boosted feature selection. In: Proceedings of Knowledge Discovery and Data Mining (2014)

An Agent-Based Model for Evaluating Post-acquisition Integration Strategies

Jing Su[1(✉)], Mohsen Jafari Songhori[1,2], Takamasa Kikuchi[1],
Masahiro Toriyama[3], and Takao Terano[1]

[1] Department of Computational Intelligence and Systems Science,
Interdisciplinary Graduate School of Science and Engineering,
Tokyo Institute of Technology, Tokyo, Japan
`sujing@trn.dis.titech.ac.jp`
[2] JSPS, Tokyo, Japan
[3] Graduate School of Management, Ritsumeikan University, Kyoto, Japan

Abstract. Mergers and acquisitions become popular means for the development of modern corporations, allowing companies to obtain quick access to new markets and source to grow. Post-acquisition integration has been recognized to be influential to the success of M&A. In this paper, we develop an agent-based model to study post-acquisition integration strategies for M&A according to the behavioral theory of the firm. Especially, the model conceptualizes firms conducting search over associated NK performance landscapes. Using this model, our simulation experiments indicate that strategies of personnel allocation, high level manager's feedback and the frequency of exchanging information could have impact on company's performance after M&A.

1 Introduction

Mergers and acquisitions (M&A) are transactions in which the ownership of companies, other business organizations or their operating units are transferred or combined. They become popular means for the development of modern corporations, because they allow companies to obtain quick access to new markets, products, technologies and source to grow. However, there is a high rate of failure after mergers and acquisitions (Christensen et al. 2011; Dauber 2012). As long as the development of research on M&A, there is a growing interest to the post-merger/post-acquisition integration in the literatures. It is found that post-merger/post-acquisition integration has great effects on the success of the M&A (Haspeslagh and Jemison 1991; Uzelac et al. 2016).

Birkinshaw et al. (2000) distinguish the post-acquisition integration between task integration and human integration. Human integration concerns generating satisfaction and shared identity among the employees, while task integration focuses on value creation and operational synergies. Some of researches on task integration discuss the value creation or company's performance are affected by the level of integration (Larsson and Finkelston 1999; Zaheer et al. 2013) while some of researches study on the influence of integration speed (Uzelac et al. 2016).

© Springer International Publishing AG 2017
S. Kurahashi et al. (Eds.): JSAI-isAI 2016, LNAI 10247, pp. 188–203, 2017.
DOI: 10.1007/978-3-319-61572-1_13

According to behavioral theory of the firm (Simon 1955; Cyert and March 1963), companies are conceptualized as bounded rational people who are doing the search. Particularly, people in a company generate alternatives by changing some decision parameters and then choose the alternatives with best performance (Knudsen and Levinthal 2007; Mihm et al. 2010). As fitness landscapes has been adopted to model human organizations, company's daily business can be seen as searching on a fitness landscape. Usually, large companies need to solve problems with complex interdependencies among each other, such as multiple relevant technologies, global markets, collaboration with external partners and so on. Thus, NK model proposed by Kauffman and Weinberger (1989) becomes a popular platform and widely used in studying organizations as complex adaptive systems as it allows researchers or modelers control over the interactions among the elements of system (Rivkin and Siggelkow 2002, 2007; Claussen et al. 2014).

In this paper, we study companies' mergers and acquisitions from the perspective of behavioral theory of the firm, especially for the case of a core company acquires a peripheral company. We conceptualize two companies' development as finding good strategies on two NK landscapes. After the acquisition, their strategies and landscapes get merged and become correlated to each other. Also, the structure of the acquiring company would be re-arranged with new comers from the target company. We define these two aspects of changes as the post-acquisition integration process. Then, we draw an agent-based model to simulate company's search process. Finally, we design several simulation scenarios and discuss how post-acquisition integration and search behaviors affect company's performance after M&A. The model is described in detail in Sect. 2, followed by the simulation result in Sect. 3 and conclusion in Sect. 4.

2 Model

In this paper, we assume a core company acquires a peripheral company. In particular, we model the core company as the acquirer with a three-level hierarchical structure. There are N front-line workers as the lowest level that equally distributed in D number of departments being managed by one CEO. The peripheral company with a flat structure of M workers and a CEO is the target company in the acquisition.

2.1 Original Development of Two Companies

In reality, company's goal is to find good strategies with high payoff. A strategy can be seen as a series of binary decisions about how to configure different activities. For instance, the company have to decide whether to develop a new product, whether to extend its market, and so forth. Thus, it can be defined as a binary string with N elements, each of which representing a decision of company's activities. We denote this N-digit string as $\mathbf{d} = \{d_1 d_2 \ldots d_N\}$, where d_i equals 0 or 1. Each strategy, that is, each configuration of the string can be evaluated by a fitness function. The value of fitness can be seen as the payoff of that strategy in the reality, while it also can be used to measure the overall company's performance in our model.

Specifically, each decision has a contribution to the fitness of strategy. The efficacy of each decision is affected not only by the choice of that decision, but also by the choices regarding other relevant decisions. Each decision i makes a contribution C_i to the fitness, and C_i depends on not only d_i but also some other decisions of $\{d_j\}$, which can be denoted by $C_i = C_i(d_i; \{d_j\})$. The exact set of $\{d_j\}$ for each d_i is determined by the relationships of the decisions. In the following part, we use "interaction matrix" to describe these relationships of the decisions for both two companies.

In acquiring company, the strategy is divided into N decisions and equally assigned to D number of departments. The decisions within each department could be highly relevant to each other, and have many interactions. However, those decisions that belong to different departments are less likely to be relevant, and have less interactions. Nonetheless, considering some of the departments may be more important than others, the decisions assigned to these departments could be more relevant to the ones of other departments. As an instance, the interaction matrix for a company with $N = 10$ decisions and $D = 2$ departments is shown in Fig. 1. According to the matrix, No. 1–5 represent the decisions belong to department 1 and No. 6–10 represent those belong to department 2. Mark "Y" represents the focal decision and mark "x" represents the interaction between the exact decision and the focal one. Consider decision No. 6, it has three interactions with decisions No. 7, 8, 9 from the same department as well as two interactions with decisions No. 2, 4 from the other department. Thus, the contribution of decision No. 6 is $C_6 = C_6(d_6; \{d_7 d_8 d_9 d_2 d_4\})$. The yellow area shows that some of the decisions of department 1 are relevant to the ones of department 2, and the former could affect the contribution of the latter. However, as shown by blank cells in matrix for columns 6–10 and rows 1–5 (i.e. the upper right of matrix area), none of decisions No. 6–10 in department 2 affect the decisions of department 1. Therefore, department 1 with more important decisions can be seen as a central department of the acquiring company.

		Department 1					Department 2				
		1	2	3	4	5	6	7	8	9	10
Department 1	1	Y	x		x	x					
	2	x	Y	x	x						
	3	x	x	Y		x					
	4	x		x	Y	x					
	5		x	x	x	Y					
Department 2	6		x		x		Y	x	x	x	
	7	x		x			x	Y	x		x
	8	x				x	x	x	Y	x	x
	9			x		x	x		x	Y	x
	10	x		x				x	x	x	Y

Fig. 1. An example of interaction matrix of acquiring company ($N = 10, D = 2$)

Unlike the acquiring company, the target company has a flat structure without department. But considering there could also be some decisions which are more influential than others, we design an interaction matrix similar to the one of acquiring company, which is shown in Fig. 2. We assume the target company's strategy contains six binary decisions. No. 1, 2, 3 are more influential, and they could affect the contributions of the other three decisions.

Fig. 2. An example of interaction matrix of target company ($M = 6$)

With the interaction matrix, the contribution of each decision can be determined. In particular, for each decision d_i, each configuration of $(d_i; \{d_j\})$ has an independent contribution value $C_i(d_i; \{d_j\})$, which is drawn at random from a uniform $U[0,1]$ distribution. Hence, changing the state of either d_i or any relevant decision d_j could result in a different contribution value C_i. Then, the overall fitness associated with a configuration of all the decisions is the average of the N (M for target company) contributions, which is shown in Eq. (1). Since the contributions are stochastic in the range of $[0,1]$, the fitness value $F(\mathbf{d})$ for each configuration is also between 0 and 1. Higher fitness value indicates better configuration of strategy. With 2^N (2^M for target company) possible strategy configurations and corresponding fitness values, the original landscapes of two companies' performance can be generated. This generating procedure is adapted from Kauffman's NK model.

$$F(\mathbf{d}) = \frac{1}{N} \sum_{i=1}^{N} C_i(d_i; \{d_j\}) \tag{1}$$

2.2 Post-acquisition Integration

We defined the original development of two companies in previous section, and this section describes their integration process after M&A. In this paper, we consider the acquisition case that the business of two companies merge together and become interdependent rather than other types of M&A. This type of acquisition may happen when a core company wants to explore some new functions on its products or to combine its own business to some other business, yet it has little knowledge in the exact fields. Then it may cover these shortages through acquiring a peripheral company that is professional in those fields. In the following paragraphs, we will define two companies' integration process from the aspects of strategy integration and structural integration.

Strategy Integration. As the business merge together, the strategies of two firms, and their performance landscapes become interdependent, too. To simplify the problem, we model new strategy after the acquisition as the simple combination of the two original ones. However, these two strategies will no longer be independent as before. Instead, they become interdependent to each other with some new interactions emerged among decisions.

We model the new interaction matrix following Claussen's work (2014) in combining two interdependent ones. Specifically, we integrate two strategies into a $(N + M)$-digit string with N digits from the acquiring company and M digits from the target. To simplify the problem, we assume original decision interactions within each company remain unchanged while some new interactions emerge among the decisions from different companies. Figure 3 shows an example of new interaction matrix after acquisition referring to the examples in Figs. 1 and 2.

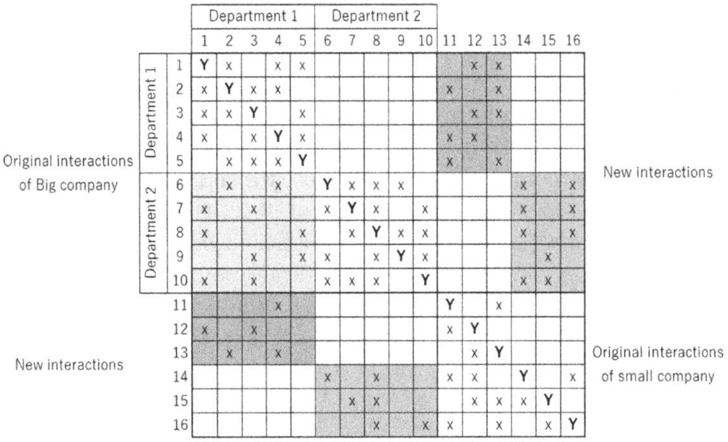

Fig. 3. Example of a new interaction matrix after acquisition (Referring to Figs. 1 and 2)

In Fig. 3, decisions No. 1–10 are the original ones from the acquiring company while No. 11–16 are from the target company. Thus, the interaction pattern of upper left and lower right areas, which represent the interactions among decisions within two companies respectively, are the same as before. The upper right and lower left areas represent new interactions emerged among decisions from different companies.

Generally, the pattern of new interactions between companies is unknown and it is likely to have particular pattern, rather than random. Nonetheless, to study how the new interactions affect company's performance after M&A, we consider an extreme case with a special type of pattern as shown in Fig. 3. We assume that more influential decisions of target company would become correlated with the ones in the central department of the acquiring company, while the less influential decisions of target company and the ones in less important department could have some interactions with each other after merging together. These new interactions are shown as the green and blue areas in the matrix.

As interaction matrix changed, the contribution of each decision to the fitness of strategy would change, too. However, the performance landscapes of both companies would not totally change. New landscape should be correlated to both of the two original ones. This correlation can be reflected by the correlations between each decision's new contributions and the original ones.

In this paper, we adapt Adner's concept (2014) on the correlation between contributions. Specifically, we define a correlation coefficient, denoted by $\rho \in [0, 1]$, to represent the correlation degree of the new contributions with the original ones. For each decision i, its new contribution C_i' could be the same as the original one C_i with probability of ρ, or could be independent from C_i, hence generated randomly following an uniform distribution with probability of $(1 - \rho)$. This process can be written as Eq. (2). For instance, the original contribution of decision No. 6 is $C_6 = C_6(d_6; \{d_7 d_8 d_9 d_2 d_4\})$ according to Fig. 1 and it becomes $C_6' = C_6'(d_6; \{d_7 d_8 d_9 d_2 d_4; d_{14} d_{16}\})$ after M&A according to Fig. 3. Thus, each configuration of $(d_6; \{d_7 d_8 d_9 d_2 d_4\})$ will derive four new configurations due to d_6's new interactions with d_{14} and d_{16}. The new contribution of each new configuration will be generated according to the original contribution value and Eq. (2).

With new interaction matrix and new contributions, we can evaluate the fitness of new strategies and generate a new landscape of company's performance.

$$C_i' = \begin{cases} C_i, & if \ \rho \\ c \sim U[0, 1], & if \ 1 - \rho \end{cases} \tag{2}$$

Structural Integration: Personnel Allocation. Although the decisions in the strategy are highly interdependent with each other, the company has to assign them to different teams or employees, because no single individual can solve all the relevant problems. In this model, lowest-level employees, that is front-line workers of each company take charge of making choices on the decisions, and each worker is assigned with one particular decision. After the acquisition, workers from the target company are allocated to different departments of the acquiring company. Assume these new comers will still work on their original tasks after allocation, then, the allocation of workers also can be seen as the allocation of decisions. Considering the complex interactions between decisions and company's search process (introduced in the next section), personnel allocation method may affect company's search performance.

Generally, the managers in either company do not know the new interactions among decisions after M&A, and they could allocate new comers in many different ways. In this paper, to study how personnel allocation affects company's performance with the influence of new interactions, we design two simple allocation methods as follows.

Allocation Method 1. The first method is to allocate employees who take charge of influential decisions to the central departments and other employees to the less central departments. Consider the case in Fig. 3, employees who take charge of decisions No. 11, 12, 13 will be allocated to department 1 and the other three employees will be allocated to department 2.

Allocation Method 2. On the contrary, the second method is to allocate employees who take charge of influential decisions to the less central departments and allocate others to the central department. That is, to allocate employees who take charge of decisions No. 11, 12, 13 to department 2 and others to department 1 for the case of Fig. 3.

Figure 4 shows the decision interaction patterns after personnel allocation. Note that these two patterns are only the examples to show the interactions among the decisions, but do not change the performance landscape. According to Fig. 4, allocation method 1 made highly relevant decisions centralized, while allocation method 2 made them decentralized. Specially, the highly relevant decisions are allocated into the same department when method 1 is practiced, while they are allocated into different department when method 2 is practiced.

Allocation-1

		Department 1								Department 2							
		1	2	3	4	5	11	12	13	6	7	8	9	10	14	15	16
Department 1	1	Y	x		x	x		x	x								
	2	x	Y	x	x		x		x								
	3	x	x	Y		x	x		x								
	4	x		x	Y	x	x	x									
	5		x	x	x	Y	x		x								
	11			x			Y		x								
	12	x		x			x	Y									
	13		x		x			x	Y								
Department 2	6		x		x					Y	x	x	x		x		x
	7	x		x						x	Y	x		x	x		x
	8	x			x						x	Y	x	x	x		x
	9		x		x					x		x	Y	x		x	
	10	x		x						x	x	x		Y	x	x	
	14					x	x			x		x			Y		x
	15					x		x	x	x	x				x	Y	
	16					x			x		x		x		x	x	Y

Allocation-2

		Department 1								Department 2							
		1	2	3	4	5	14	15	16	6	7	8	9	10	11	12	13
Department 1	1	Y	x		x	x										x	x
	2	x	Y	x	x										x		x
	3	x	x	Y		x										x	x
	4	x		x	Y	x									x	x	
	5		x	x	x	Y										x	
	14						Y		x	x		x			x	x	
	15						x	Y		x	x				x		x
	16						x	x	Y		x		x		x	x	
Department 2	6		x		x		x		x	Y	x	x	x				
	7	x		x			x	x	x	x	Y	x		x			
	8	x			x	x	x		x	x	x	Y	x	x			
	9		x		x		x		x	x	x	x	Y	x			
	10	x		x			x	x		x	x	x	x	Y			
	11			x											Y		x
	12	x		x											x	Y	
	13	x		x												x	Y

Fig. 4. Example of interaction patterns after personnel allocation (According to Fig. 3)

2.3   Search Process

In reality, company's goal is to find good strategies to get high payoff. According to the behavior theory of the firm, this process is conceptualized as search process. It also can be seen as company climbing on the landscape from the perspective of NK model. Nonetheless, employees and managers in company have no idea of the whole landscape and no one can finish this task by himself. Thus, a three-level vertical search process is proposed in this paper. Note that the personnel allocation after the acquisition do not change the vertical structure of acquiring company and thus has no influence of search process, we do not differentiate workers from two companies in this section.

Figure 5 shows an example of the three-level hierarchical structure of the acquiring company. We assume there are N workers and D departments managed by CEO, and workers equally distributed in each department. The detailed process is described in the following subsections.

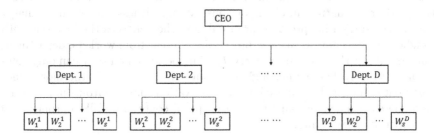

Fig. 5. Hierarchical structure of the acquiring company

Front-Line Workers' Local Search. As mentioned in the previous section, each front-line worker is assigned with one particular decision of the strategy, and he can make choices (choosing 1 or 0) only on that decision. However, he has information of other decisions either through a guidance from high levels or through an information exchange with his colleagues (introduced in the later subsections). Each worker has to make a proper choice on his own decision with his information of other decisions to make the whole strategy get a higher fitness.

For instance, worker i in department 1 is denoted by W_i^1. He has a set of information on other decisions, denoted by $\{\overline{d_1} \ldots \overline{d_{i-1}}, \overline{d_{i+1}} \ldots \overline{d_N}\}$. Hence, he has two alternative configurations of the whole strategy with that information set and his own decision d_i as 1 or 0. Then, he evaluates these two alternatives by their fitness according to Eq.(1), and chooses a better one with a higher fitness value. This process of making a choice can be written as

$$d_i^* = arg\ max\ F(d_i, \{\overline{d_j}\}), \text{where } j \in [1, N], j \neq i \qquad (3)$$

In Eq. (3), $F(\cdot)$ indicates the fitness of the strategy configuration according to Eq. (1), and $\overline{d_j}$ indicates the information of the state on decision d_j that worker W_i^1 has.

Periodically, that is every T_W time periods in this paper, workers submit their proposals to their department managers (Rivkin and Siggelkow 2003; Siggelkow and Rivkin 2005). Considering worker's observation is bounded, we assume each worker can submit a proposal containing only the information of decisions within his own department. For instance, worker W_i^1 of department 1 in Fig. 5 can submit a proposal with s digits information of $\{\overline{d_1} \ldots \overline{d_{i-1}}, d_i, \overline{d_{i+1}} \ldots \overline{d_s}\}$. In this proposal, d_i indicates the choice of W_i^1's own decision, and $\overline{d_j}$ $(j \in [1, s], j \neq i)$ indicates the information of his colleagues' decisions within department 1.

Department Managers' Decision-Making. Every T_W time periods, department managers gather the proposals submitted by their subordinates. Similar to workers, each manager has not only the s-digit information submitted by the workers, but also $(N - s)$-digit information on other decisions through the guidance from high level or the information exchange with his colleagues. Then he combines each proposal to the other $(N - s)$-digit information as different options. After evaluating all of the options by their fitness values, the manager compares those with his previous proposal, and then chooses the best one with the highest fitness value as his new proposal. Different from workers, department managers submit their proposals every T_D time periods, and they can only compare the options rather than changing the state of any digit. Furthermore, since department managers have more holistic observation than the front-line workers, we assume each manager can submit a proposal containing full information of the N-digit strategy string.

Meeting Colleagues and Exchanging Information. During the search process, workers and department managers may have chances to meet their colleagues (of the same level) by either regular meetings or "random" encounters. Through these meetings, workers or department managers can exchange and update their own information sets. Consider worker W_i and W_j, each of them has a set of information about the states of all other decisions besides his own one, denoted by $I_i = \{\overline{d_k}\}$, $(k \in [1, N], k \neq i)$ and $I_j = \{\overline{d_l}\}$, $(l \in [1, N], l \neq j)$ respectively. When they meet each other, they exchange the information of their own decisions and update their information sets, which become $I_i' = \{\overline{d_j'}, \{\overline{d_k}\}\}$, $(k \in [1, N], k \neq i, j)$ and $I_j' = \{\overline{d_i'}, \{\overline{d_l}\}\}$, $(l \in [1, N], l \neq j, i)$. In the equations, $\overline{d_j'}$ and $\overline{d_i'}$ represent the up-to-date information that worker W_i and W_j obtained. With these updated information, workers can do the local search again according to Eq. (3). Similar to workers, department managers can also meet each other. But instead of exchanging the information of a particular digit, they exchange all information of decisions within their departments.

We assume each worker or department manager meets his colleagues following a Poisson Process (Mihm et al. 2003, 2010), including the cases of company's regular meetings and the "random" encounters. Hence, the time interval of two employees' meeting follows an exponential distribution with a scale parameter as the mean of meeting time interval. It is also plausible to assume that workers who are working in the same department meet more frequently than workers from different departments do. Thus, we denote $scale_w$ as the average time interval that two workers within the same department meeting each other, and $scale_b$ as the average time interval that two workers from different departments meeting each other, as well as the average time interval two department managers meeting each other. Besides, each employee's meeting process is independent process.

CEO's Decision-Making and Feedback. Every T_D time periods, CEO gathers the proposals that contain N-digit information from department managers. Then she compares these proposals besides company's previous strategy (her last choice) by evaluating their fitness values according to Eq. (1), and chooses the best one as company's new strategy. The fitness value of this strategy can be seen as the measurement of company's performance. This procedure represents CEO's decision making process. In addition, in some cases, CEO may give her choice of the best strategy back to the lower levels as a developing guidance, hence we assume three types of information feedback process.

Full Information Feedback. In this case, CEO always gives back her choice of the best strategy to the department managers every time after she making a decision. Then the department managers give this feedback that contains full information to the workers. Thus, everyone can update his own information set and do his local search with this new information in the next time period.

Partial Information Feedback. Similar to the full information feedback case, CEO always gives back her choice to the department managers. However, in this case, each department manager only cares about the decisions of his own department. Thus, instead of the full feedback that he obtained, each manager picks a particular part of it which contains only the information of his own department and give this partial feedback to the workers. Therefore, the department managers can update their whole information sets, while the workers can get only part of their information sets updated.

No Feedback. In this case, CEO do not give back her choice as the developing guidance at all. Department managers and workers can update their information only through meeting their colleagues.

3   Simulation and Results

In this paper, we discuss the organization's behavior after M&A from two aspects: (i) the company's search behavior, and, (ii) the influence of post-acquisition integration with particular integrated performance landscape and personnel allocation methods. Specifically, we test the influence of different types of information feedback, different sets of employees' meeting frequency, different personnel allocation methods, and different complexity of landscapes.

To simplify the problem, we assume the acquiring company has $N = 10$ workers who are equally assigned to $D = 2$ departments, and the target company has $M = 6$ workers. Hence, the scale of strategies of two companies can also be determined as the same. The original landscapes of two companies as well as the new landscape are generated according to the process previously described in the model section with the correlation coefficient $\rho = 0.6$. Regarding the search process, we set a total simulation time of $T = 3000$ time steps. In addition, workers submit their proposals every $T_W = 5$ steps and department managers submit theirs every $T_D = 23$ steps. In each time step, each person can only focus on one activity, that is, either (i) conducting search, (ii) submitting proposals (gathering submissions), or (iii) meeting with one colleague.

To study company's behavior, we design 90 simulation scenarios with 3 types of information feedback (details in model section), 5 sets of meeting frequency, 2 personnel allocation methods (details in model section), as well as 3 levels of landscape complexity. As introduced in the model section, we can use the scale parameter of exponential distribution to measure employees' meeting frequency. Specially, we set $scale_w = \{4, 6, 8, 10, 12\}$ and $scale_b = 4 \cdot scale_w$, where smaller values represent that employees meet each other more frequently. In this paper, we measure the complexity of NK landscape as K/N, where N represents the number of decisions, and K represents that each decision d_i has in average K number of relevant decisions $\{d_j\}$ which have impact on d_i's contributions. Due to the special pattern of interaction matrix, the maximum complexity of the landscape could be around 0.5. Thus, we set 0.16, 0.35, and 0.49 as low, medium and high level respectively. For each complexity level, we generate 100 different landscapes and simulation runs 5 times on each landscape. Due to the limited space, only the results of low complexity and high complexity are shown in this paper. The results of medium complexity are similar to the high level ones (and available from the authors).

Figures 6, 7, 8 and 9 show the simulation results of different scenarios by box plots. Figures 6 and 7 show the results of simulations with low level complexity, while Figs. 8 and 9 show the results of high level complexity. The three panels of each figure shows the results of three types of information feedback. In each panel, x-axis shows different value of $scale_w$ which represents different meeting frequencies, and y-axis shows either the normalized performance or the convergence time of search. For each value of $scale_w$, there is a blue box showing the simulation results of personnel allocation method 1 and a red box illustrating the results of allocation method 2. The red mark in each box represents the statistical average of the exact set of data.

Influence of Different Types of Feedback. In the low complexity case in Figs. 6 and 7, the landscape is very smooth with less local peaks, and it is easy for company to get high performance via the search process. Thus, different scenarios show similar results. Figures 8 and 9 that represent the results of high complexity scenarios show the obvious difference between different types of feedback. In full feedback case, everyone gets a feedback from CEO periodically, and then conducts the search with the information of this feedback. Since the feedback was chosen as the best strategy among all the options in the mean time, it could lead everyone quickly reach to a higher position on the landscape. Hence, this type of feedback process may accelerate the search process and make it converge

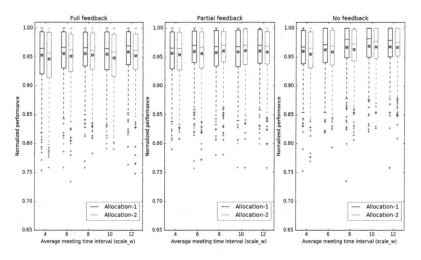

Fig. 6. Search performance of different simulation scenarios with complexity = 0.16

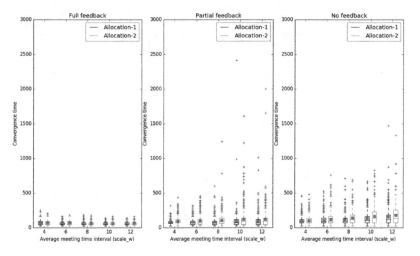

Fig. 7. Convergence time of different simulation scenarios with complexity = 0.16

in a very short time. Nonetheless, it also makes everyone's configuration same, hence will cause everyone getting stuck once the feedback reaches to a local optimum.

On the contrary, in no feedback case, CEO has no influence to the lower levels' search process. Everyone updates his information through meeting colleagues. Hence, company's search process could be much slower than those in the feedback cases. However, search without feedback somehow keeps the strategy options highly diverse, so the company could have opportunities to jump out of a local optimum and find better strategies. The partial feedback is the case between the other two. Periodical feedback reduces the search time while partial feedback

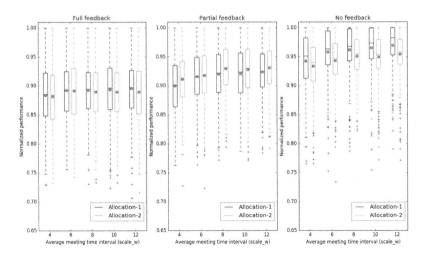

Fig. 8. Search performance of different simulation scenarios with complexity $= 0.49$

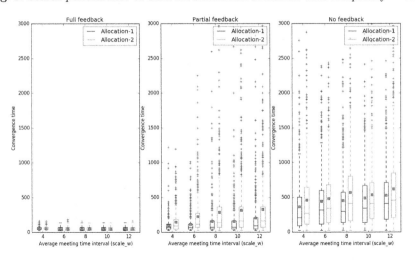

Fig. 9. Convergence time of different simulation scenarios with complexity $= 0.49$

from department managers retains some of the diversity of the strategy options. Therefore, no feedback case takes the longest time to converge but obtains the highest performance, while the full feedback case shows the other way around.

Influence of Different Meeting Frequencies. The high complexity simulation results shows the difference between different meeting frequencies especially in the partial feedback and no feedback cases. In full feedback case, periodical feedback coming from high levels affects the search process more than employees' meeting does. Thus, there is little difference between the results of different meeting frequencies. However, in partial feedback and no feedback cases, both the search performance and convergence time increase when employees' meeting frequency decreases. When $scale_w$ gets smaller, employees meet each other and hence get their information updated more frequently. Consequently, employees' configurations become the same quickly, which could cause everyone rapidly getting stuck at the same point. In addition, spending lots of time to meet colleagues would occupy the search time. Therefore, too much meeting may lead to a lower performance but less convergence time.

Influence of Different Personnel Allocation Methods. In partial feedback and no feedback cases of Figs. 8 and 9, the difference between red boxes and the blue ones appears to be significant, which represent the different personnel allocation methods would affect the performance and convergence time of the search. Allocation method 1 and 2 perform similar in the full feedback and partial feedback cases, yet method 1 performs better than method 2 in the no feedback case. However, method 2 always takes longer time to converge than method 1.

In no feedback case, workers and department managers update informations only through meeting colleagues. Workers in the same department meet each other more frequently than the ones from different departments. Department managers meet each other infrequently as well. As a result, everyone focuses on the decisions in his own department, yet, to some extent, ignore the ones of the other department. As for the case of allocation method 1, highly relevant decisions are in the same department, thus workers can promptly obtain the up-to-date information of these decisions and make a proper choice. However, with allocation method 2, workers cannot promptly obtain the up-to-date information which is highly relevant to their own decisions. Thus, allocation method 1 performs better than method 2, and takes shorter time to converge.

Partial feedback and full feedback cases are quite different from no feedback case. As mentioned in the previous sections, the feedback from high levels rather than the meetings among the employees dominates the information update. Employees can obtain the up-to-date information of the other department from this feedback. Thus, there is little difference between the two allocation methods. Specially, in the partial feedback case, workers cannot promptly update the information of the other department because of the partial feedback from their managers. Thus, it may take longer time to converge when allocation method 2 put into practice.

4   Conclusion and Future Work

In this paper, we propose an agent-based model to study the effect of post-acquisition integration from the perspective of behavioral theory of the firm. In particular, we define the two companies' original developments by generating two NK landscapes. We also define the post-acquisition integration process on both the landscapes and the structures of two companies. Then we elaborate a search process to simulate the organization's behavior in finding good strategies.

According to the simulation result, when problem complexity is low, that is, decisions of company's strategy are less interdependent to each other, it is easy for employees to find superior strategies with high performance even after mergers and acquisitions. However, when the complexity becoming high, that is, decisions of strategy become highly interdependent with each other, it becomes difficult for employees to find superior strategies with high performance. Many factors could influence the search process. Specially, excessive feedback from high levels may help the company quickly find some good strategies, but it may restrict employees' and managers' cognition to search for other possible strategies. Thus, it could easily make company's search get stuck with low performance. Without feedback, company's search process is dominated by low level employees' cooperation. In this case, frequent meetings among the employees may do harm to the search performance by occupying employees' search time as well as make their cognition quickly converge. As for re-arranging employees after mergers or acquisitions, employees who take charge of highly relevant tasks working together could help the company get high performance, especially when there is no information feedback from high levels.

In this paper, there are some limits to our model. Specifically, we modeled the company's post-acquisition integration with some simple and particular definitions. For instance, we defined the landscapes of two companies' original development by special interaction patterns of decisions. We also defined the post-acquisition interaction pattern in a particular way. These can be released in the future works. Besides, the personnel allocation methods and some other settings of search process can also be released with more scenarios.

References

Adner, R., Csaszar, F.A., Zemsky, P.B.: Positioning on a multiattribute landscape. Manage. Sci. **60**(11), 2794–2815 (2014)

Birkinshaw, J., Bresman, H., Hkanson, L.: Managing the postacquisition integration process: how the human integration and task integration processes interact to foster value creation. J. Manag. Stud. **37**(3), 395–425 (2000)

Christensen, C.M., Alton, R., Rising, C., Waldeck, A.: The big idea: the new M&A playbook. Harv. Bus. Rev. **89**(3), 48–57 (2011)

Claussen, J., Kretschmer, T., Stieglitz, N.: Vertical scope, turbulence, and the benefits of commitment and flexibility. Manag. Sci. **61**(4), 915–929 (2014)

Cyert, R.M., March, J.G.: A Behavioral Theory of the Firm. Prentice-Hall Inc., Englewood Cliffs (1963)

Dauber, D.: Opposing positions in M&A research: culture, integration and performance. Cross Cult. Manag.: Int. J. **19**(3), 375–398 (2012)

Haspeslagh, P.C., Jemison, D.B.: Managing Acquisitions: Creating Value Through Corporate Renewal, vol. 416. Free Press, New York (1991)

Kauffman, S.A., Weinberger, E.D.: The NK model of rugged fitness landscapes and its application to maturation of the immune response. J. Theor. Biol. **141**(2), 211–245 (1989)

Knudsen, T., Levinthal, D.A.: Two faces of search: alternative generation and alternative evaluation. Organ. Sci. **18**(1), 39–54 (2007)

Larsson, R., Finkelstein, S.: Integrating strategic, organizational, and human resource perspectives on mergers and acquisitions: a case survey of synergy realization. Organ. Sci. **10**(1), 1–26 (1999)

Mihm, J., Loch, C., Huchzermeier, A.: Problem solving oscillations in complex engineering projects. Manag. Sci. **49**(6), 733–750 (2003)

Mihm, J., Loch, C.H., Wilkinson, D., Huberman, B.A.: Hierarchical structure and search in complex organizations. Manag. Sci. **56**(5), 831–848 (2010)

Rivkin, J.W., Siggelkow, N.: Organizational sticking points on NK landscapes. Complexity **7**(5), 31–43 (2002)

Rivkin, J.W., Siggelkow, N.: Balancing search and stability: interdependencies among elements of organizational design. Manag. Sci. **49**(3), 290–311 (2003)

Rivkin, J.W., Siggelkow, N.: Patterned interactions in complex systems: implications for exploration. Manag. Sci. **53**(7), 1068–1085 (2007)

Siggelkow, N., Rivkin, J.W.: Speed and search: designing organizations for turbulence and complexity. Organ. Sci. **16**(2), 101–122 (2005)

Simon, H.A.: A behavioral model of rational choice. Q. J. Econ. **69**, 99–118 (1955)

Uzelac, B., Bauer, F., Matzler, K., Waschak, M.: The moderating effects of decision-making preferences on M&A integration speed and performance. Int. J. Hum. Resour. Manag. **27**(20), 2436–2460 (2016)

Zaheer, A., Castaer, X., Souder, D.: Synergy sources, target autonomy, and integration in acquisitions. J. Manag. **39**(3), 604–632 (2013)

Assessing Long-Term Care Resource Distribution in China by Simulating Care-Seeking Behaviors

Shuang Chang[1(✉)], Wei Yang[2], and Hiroshi Deguchi[1]

[1] Tokyo Institute of Technology, Tokyo, Japan
chang@cs.dis.titech.ac.jp, deguchi@dis.titech.ac.jp
[2] King's College London, London, UK
wei.yang@kcl.ac.uk

Abstract. Over the past decades, there has been an increasing attention to develop systems of ageing-related services in China to meet the swelling long-term care (LTC) needs along with a rapid aging population. However, it is challenging to design and evaluate integrated LTC service systems, especially of the resource distribution among various care service types, i.e. home-based care, hospital care and nursing house care. Furthermore, the divergent LTC needs and biased LTC service supply lead to unclear implications of any policy changes on distributional impacts. Therefore, by deploying an agent-based simulation approach, this paper aims to evaluate LTC service resource distribution, in terms of equity and effectiveness, among types of care provided in China. We first estimate LTC needs across different age groups and their preference on different types of LTC services. Subsequently, we evaluate service distribution plans by simulating individuals' care-seeking behaviors towards corresponding service providers. Simulation results indicate the co-existence of a waste of LTC service resource and unmet needs of LTC service in certain districts. The results also suggest disparities in unsettled rate among different social groups. This work is expected to further facilitate policy-makers on LTC resource distributive plans.

Keywords: Long-term care · Agent-based simulation

1 Introduction

Long-Term Care (LTC) services are primarily provided to support people who may require assistance with daily living activities, such as eating, bathing, dressing, personal hygiene, incontinence, and in-door moving [19]. Over the past decades, there has been an increasing attention to develop systems of ageing-related services in China to meet the swelling needs, attributed to an increase in longevity and decline in fertility [6,13,17]. On the other side, the family structure change due to rapid demographic shifts and profound socioeconomic situations lead to a shrink of family caregivers, thus eroding the tradition of filial piety and fostering emerging needs of institutional LTC services [4].

© Springer International Publishing AG 2017
S. Kurahashi et al. (Eds.): JSAI-isAI 2016, LNAI 10247, pp. 204–219, 2017.
DOI: 10.1007/978-3-319-61572-1_14

In response to the escalating needs, the twelfth Five-year Plan (2011–2015) released by the State Council outlined three tiers of LTC service provision: family care as the base; community-based service as the support, and institutional care as the supplementation [13]. Not limited to actively promoting family care and developing community-based services, the Chinese government has been putting concerted efforts to construct institutional LTC facilities to handle the beds shortage. In addition, incentives and subsidies in compliance with national-level policies at municipal level, though varied substantially across municipals, have been experimented to accelerate the development of private-sector facilities to enhance the capacity of institutional LTC [4].

However, it is challenging to design and assess integrated LTC service systems, especially of the resource distribution among various service types, i.e. home-based care, hospital care and nursing house care. The compound situation stemmed from both stratified LTC needs and biased LTC service supply lead to unclear implications of policy changes for the elderly welfare [18]. Moreover, relevant policies and regulations initiated from the government may have various impact on different social groups, thus further increase the complexity of analyzing the service distributive effect [3]. Therefore, by simulating care-seeking behaviors among social groups, we aim to assess the equity and effectiveness of LTC service resource distribution, in terms of bed provision, among types of care provided in China. More specifically, we aim to examine how different types of LTC resources should be distributed in the presence of individual heterogeneity to achieve effectiveness and equity.

1.1 Related Works

There have been extensive works on funding LTC and distributing relevant resources in both developed and developing countries [4,8,12,15,17]. Some of them rooted in projecting unmet needs or demands of LTC services based on the exploration of influential factors within particular contexts; micro-simulation, macro-simulation and statistic models are widely adopted [2,6,7,16,19]. Another stream of works focused on modeling health-seeking behaviors by analyzing the influence of both design variations and individual heterogeneity, though not applied to China's LTC scheme yet [1,2]. However, few of them examined the dynamics between individuals' LTC service-seeking behaviors and resource distributive effects upon regulations, especially of the Chinese case. In addition, due to the unavailability of relevant empirical data on China's LTC schemes, especially of the usage of LTC services, conventional statistical methods may not be applicable. Therefore, a simulation approach is deployed to link dynamically the micro-level care seeking behaviors of stratified individuals and macro-level resource distribution assessment. Further, the simulation approach enables large scale simulation and "to-be" scenario analysis to capture the heterogeneity of situations, which may provide insight for policy-makers on future plan.

This paper is thus organized as follows. The conceptual framework is proposed and explained in Sect. 2. The formal modeling of agents are defined in

Sect. 3 and the simulation work of a Chinese case is discussed in Sect. 4. Conclusion and a discussion of future works are provided in the last section.

2   Conceptual Framework

The framework comprises two entities: LTC service users (individuals with LTC needs) and LTC service providers. We follow Zhu's work [19] to identify influential factors on LTC needs and adopt Andersen's behavioral model of health services use [1] to simulate individuals' selection behaviors of different types of LTC services.

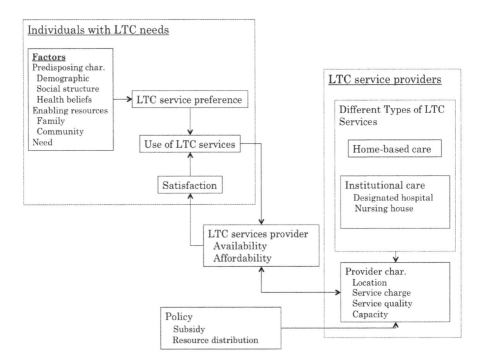

Fig. 1. Conceptual framework capturing the major stakeholders

These two entities are closely related with each other and should not be examined separately when framing LTC-related policies on developing and distributing LTC services. On one side, there are unmet needs of LTC services primarily due to (1) divergent needs of heterogeneous individuals associated with their socioeconomic status, social security condition (enrolled insurance types), and availability of family care provided by informal caregivers, such as family members [19] and (2) insufficient LTC resources and the disparity of resource allocation among institutional facilities [4]. On the other side, LTC resources might not be efficiently and effectively utilized due to (1) their inherit properties, such as amenities, professions and caregivers, location and service charges and (2) a bias plan and distribution of LTC services not in tune with divergent needs of individuals [13]. These two aspects are integrated in a recursive manner that any policy change may lead to behavioral changes, which may further influence the future policy-making. Therefore, we propose a holistic framework as shown in Fig. 1 to integrate these two aspects and capture their influence to each other.

With respect to LTC service users, Andersen defined three factors which may influence individuals' use of health services: predisposing characteristics, enabling resources and actual needs [1]. Upon determining the type of LTC services taking into account their socioeconomic status (predisposing char.), availability of caregivers (enabling resources), and level of ADL (Activities of Daily Living scales) limitations (actual needs), individuals would choose a particular service provider depending on key measurements of LTC service providers: availability, affordability and service quality [9]. Individuals might repeat the selection behaviors for several times until they are satisfied with the provided services [1].

With respect to LTC service providers, we assume there are three types of LTC services: institutional care provided by designated hospitals and nursing houses, and home-based care. Designated hospitals and nursing houses provide medical and nursing cares to people in critical LTC needs, whereas home-based care includes regular home-visit by professions. In compliance with municipal level subsidies and health resource distributive plan in terms of LTC beds, the characteristics of LTC service providers vary, which would further influence individuals' LTC service use in a recursive manner.

3   Modeling

We assume a long-term care insurance scheme launched in an artificial city of China. The city is divided into four districts where LTC resources are distributed in terms of bed provision. There is no district-level restriction on hospital care and nursing house care. In contrast, home-based care is only available to residences living within a pre-defined distance from the care center.

Based on the proposed conceptual framework, we define two types of agents: individuals with LTC needs and LTC service providers. We assume rational individuals who make decisions based on an utility evaluation [5] of each service provider.

3.1   LTC Service Provider

We define a set of service providers $S = \{s_1, s_2, \ldots, s_n\}$, $n \in N$. Each $s_i \in S$ is defined by its type, capacity, service quality, service charge and location. Depending on the type of LTC service providers, eligible individuals would be different. For instance, people with critical LTC needs could enroll in designated hospitals and nursing houses whereas those with less LTC needs could only enroll in home-based cares; once a service provider is fully occupied in terms of beds, it could no longer enroll any new individuals; the service quality is assigned randomly and a larger value indicates better services, such as well-equipped amenities, spacious environment, sufficient trained professions, and so on; the location of each service provider is the district where they locate. The description and definition of variables of service provider s_i are summarized in Table 1.

Table 1. Explanation of service provider agent variables

Variable	Definition	Description
Type	$TP_i \in \{0, 1, 2\}$	Type of service providers: 0 as designated hospital care, 1 as nursing house care, and 2 as home-based care
Capacity	$SA_i \in N$	Number of available beds for LTC services
Quality	$SQ_i \in (0, 3]$	Service quality; a larger value indicates better services
Charge	$SC_i \in (0, 5000]$	Monthly charge of LTC services
Location	$Loc_i \in \{d_i\}$	District where the service provider locates For home-based care, the service is only available to people living in the same district

Regarding nursing house care, we further categorize the service providers into four groups as follows: Type I could be treated as private-owned ones that aiming at people with high income; Type II as government-owned ones with high reputation and low charge; Type III as outdated ones with low quality and charge; and Type IV as ordinary ones varied in quality and service charge (Table 2).

Table 2. Nursing home care providers

Type	Character	Definition
Type I	High quality, High charge	$SQ \in [2, 3]$, $SC \in [4000, 5000]$
Type II	High quality, Low charge	$SQ \in [2, 3]$, $SC \in (0, 2000)$
Type III	Low quality, Low charge	$SQ \in (0, 1)$, $SC \in (0, 2000)$
Type IV	Fair quality, Fair charge	$SQ \in [1, 2)$, $SC \in [2000, 4000)$

Service quality $SQ_j(t) \in [\nu, \upsilon]$ partially depends on the number of individuals currently enrolled, denoted as $cap_j(t)$, and is updated accordingly as below; α is a control parameter either be positive or negative to reflect whether an increase of bed occupation could improve the service quality or not.

$$SQ_j(t) = \begin{cases} SQ_j(t-1) * (1 + \alpha * \frac{\Delta cap_j(t)}{SA_j}) \\ \upsilon & \text{if } SQ_j(t-1) > \upsilon \\ \nu & \text{if } SQ_j(t-1) < \nu \end{cases} \qquad (1)$$

3.2 Individuals with LTC Needs

We first estimate LTC needs by studying the sample of the first wave of China Health and Retirement Longitudinal Study (CHARLS) conducted between June 2011 and March 2012 [11]. This national-level survey interviewed 10069 individuals older than 45 and their spouses (7639 individuals) focusing on their social, economic and health-related issues [18]. We divide respondents into four age groups: 45–54, 55–64, 65–79, and 80–101; the sample characteristics of each age group, including marital status, number of children (male/female) living together, enrolled health insurance, household expenditure on non-food, number of ADL limitations and caregivers, are summarized in the following Tables 3, 4 and 5.

We target urban residences who had required assistance in at least 1 of six ADLs-related questions in CHARLS. It asked the respondents whether they have difficulty with any of the six everyday activities for at least more than 3 months, including dressing, showering, eating, using the toilet, incontinence, and in-door moving. Four answers are provided indicating the degree of seriousness, from the lowest to highest: no difficulty, have difficulty but can still do it, need help, and cannot do it. Depending on the number of ADL limitations, they may choose different kinds of services.

Formal Definition. We define a set of individuals with LTC needs as $Ind = \{I_1, I_2, \ldots, I_n\}$, $n \in N$. Each individual $I_i \in Ind$ is defined by sex, age, economic status, marital status, number of children (male and female separately), ADL level and insurance type. The description and definition of each variable of individual I_i is summarized in Table 6.

3.3 Individuals' Care-Seeking Behaviors

The care provider selection process is depicted in Fig. 2. It is a two-stage decision-making process. First, individuals with LTC needs determine the type of LTC services. Second, upon the determination of care type, they may choose a particular service provider based on the corresponding utility estimation [5]. We also assume that individuals hold full information of service providers from the same district and of those located at other districts with certain probability ($\approx 80\%$). For the current work, we make this strong assumption about the probability due to a lack of empirical data. Different probability may influence the service provider selection results. For future work, we would examine the influence of service provider promotion with varied probabilities across service providers. For instance, a service provider with a better reputation or having been massively promoted would have a higher probability of being known. If a service provider

Table 3. Sample distribution of respondents aged 45–64 by sex

Age: 45–54 (n = 1374)	Number (F)	Percentage (F) (%)		Number (F)	Percentage (F) (%)
Sex			ADL		
Male	644	46.87	Yes	28 (27)	4.37 (3.72)
Female	730	53.13	No	612 (699)	5.63 (96.28)
Marital status			No. of ADL items		
Single	4 (3)	0.63 (0.41)	1	11 (16)	39 (59.26)
Married	614 (677)	95.94 (93.51)	2	7 (3)	25 (11.11)
Separated	22 (44)	3.44 (3.85)	3 and above	10 (8)	36 (29.63)
No. of children (M)			Insurance type		
0	373 (425)	58.28 (58.54)	URI	110 (155)	20.3 (25.24)
1	237 (273)	37.03 (37.60)	UEI	269 (281)	49.7 (45.77)
2 and above	30 (28)	4.69 (58.28)	Others	81 (105)	15 (17.1)
No. of children (F)			No	81 (73)	15 (11.89)
0	500 (569)	78.13 (78.37)			
1	126 (139)	19.69 (19.15)			
2 and above	14 (18)	2.19 (2.48)			
Type of caregiver					
Spouse	33 (32)	47.14 (50)	Others	1 (4)	1.43 (6.26)
Children	15 (13)	21.43 (20.31)	No	9 (9)	12.86 (14.06)
Relatives	12 (6)	17.14 (9.38)			

Age: 55–64 (n = 1499)	Number (F)	Percentage (F) (%)		Number (F)	Percentage (F) (%)
Sex			ADL		
Male	768	48.77	Yes	39 (43)	5.1 (5.91)
Female	731	51.23	No	726 (685)	94.9 (94.09)
Marital status			No. of ADL items		
Single	3 (2)	0.39 (0.28)	1	21 (25)	53.8 (58.14)
Married	729 (621)	95.54 (85.54)	2	8 (2)	20.5 (4.65)
Separated	31 (103)	4.06 (14.19)	3 and above	10 (16)	25.6 (37.21)
No. of children (M)			Insurance type		
0	470 (443)	61.44 (60.85)	URI	136 (158)	20.82 (25.04)
1	255 (254)	33.33 (34.89)	UEI	357 (306)	54.67 (48.5)
2 and above	40 (31)	5.23 (4.35)	Others	96 (74)	14.7 (11.73)
No. of children (F)			No	64 (93)	9.8 (14.74)
0	589 (563)	76.99 (77.34)			
1	148 (152)	19.35 (20.88)			
2 and above	28 (13)	3.66 (1.79)			
Type of caregiver					
Spouse	44 (49)	51.16 (53.85)	Others	3 (1)	3.49 (1.10)
Children	15 (20)	17.44 (21.98)	No	17 (8)	19.77 (8.79)
Relatives	7 (10)	8.14 (10.99)			

Table 4. Sample distribution of respondents aged 65–101 by sex

Age: 65–79 (n = 1172)	Number (F)	Percentage (F) (%)		Number (F)	Percentage (F) (%)
Sex			ADL		
Male	649	55.38	Yes	39 (30)	6.06 (5.76)
Female	523	44.62	No	605 (491)	93.94 (94.24)
Marital status			No. of ADL items		
Single	2 (0)	0.31 (0)	1	21 (23)	53.85 (76.67)
Married	589 (357)	91.46 (68.92)	2	6 (6)	15.38 (20)
Separated	53 (161)	8.23 (31.08)	3 and above	12 (1)	30.77 (3.33)
No. of children (M)			Insurance type		
0	415 (339)	64.44 (65.07)	URI	70 (127)	12.05 (27.19)
1	206 (160)	31.99 (30.71)	UEI	316 (207)	54.39 (44.33)
2 and above	23 (22)	3.57 (4.22)	Others	153 (82)	26.33 (17.56)
No. of children (F)			No	42 (51)	7.22 (10.92)
0	499 (419)	77.48 (80.42)			
1	137 (99)	21.27 (19)			
2 and above	8 (3)	1.24 (0.58)			
Type of caregiver					
Spouse	36 (29)	46.15 (44.62)	Others	3 (6)	3.84 (9.32)
Children	17 (14)	21.79 (21.54)	No	11 (8)	14.10 (12.31)
Relatives	11 (8)	14.10 (12.31)			

Age: 80–101 (n = 174)	Number (F)	Percentage (F) (%)		Number (F)	Percentage (F) (%)
Sex			ADL		
Male	92	47.13	Yes	10 (5)	11.11 (6.1)
Female	82	52.87	No	80 (77)	88.89 (93.9)
Marital status			No. of ADL items		
Single	1 (0)	1.11 (0)	1	6 (4)	60 (80)
Married	68 (26)	75.56 (31.71)	2	1 (0)	10 (0)
Separated	21 (56)	23.33 (68.29)	3 and above	3 (1)	30 (20)
No. of children (M)			Insurance type		
0	63 (52)	70 (63.41)	URI	13 (16)	15.66 (22.22)
1	20 (25)	22.22 (30.49)	UEI	30 (34)	36.14 (47.22)
2 and above	7 (5)	7.78 (6.1)	Others	32 (9)	38.55 (40.27)
No. of children (F)			No	8 (13)	9.63 (18.05)
0	70 (64)	77.78 (78.05)			
1	18 (15)	20 (18.29)			
2 and above	2 (3)	2.22 (3.66)			
Type of caregiver					
Spouse	6 (4)	42.86 (36.36)	Others	2 (0)	14.28 (0)
Children	2 (3)	14.29 (27.27)	No	3 (2)	21.43 (18.18)
Relatives	1 (2)	7.14 (18.18)			

Table 5. Sample characteristics of yearly expenditure on non-food consumption

Age: 45–54					Age: 55–64				
Obs	Max	Min	Mean	Std. dev.	Obs	Max	Min	Mean	Std. dev.
1228	753402.6	379.0857	28768.89	47943.45	1318	753402.6	0	24862.48	39050.39
Age: 65–79					Age: 80–101				
Obs	Max	Min	Mean	Std. dev.	Obs	Max	Min	Mean	Std. dev.
1035	913368.9	0	29236.58	56619.7	159	320211.5	1634.979	27975.67	32032.29

Table 6. Explanation of individual agent variables

Variable	Definition	Description
Sex	$sex_i \in \{0,1\}$	0 for male and 1 for female
Age	$age_i \in \{0,1,2,3\}$	0: aged 45–54; 1: aged 55–64; 2: aged 65–79; 3 aged 80–101
Economic status	$inc_i \in R^+$	Monthly expenditure on non-food
Location	$L_i \in \{d_i\}$	The district where the individual reside
Number of children	$nc_i^f, nc_i^m \in N$	Number of female children (nc_f) and male children (nc_m)
Insurance type	$ins_i \in \{0,1,2\}$	0 for no health insurance, 1 for URI and 2 for UEI
ADL level	$ADL_i \in \{0,1,2\}$	The level of disability depending on ADL score. A larger value indicates a more serious condition
Service type	$st_i \in \{0,1,2\}$	Chosen type of LTC service
Probability of service type	$P_{st,i} \in (0,1)$	The probability of choosing each type of LTC services
Probability of service provider	$P_{s_j,i}^{st} \in (0,1)$	The probability of choosing each service provider s_j with the determined service type st
Utility value	$Util_{s_j,i}^{st} \in R^+$	The utility value of service provider $s_j \in S$ with type st
Time of service	$ts_{s_j,i} \in N$	Time unit of staying at one particular service provider $s_j \in S$

is fully occupied, it will be removed from the candidate list, which is updated at each iteration. In addition, individuals decide whether or not they would stay at the same service provider depending on their satisfaction degree. We assume that the satisfaction rate only relates to the service quality: if the service quality is extremely low (<0.2), with probability 50% individuals may seek for new providers.

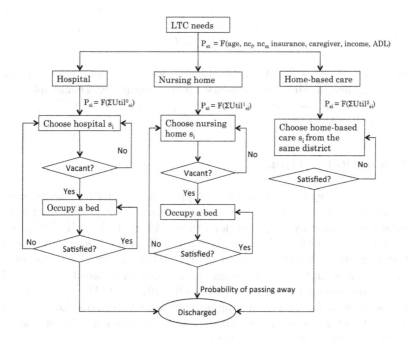

Fig. 2. Individuals' behavior of service provider selection

LTC policies and regulations learned from an LTC scheme in China are considered for modeling [10]. We assume only individuals with the number of ADL limitations larger than 3 could choose hospitals and nursing houses for medical services; individuals enrolled in different insurance schemes prefer different service types since the reimburse rate varies; only individuals with income higher than the service charge could afford nursing houses; individuals with informal caregivers at home will possess a higher possibility of utilizing home-based care.

For each individual $I_i \in Ind$, the utility value $Util^{st}_{s_j}$ of service provider $s_j \in S$ and the corresponding probability P_{s_j} are updated as follows.

$$Util^{st}_{s_j}(t) = w_1 * Eva_{SQ} + w_2 * Eva_{SC} + w_3 * Eva_{TS} \tag{2}$$

$$P_{s_j}(t) = \frac{e^{(Util^{st}_{s_j}(t))}}{\sum_{s_j, TP_j == st} e^{(Util^{st}_{s_j}(t))}} \tag{3}$$

The affordability Eva_{SC} is updated as follows. We assume that if the service charge is higher than household expenditure on non-food items, it is then not affordable.

$$Eva_{SC} = \begin{cases} 1 - e^{-(inc_i - SC_j)} & \text{if } inc_i > SC_j \\ 0 & \text{otherwise} \end{cases} \tag{4}$$

The evaluation of usability depends on two criteria, service quality Eva_{SQ} and utilization duration Eva_{TS}. We assume individuals may avoid frequent switch of service providers for stable services.

$$Eva_{SQ} = 1 - e^{-SQ_j(t)} \tag{5}$$

$$Eva_{TS} = 1 - e^{-ts_{s_j,i}} \tag{6}$$

We require $st == TP_j$. When st equals 2, we require one more constraint $Loc_j == L_i$ indicating that for home-based service, individuals could only choose the service provider from the same district; w_i, $i \in \{1, 2, 3\}$ and $\sum_i w_i = 1$, are weights to adjust the proportion of each indicator.

4   Simulation Setting

We assume there are 700000 individuals (approximately 1:10 to the population of a middle-size city in China) and allocate them to age groups according to UN's population estimation by age group for China [14]: aged 45–54 (13.58%), 55–64 (10.25%), 65–79 (6.92%) and above 80 (1.32%). For each age group, we create individual agents with LTC needs (at least 1 ADL) and set the characteristics referring to the sample characteristics summarized in Tables 3, 4 and 5. The agents are distributed to 4 districts proportionally (2:4:2:1). w_i, $i \in \{1, 2, 3\}$ are assigned evenly (=0.33) indicating no preference of a particular factor. The preference of certain factors across different social groups could be simulated by setting varied value of w_i in future work.

For service providers, we set 3 designated hospitals with 60 beds, 10 nursing houses with 230 beds and 55 home-based care with 2000 beds. Among 10 nursing houses, we assume one of them is Type I, two of them are Type II, the same to Type III, and the rest five are Type IV. We simulate three scenarios of home-based care distribution. In Scenario 0, home-based care providers are distributed randomly to 4 districts. In Scenario 1, service providers are distributed proportional to the population whereas in Scenario 2 are proportional to the population with at least 1 ADL. The number of beds of each provider are randomly assigned.

For each scenario, we run the models for 20 times and analyze the average value.

4.1   Simulation Results

Around 21% of 11432 individuals with LTC needs choose LTC services, among whom approximately 4% individuals choose designated hospital care, 13% individuals choose nursing house care and 83% individuals choose home-based care.

Figure 3 summaries the occupation information of nursing house care by type (Fig. 3(a) and (b)), and the capacity, occupation and number of individuals not accepted in any nursing house care by district (Fig. 3(c)). From Fig. 3(a) we could observe that Type II (low charge, high quality) nursing house care is the most popular one that the occupation rate is always 100%, followed by Type III and Type I ones. Occupation rate of Type III care providers shows some turbulence which indicate a frequent opt-out rate. Individuals in need may choose Type III care at the outset due to a relatively low charge, but may finally change to

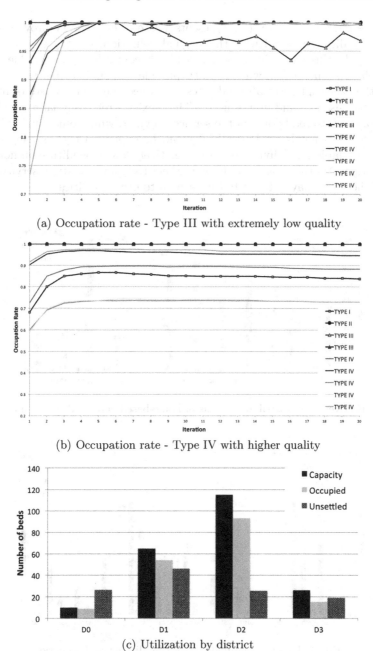

(a) Occupation rate - Type III with extremely low quality

(b) Occupation rate - Type IV with higher quality

(c) Utilization by district

Fig. 3. Effectiveness of distributive plans of nursing house care by district and type

others because of the low care quality provided (≈ 0.1). Variations in occupation rate exist in Type IV providers. By scrutinizing into the detailed record, it shows that the service providers with either relatively high quality or low charge are the most preferred ones. Figure 3(b) represents a case when Type IV providers have a higher quality compared with the original definition, i.e. $SQ \in [2,3]$ rather than $[1,2)$. In this case, the Type I provider becomes less popular due to the service charge competition with Type IV providers in similar quality.

On the other side, Fig. 3(c) indicates the co-existence of a waste of nursing care service resource and unmet needs. There exist a relatively large number of individuals not served by any of the service provider, while some nursing houses are still have vacant beds. Partially it could be explained that nursing houses are not affordable for individuals with a relatively low expenditure on non-food consumption. Another reason could be the relatively low quality of services that most of the users may not satisfy thus switch to other facilities.

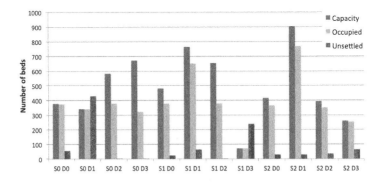

Fig. 4. Effectiveness of distributive plans of home-based care by district and scenario

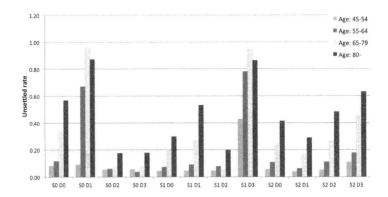

Fig. 5. Unsettled individuals by age, district and scenario

Figure 4 presents the capacity, occupation rate and unsettled individuals of home-based care by district and scenario. $D0S0$ indicates the case for district 0 and scenario 0. We could observe that the number of individuals not served is relatively large at district 1 in Scenario 0, and at district 3 in Scenario 1.

It could be explained by the lack of sufficient home-based care centers distributed to the districts. In general, Scenario 2 performs better than other scenarios in terms of minimizing unsettled individuals. Although allocating service resources proportional to populations in need is the most effective strategy, it might be difficult to implement in reality because of the lack of data at district level. Alternatively, although allocation plans proportional to the population may lead to imbalance across districts, it still outperforms the random allocation.

Figure 5 presents the unsettled rate of individuals by age group, district and scenario. We could identify individuals older than 80 as the most vulnerable group who has a higher unsettled rate compared with other age groups. It could be partially due to a relatively lower household income for the elderly, as well as to the fact that neither of the three proposed scenarios favors a particular social group specifically. In order to guarantee the equity of access, policies favoring vulnerable social groups are expected.

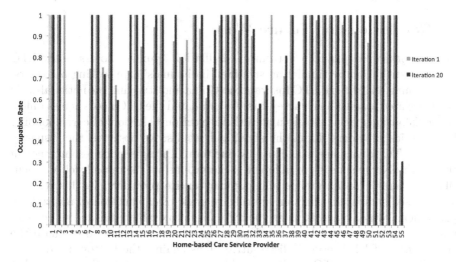

Fig. 6. Occupation rate of each home-based care provider

Figure 6 presents the occupation rate of each home-based care service provider at iteration 1 and 20 of Scenario 2. x-axis represents the service provider of home-based care and y-axis represents the corresponding occupation rate at iteration 1 (in dark grey) and iteration 20 (in light grey) of each service provider. Through this figure, we could identify service providers with a larger difference in occupation rate along with the simulation runs and scrutinize into the micro-level record of individuals and service providers to trace their properties. Generally service providers with a relatively high quality could attract more individuals while the ones with a lower quality have a higher opt-out rate. In order to avoid this situation, regular assessment of service providers may be necessary to guarantee the quality. We will not conduct the detailed micro-level analysis in this paper but leave it for future works.

In summary, all the simulation results analyzed are under the assumptions and limitations embedded in the models. At the district level, distribution plan proportional to population in LTC needs, in terms of ADL assessment, is the most effective one to reduce the disparity between allocated LTC resources and unsettled individuals; at the service provider level, good quality and fair service charge are key factors to guarantee a stable occupation rate; at the individual level, individuals older than 80 are identified as the most vulnerable social groups that policies in favor of these groups should be considered to guarantee the equity.

Regarding the validation process, since empirical data is not available at the service provider level and district level, we could only conduct scenario analysis and interpret the simulation results strictly under simulation assumptions. Qualitatively the simulation results reflect certain aspects of the reality that there is a long waiting queue for service providers with higher service quality and vacant LTC beds in other service providers, though more rigorous quantitative validation is required for future works.

5   Conclusion and Future Work

In this work, we deployed a bottom-up simulation approach to assess the LTC service distribution by simulating stratified individuals' care-seeking behaviors. By assuming an artificial LTC Insurance in China and studying the national-wide health-related longitudinal survey, first we estimated the LTC needs across different social groups, and then simulated their selection behaviors of corresponding LTC service providers. From the simulated results, we observed the disparity between LTC service needs and supply due to allocating service resource not in tune with divergent needs of different social groups; at the meso level, we identified vulnerable social groups and popular service providers. In this respect, agent-based simulation offers an alternative angle of explanation of macro-level phenomena, which could facilitate policy-makers to evaluate future "to be" scenarios. This work could be treated as an analytic framework to evaluate the effectiveness of LTC services distributive plans against the bottom-up dynamics of heterogeneous LTC needs. It could also be served as the base to predict future trends of LTC needs and selection dynamics as leveraging the advantages of agent-based simulation. Policies and regulations of particular LTC schemes in China could be further integrated into the model.

This work is still at the preliminary stage that many future works are expected. Firstly, this work assumes a constant population with LTC needs when simulating their selection behaviors of LTC services. For future work, with more available empirical data, such as household dynamics and individuals' disability transition rate, the way of estimating LTC needs of different care could be improved. Secondly, as leveraging the advantages of agent-based simulation, interactions among individuals such as information diffusion and knowledge sharing could be integrated to individuals' behaviors to capture more aspects of the complex reality. Thirdly, it might be interesting to analyze the operation cost of each service provider when they have different proportion of resources devoted to LTC services, since the public subsidy to LTC services might vary to different types of services.

References

1. Andersen, R.M.: Revisiting the behavioral model and access to medical care: does it matter? J. Health Soc. Behav. **36**, 1–10 (1995)
2. Brown, P.H., Theoharides, C.: Health-seeking behavior and hospital choice in China's new cooperative medical system. Health Econ. **18**(S2), S47–S64 (2009)
3. Colander, D., Kupers, R.: Complexity and the Art of Public Policy: Solving Society's Problems from the Bottom Up. Princeton University Press, Princeton (2014)
4. Feng, Z., Liu, C., Guan, X., Mor, V.: China's rapidly aging population creates policy challenges in shaping a viable long-term care system. Health Aff. **31**, 2764–2773 (2012)
5. Grossman, M.: On the concept of health capital and the demand for health. J. Polit. Econ. **80**(2), 223–255 (1972)
6. Gu, D., Vlosky, D.A.: Long-term care needs and related issues in china (2008)
7. Hancock, R., Comas-Herrera, A., Wittenberg, R., Pickard, L.: Who will pay for long-term care in the UK? Projections linking macro- and micro-simulation models. Fisc. Stud. **24**(4), 387–426 (2003)
8. Karlsson, M., Mayhew, L., Rickayzen, B.: Long term care financing in four OECD countries: fiscal burden and distributive effects. Health Policy **80**(1), 107–134 (2007)
9. Lei, P., Feng, Z., Wu, Z.: The availability and affordability of long-term care for disabled older people in China: the issues related to inequalities in social security benefits. Arch. Gerontol. Geriatr. **67**, 21–27 (2016)
10. Municipal Government of Qingdao P.R. China: The policy document of long-term nursing insurance in Qingdao. 2012(52) (2012). https://esa.un.org/unpd/wpp/
11. Peking University: China health and retirement longitudinal study (CHARLS) (2011). http://charls.ccer.edu.cn/en
12. Rhee, J.C., Done, N., Anderson, G.F.: Considering long-term care insurance for middle-income countries: comparing South Korea with Japan and Germany. Health Policy **119**(10), 1319–1329 (2015)
13. UNESCAP: Long-term care for older persons in China. SDD-SPPS PROJECT Working Papers Series: Long-Term Care for Older Persons in Asia and the Pacific (2015)
14. United Nations Department of Economic and Social Affairs Population Division: World Population Prospect: The 2015 Revision (2015). https://esa.un.org/unpd/wpp/
15. Wittenberg, R., Sandhu, B., Knapp, M.: Funding long-term care: the public and private options. In: Mossialos, E., Dixon, A., Figueras, J., Kutzin, J. (eds.) Funding Health Care: Options for Europe, pp. 226–249. The European Observatory on Health Care Systems (2002). (Chap. 10)
16. Worrall, P., Chaussalet, T.J.: A structured review of long-term care demand modelling. Health Care Manag. Sci. **18**(2), 173–194 (2015)
17. Yang, W., He, A.J., Fang, L., Mossialos, E.: Financing institutional long-term care for the elderly in China: a policy evaluation of new models. Health Policy Plan. **31**, 21–27 (2016)
18. Zhao, Y., Hu, Y., Smith, J.P., Strauss, J., Yang, G.: Cohort profile: the China health and retirement longitudinal study. Int. J. Epidemiol. **43**, 61–68 (2014)
19. Zhu, H.: Unmet needs in long-term care and their associated factors among the oldest old in China. BMC Geriatr. **15**(1), 1–11 (2015)

The Research of Bankruptcies' Succession by Systemic Risk Index

Morito Hashimoto$^{(\boxtimes)}$ and Setsuya Kurahashi

Business Science, Tsukuba University, Bunkyo-ku, Tokyo, Otsuka 3-29-1, Japan
hasimoto@gssm.otsuka.tsukuba.ac.jp, kurahashi.setsuya.gf@u.tsukuba.ac.jp
http://www.gssm.otsuka.tsukuba.ac.jp/

Abstract. To lower the risk of a chain reaction of bank failures, active studies of fund transaction networks that are related to systemic risks have been conducted globally, and they are centered in Europe. In this study, we propose a new systemic risk index that reduces the risk of a chain reaction of failures at minimum cost by building a model of inter-bank fund transaction networks. This model's structure has as its basis on the Erdos-Renyi network and considers the network's characteristics. Our verification, using an agent-based modeling method, confirms that financial assistance given to stop chain reaction failures could increase the possibility of a chain reaction and that the systemic risk index can be used as a reference to determine financial institutions that should be given financial assistance.

Keywords: Agent-based modeling · Systemic risk · Network theory · Interbank transaction

1 Introduction

Fund transactions performed between financial institutions through various securities could cause a chain reaction of financial institutions' failures that would destabilize financial markets. Financial institutions are generally under regulations that use net worth ratios that are stipulated by Basel III, which is the uniform international standard for international banks that are published by the Basel Committee on Banking Supervision. This standard serves as one of the measures of financial-market stability. However, financial institutions compete intensely for interest rates and various charges because of financial liberalization. They, hence, can have problems with their operating foundations. Amid such circumstances, we must take necessary measures, that assume the failures of financial institutions could happen unexpectedly.

The mechanism of fund transactions conducted between financial institutions comprises of a network of multiple financial institutions. This financial institution network consists of individual interbank transactions. However, an interbank transaction is actually a market that is limited to financial institutions, with banks as market participants. These markets include the call market in Japan, which is a financial market for borrowing and lending short-term

© Springer International Publishing AG 2017
S. Kurahashi et al. (Eds.): JSAI-isAI 2016, LNAI 10247, pp. 220–236, 2017.
DOI: 10.1007/978-3-319-61572-1_15

funds. This call market serves to adjust financial surpluses or deficits on a daily basis. Other markets, which exist all over the world, include the foreign spot and futures exchange markets. An interbank transaction market consists of financial institutions, money market brokers, and financial authorities. In such a market, transactions are conducted mainly by direct dealings or through money market brokers.

Taking a top-down view of the overall systemic risk, we can find some undiscovered points that are related to the risk of chain-reaction failures in fund transaction networks. These have occurred in financial crises such as the savings and loan crisis and the Lehman collapse. In Fig. 1, the vertical axis indicates the number of financial-institution failures, and the horizontal axis indicates the year. When comparing Japan and the United States, we see that there is no relationship between the peaks of financial institution failures.

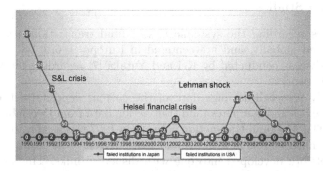

Fig. 1. Comparison of the number of failed financial institutions in Japan and the United States. The vertical axis is the number of failed financial institutions, and the horizontal axis is the year. There is a difference in the peaks of the number of failed financial institutions in Japan and the United States.

The business suspension of one financial institution risks halting fund transactions and causing a chain reaction. The suspension of bank-fund transactions directly causes the collapse of their means of support. The government establishes safety-net measures, such as the injection of public funds, formulation of deposit-insurance systems, and lending by central banks. This subject attracts much attention from the nation because of the possibility of an increase in the national burden if public funds are injected into financial institutions in the event of a financial crisis. To rescue failed financial institutions, the country's Prime Minister must recognize the need for a well-ordered treatment of financial institutions through discussion at the Financial Crisis Response Council. However, it is necessary to carry out careful discussions to determine financial institutions that need rescue and the amount of public funds to inject in them. To do so, the mechanism of interbank networks' chain reaction failure needs to be clarified. Furthermore, some specific matters need to verified: financial institutions into which the public funds should be injected and how propagation can be prevented at a minimum cost.

In this paper, Sect. 2 overviews related systemic risk model studies, and Sect. 3 describes the proposed model. Section 4 analyzes the impact of a chain reaction of failures and node characteristics, Sect. 5 proposes the systemic risk index (SRI), and Sect. 6 analyzes and discusses this index. Finally, Sect. 7 summarizes this paper and describes the future research.

2   Studies of Systemic Risk Using Network Theory

This section overviews studies of systemic risk using network theory, and describes related studies.

2.1   Related Studies

Active studies regarding the systemic risk of fund transaction networks have been performed globally, and are centered in Europe [1–6,9–14]. In Japan, a previous study was conducted by Kei and Yutaka [7] regarding modeling fund transaction networks.

This study uses call-dealings data of the current deposits of the Bank of Japan for analysis and compares fund transaction data from December 1997 and December 2005. The fund transaction network in 1997 was a centralized network with money market brokers as hubs and was referred to as a star network. In 2005, the network developed into one with dispersed links. This network was close to a perfect network; however, node degrees tended to follow an exponential distribution. This resulted in a scale-free network. Here the term degree indicates the number of nodes that connect to other nodes.

Those financial institutions that had a large number of frequent transactions (high flow) by dealing with a number of other financial institutions (high degree) played important roles in the market. As for the network structure, there was a risk of the immediate expansion of a chain reaction of payment defaults if it was caused within the network core because of the short average distance. This previous study showed that it would be efficient and effective to intensively provide the core members with liquidity.

2.2   Basic Systemic Risk Model

Figure 2 shows an interbank transaction network. The nodes in this figure indicate financial institutions while the links indicate directed borrowing and lending between banks. When a certain financial institution fails in this network, all the lending financial institutions that are linked destinations suffer losses. Among the financial institutions indicated by the arrows in Fig. 2, financial institution A is the one that suffers a loss.

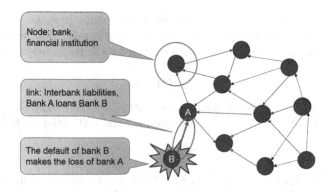

Fig. 2. Modeling of an interbank network.

2.3   Verification of the May and Arinaminpathy Model

In this study, we used a model proposed by May and Arinaminpathy [5] in order to model the propagation of failures of financial institutions. The model of May and Arinaminpathy uses a balance sheet as shown in Fig. 3 for modeling, which includes the following items below. Since these parameters are the same for all nodes, all financial institutions will have the same balance sheet. This research can, hence focus on network characteristics.

f: Shock percentage for external assets
θ: Percentage of interbank financing for assets
S(I): Shock of phase I
γ: Net worth ratio.

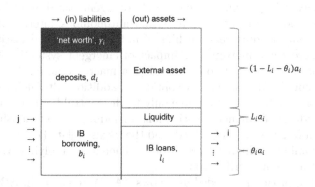

Fig. 3. Balance sheet using the model of May and Arinaminpathy. The left side shows debts, and the right side shows assets. Absorb with capital against the loss of external assets.

The model of May and Arinaminpathy defines three propagation phases of financial institution failures: phase I, phase II, and phase III. The shock of phase I is defined as Eq. (1). However, the same values for the items below are used for all banks.

$$S(I) = f(1 - \theta) \tag{1}$$

Here, the shock of phase I, which propagates into the conditions defined by Eq. (2), exceeds net worth. This causes the financial institution to fail.

$$S(I) > \gamma \tag{2}$$

Next, in the shock of phase II, the debt defaults caused by individual banks in phase I cause phase II to shock the banks as debtors. The shock of phase II is expressed as Eq. (3), and it is mitigated when the number of debtor banks increases. Here, the subscript "MIN" indicates that a smaller value is used.

z: The number of interbank loans (the average degree of the interbank network nodes).

$$S(II) = \frac{[\theta, s(I) - \gamma]_{MIN}}{z} \tag{3}$$

Additionally, based on similar considerations of the shock of phase II, the shock of phase III can be defined as Eq. (4) by nesting.

$$S(III) = \frac{\{\theta, \frac{[\theta, f(1-\theta)-\gamma]_{MIN}}{z} - \gamma\}_{MIN}}{z} \tag{4}$$

Here, N, the number of financial institutions, is sufficiently larger than z^2, the shock continues to propagate on the network.

3   Extension to Agent-Based Modeling

The advantages of applying systemic risk to agent-based modeling (ABM) include the efficiency of verification. While autonomous interactions such as the borrowing and lending relationships of individual financial institutions are considered microscopic matters, the impact of emergent systemic risk on fund transaction networks has yet to be solved as a macroscopic matter. Therefore, verification using ABM is an effective option. In addition, although the raw data of interbank transactions are available only to the central bank, the transaction behavior of individual financial institutions including the relationships of borrowing and lending have been clarified, and there exist related external data that include the number of financial institutions. Therefore, applying ABM enables us to conduct adequate verification.

Our ABM approach proceeded as follows, (A) We created an ABM verification environment using the model of May and Arinaminpathy. (B) By comparing the mean field approximation result of the model of May and Arinaminpathy with that of ABM verification, we verified the validity of our proposed model. (C) In this ABM environment, we identified the network characteristics that reduce systemic risk, which is the purpose of this study.

3.1 Comparison with the Verification Result of the Model of May and Arinaminpathy

Next, in the same way as May and Arinaminpathy, we verified the model using the following parameters for our ABM environment.

- Network: Erdos-Renyi network
- Net worth ratio: $\gamma = 0.8$ (Phase I), 0.042 (Phase II), 0.016 (Phase III)
- Asset damage ratio: $f = 1$
- Ratio of interbank loans for assets: $\theta = 0.20$
- Probability of bank A to provide a loan to bank B: $p = 0.2$.

May and Arinaminpathy also compared the validity using simulation results of NYYA [6]. Here, May and Arinaminpathy used the value 4.8 for the number of interbank loans z by mean field approximation. In this research, however, we use the number of inward links of actual nodes. Naturally, the same verification results are always obtained when the same parameters are used on the same network. For $N = 25$ nodes, the average number of banks that failed when failures were caused by each node was 18.796 in phase III and 3.408 in phase II see Table 1. These values are indicated by '★' in Fig. 4 and are the same as the results obtained by the mean-field approximation., We, hence, determined that the modeling was valid. Here, the column of Table 1 contains the items below.

- count links: The total number of network links
- mean path length: The average network path
- phase II/III mean defaults: The average number of financial institutions that fail because of the shocks of phase II or phase III.

Table 1. Simulation results of ABM using the model of May and Arinaminpathy ($N = 25$).

Count links	Mean path length	Phase III number of defaults	Phase II number of defaults
121.6	2.138	18.796	3.408

4 Impact of a Chain Reaction of Failures and Node Characteristics

This section describes the impact of node characteristics on chain reaction of failures, which is the major topic of this study. The node characteristic discussed in this study is the index of each node organized in a fund-transaction network.

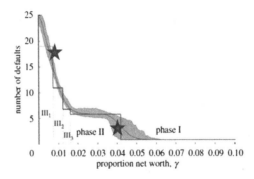

Fig. 4. Comparison of the results of the model of May and Arinaminpathy and ABM.

4.1 Fund-Transaction Networks and Fund Assistance

To verify the fund-transaction network, we created a similar Erdos-Renyi network with the following parameters: 500 nodes and a connection probability of 0.02. The target of this research is Japanese financial-institution networks. Currently in Japan, there are 548 financial institutions; therefore, we used 500 as the number of nodes. Additionally, Kei and Yutaka [7] measured the mean path of the top 200 financial institutions on the interbank fund-transaction network in 2005. This path had three steps. Hence, when we measured the mean path by adjusting the connection probability, we confirmed that 0.02 obtained a mean path value close to 3. These parameters are organized as shown below.

- Asset damage ratio: $f = 1$
- Ratio of interbank loans for assets: $\theta = 0.20$
- Probability of bank A to provide a loan to bank B: $p = 0.02$
- Total number of banks: $N = 500$

Here, the degree of the actual network was z (the number of interbank loans). The shock propagates on the network because it is expected that N will become substantially greater than z^2. When a bank survives by receiving fund assistance, the node exists. By contrast, when the bank goes bankrupt, the node is unlinked from the network. In other words, all links of the targeted node must be deleted.

4.2 The Verification Model

In this research, we suppose that when the business condition of a financial institution worsens, we calculate the differences between the number of chain-reaction bankruptcies that result if the bank gets financial assistance and survives ad that if the bank goes bankrupt and all its links are deleted. In general, giving fund assistance to a bank is better for the stability of financial systems. However there is the moral hazard of lack of management efficiency, when financial institutions at risk are rescued. Furthermore, the study of cascade failure in network theory suggests that deleting nodes from financial transaction networks, (in other words, the bankruptcy of financial institutions) can reduce the risk of successive bankruptcies. Hence, in this study, we implement the transaction network to determine the effect of removing all node links on the number of successive bankruptcies.

Specifically, as shown in Fig. 5 we verified how much the chain reaction of failures could be reduced when a certain financial institution i fails while another financial institution j is assisted financially (case A), by comparing it with the case where no financial assistance is given (case B). Specifically, we compared the number of chain-reaction failures of the entire network, when all the links of a certain node j are not unlinked on the Erdos-Renyi network when another node i fails (case A), with the case where the links of node j are unlinked (case B). By measuring 500 patterns for each of nodes j and node i, we measured a total of 250,000 patterns for this verification. Figure 6 shows the scatter diagram of the verification results, where the vertical axis indicates the difference in the average number of chain-reaction failures before and after the node links were unlinked, while the horizontal axis indicates the node number.

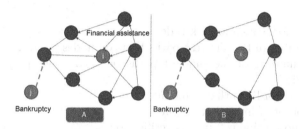

Fig. 5. Agent based modeling of systemic risks. Assuming the default of node i in 2 cases, we measured the differences of number of chain-reaction failures started from another node j. (A) all the links of node i was deleted. (B) Not deleting the link of node i because of financial assistance.

The scatter diagram of Fig. 6 shows that removing the node links increases the chain reaction of failures. This suggests that financial assistance, which should prevent a chain reaction of failures, could amplify it. Careful selection of the institutions to receive financial assistance is needed.

5   Systemic Risk Index

Because the financial institutions to which assistance is given need to be selected carefully, we examine an index which is called the systemic risk index(SRI), that indicate the impact of financial assistance on a chain reaction of failures. This SRI is expressed with Eq. (5).

$$SRI = \sum_{i \neq j \neq k} \frac{g_{jk}(i)}{g_{jk}} + [\frac{D_{out} - D_{in}}{D_{in} + D_{out}} * \alpha N]_{cond} \tag{5}$$

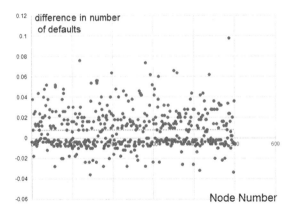

Fig. 6. Differences in the chain reaction of failures caused by node deletion. The vertical axis shows the difference in the average number of failed chains, and the horizontal axis shows the node number. There are some cases for which the average number of failed chain sequences decreases.

- SRI: Stands for Systemic Risk Index
- g_{jk}: The number of the shortest paths between nodes j and k
- $g_{jk}(i)$: The number of the shortest paths between nodes j and k which go through node i
- D_{in}: The order of links that come in
- D_{out}: The order of links that go out
- α: Factor whose value is 2 in this verification
- N: The number of nodes
- $cond$: Considered only when both degrees of links that come in and go out are greater than the average number of degree.

For the same results, as shown in Fig. 6. Figure 7 shows a scatter diagram whose horizontal axis is the SRI.

Table 2 shows the results of conducting the same verification 10 times under the same conditions on another network. We confirmed the correlation for all trials, which was strongly statistically significant.

6   Effects of the SRI and Consideration

The Eq. (5) of the SRI has the betweenness centrality value in the first term. The second term expresses all the links that indicate financial institutions that could provide financial assistance as the difference of the in- and out-degrees of the node. This is because betweenness centrality is the most explanatory variable of the SRI. We verified the effects of the following explanatory variables using the random-forest method: betweenness centrality (bc), closeness, degreeness, evcentness, graphcent, inward degrees (in.degree), outward degrees (out.degree),

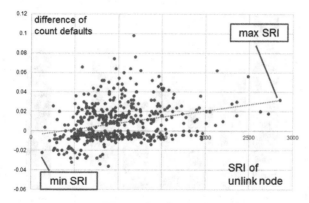

Fig. 7. Scatter diagram on number of chain-reaction of failures and the SRI. The vertical axis shows the difference in the average number of chain-reaction of failures, and the horizontal axis shows SRI. From the lower left to the upper right we can see a form that spreads like a fan.

Table 2. Coefficient of correlation between the number of defaults and the SRI.

No.	Coefficient of correlation using the SRI	t value	p value
1	0.285	6.627	8.925E-11
2	0.286	6.667	6.968E-11
3	0.319	7.499	2.962E-13
4	0.356	8.496	2.220E-16
5	0.269	6.233	9.784E-10
6	0.229	5.241	2.368E-07
7	0.306	7.174	2.662E-12
8	0.319	7.512	2.718E-13
9	0.291	6.776	3.497E-11
10	0.330	7.803	3.575E-14
Average	0.299	7.003	2.380E-08

and PageRank (PR). Figure 8 shows the measurement results, while Fig. 9 indicates the partial subordinate plot of each individual explanatory variable.

These measurements confirmed that betweenness centrality had the highest effect and was the most appropriate for the SRI. Kei and Yutaka (2008) [7] referred to betweenness centrality. They noted that although the degrees of banks A and B may be the same, if bank A has a higher betweenness centrality, bank A is located on a more important path of the network. Therefore, this means that bank A could easily become involved in the propagation of a liquidity shock, or could easily have a more significant impact on the network than that of bank B. Betweenness centrality is a centrality that uses the concept that the important nodes are those nodes that serve as bridges for many nodes. Of the

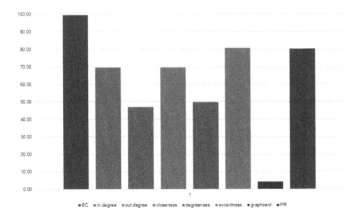

Fig. 8. Validation of explanatory variable effects using a random forest. The vertical axis is variable importance. The most important variable is betweenness centrality.

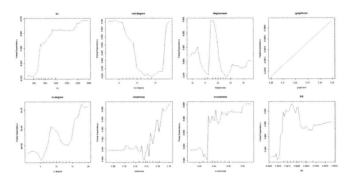

Fig. 9. Partial dependencies of the explanatory variables.

shortest paths between arbitrary node pairs, nodes are rated depending on the percentage of intermediate paths. With regard to betweenness centrality, Norio and Naoki [8] referred to countermeasures for cascade failures and stated that when points with small betweenness centrality are removed, the peaks that fail in the ultimate condition decrease. This is because these peaks are the path of the flow, and, at the same time, they are the starting points of the flow making them edge points. Additionally, the rate of flow starting from their edge points has the same independence with respect to the peak. The points with small betweenness centrality contribute little to the flow, but they generate a certain rate of flow starting from them. Therefore, it is better to remove these points. By contrast, removing the points with large betweenness centrality results in the flows passing through these points taking detours through other paths. This should affect a chain reaction. Norio and Naoki [8] have concluded the removal of points with large betweenness centrality has a very little impact.

When this concept is applied to the fund-transaction network, shown in Fig. 10, removing nodes with small betweenness centrality removed flows that start from the relevant nodes. This decreases the risk of a chain reaction of failures. When nodes with large betweenness centrality were removed, our verification confirmed that the network flow was disturbed, while the flow of neighboring nodes increased, which increased the number of chain-reaction failures. However, the contrasting diagram in Fig. 11 does not indicate the risk of causing a chain reaction of failures when compared with the case of nodes with small betweenness centrality. In Fig. 10, dashed lines from nodes with minimum betweenness centrality indicate the flow to be removed, while the bold lines to nodes with maximum betweenness centrality indicate flows that are to be increased. In Fig. 11, the vertical axis is the difference in the average number of chain-reaction failures before and after the node links that were unlinked, and the horizontal axis indicates betweenness centrality.

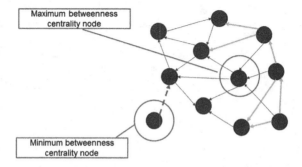

Fig. 10. Node betweenness centrality.

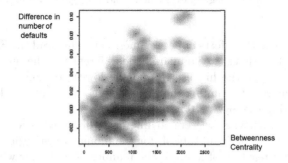

Fig. 11. Contrasting scatter diagram of default numbers and betweenness centrality.

As described earlier, betweenness centrality has limitations when indicating a chain reaction of failures. Therefore, the second term on the right side was added to Eq. (5). The purpose of this addition is to correct patterns with small betweenness centrality for the unlinked nodes, although these nodes are unlinked when a chain reaction of failures increases. Because the difference between the number of in- and out-degrees can vary, we consider the total degree and its proportion with respect to the nodes. In other words, the final value is obtained by dividing the difference of the degrees by the sum of the degrees, and this value is corrected for the scale of the network by multiplying this coefficient by the number of nodes. However, those nodes with fewer degrees tend to have a value with less betweenness centrality depending on their characteristics. Therefore, we decided to consider only those with in- and out- degrees that are greater than the average, and we added this to the conditions. We verified this choice using the CART tree algorithm Fig. 12. Our verification confirmed that for cases where chain-reaction failures increase frequently, the in- and out- degrees are effective for classifying these cases.

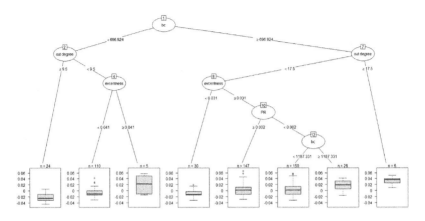

Fig. 12. Classification and regression trees for the CART algorithm. The root is betweenness centrality. Out-degree, evcentness, and PageRank are under the root.

Next, to compare the effects of betweenness centrality and the SRI, using one networks verified 10 times, we compared the difference in the number of chain-reaction failures where the highest 10 and the lowest 10 nodes were unlinked. Table 3 shows the result of unlink the highest 10 nodes, while Table 4 shows the results of unlinking the lowest 10 nodes.

As there results show, there were no differences observed betweenness centrality and the SRI in Table 3. In Table 4, for the financial institutions that should increase the chain reaction of failures, the number of nodes for which the chain reaction of failures decreased was four when betweenness centrality was used, but it decreased to one when the SRI was used.

To verify the effectiveness of ranking, we validated the rank correlation of Spearman and Kendall. Furthermore, to verify the effect of the SRI on the number of nodes with high values, we sorted nodes in ascending order of betweenness

Table 3. Difference in the number of defaults ordered by the minimum betweenness centrality and the SRI.

No.	BC	Differences	SRI	Differences
1	0	0.002	0	0.002
2	110.4802	−0.006	110.4802	−0.006
3	179.5253	−0.008	179.5253	−0.008
4	191.2865	0.001	191.2865	0.01
5	196.8176	0	196.8176	0
6	200.5888	−0.002	200.5888	−0.002
7	234.8976	−0.012	234.8976	−0.012
8	239.4583	0	239.4583	0
9	243.8241	−0.014	243.8241	−0.014
10	246.0996	−0.016	246.0996	−0.016

Table 4. Difference in the number of defaults ordered by maximum betweenness centrality and the SRI.

No.	BC	Differences	SRI	Differences
1	2783.172	0.004	2661.996	0.054
2	2758.77	0.054	2631.656	0.004
3	2554.198	0.054	2625.626	0.054
4	2540.356	0.014	2438.516	−0.01
5	2471.129	0.002	2309.839	0.002
6	2458.212	−0.002	2299.403	0.016
7	2455.867	0.02	2290.356	0.014
8	2341.742	−0.01	2230.06	0.02
9	2261.595	−0.004	2200.889	0.102
10	2209.038	−0.008	2194.838	0.048

centrality and divided them by 250 nodes in the first half and the last half as shown in Fig. 13. We then validated the correlation coefficients and rank correlation. The results are shown in Table 5.

Table 5 shows that the value decreases because the range of betweenness centrality is halved, and the overall values are improved from 3% to 5%. However the value for the last half improved from 28% to 31%, and we can hence, greatly improve the explanation for the betweenness centrality, although we regarded it as a challenge.

Table 5. Result for the Spearman and Kendall test.

	Methods	All	First half	Last half
BC	Spearman	0.265	0.168	0.111
	Kendall	0.183	0.118	0.075
	Pearson	0.272	0.196	0.113
SRI	Spearman	0.273	0.170	0.142
	Kendall	0.189	0.119	0.096
	Pearson	0.286	0.197	0.148
Improvement	Spearman	1.030	1.013	1.280
	Kendall	1.031	1.012	1.283
	Pearson	1.054	1.005	1.315

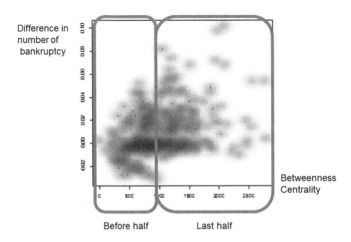

Fig. 13. Classification by betweenness centrality. Each class has 250 nodes.

7   Conclusion and Future Development

In this research, we built an interbank fund transaction network model using the Erdos-Renyi network. By considering the network characteristics, we proposed a new SRI that reduces chain-reaction failures at minimum cost. Moreover, we verified that a chain reaction of failures and the proposed index are related using ABM and obtained the following results.

(1) Financial assistance can increase chain-reaction failures.
(2) Adding the degree factor to the betweenness centrality of nodes to be unlinked in the SRI which enhances its correlation with the average number of chain-reaction failures. This is effective for selecting financial institutions for which financial assistance should be provided.

Regarding the first conclusion, we verified that financial assistance, whose purpose is to reduce chain-reaction failures, has the risk of increasing such failures, which is contrary to expectations. In addition, we proposed the SRI as an index to avert such risks, and verified the effectiveness of this index. If a chain reaction is increased because of the survival of financial institutions that use financial injections, the systemic risk is extended, the number of failed financial institutions increases, and there is a risk that the public fund injection might have to increase. Finally, the cost, which depends on the national burden, increases.

As for our future research, we will conduct verifications based on links. The analysis ind this research was only based on the unlinking of nodes. However, considering the difference in chain-reaction failures caused by specific link deletion is effective. Rather than unlinking all node destinations of a particular financial institution, it is more realistic to eliminate financial institutions or set loans between them according to each link. However, an index for links can easily be unlinked by considering the current node's characteristics. However, we need to consider an index in which new links are added more adequately. For example, nodes with higher degrees are highly likely to be identified. But, although nodes with smaller degrees are linked, the effect of risk reduction could vary because multiple nodes with smaller degrees could exist.

References

1. Acharya, V., Engle, R., Richardson, M.: Capital shortfall: a new approach to ranking and regulating systemic risks. Am. Econ. Rev. **102**(3), 59–64 (2012)
2. Benoit, S., Colliard, J.-E., Hurlin, C., Perignon, C.: Where the risks lie: a survey on systemic risk. HEC Paris Research Paper (2015)
3. Eisenberg, L., Noe, T.H.: Systemic risk in financial systems. Manag. Sci. **47**(2), 236–249 (2001)
4. Gai, P., Kapadia, S.: Contagion in financial networks. Bank Engl. Q. Bull. **50**(2), 124 (2010)
5. May, R.M., Arinaminpathy, N.: Systemic risk: the dynamics of model banking systems. J. Roy. Soc. Interface/Roy. Soc. **7**(46), 823–838 (2010)
6. Nier, E., Yang, J., Yorulmazer, T., Alentorn, A.: Network models and financial stability. J. Econ. Dyn. Control **31**(6), 2033–2060 (2007)
7. Kei, I., Yutaka, S.: Funds trading network of the call market. Stud. Financ. **27**, 47–99 (2008)
8. Norio, K., Naoki, M.: Complex Network. Kindai Kagaku-Sya, Tokyo (2010)
9. Cifuentes, R., Ferrucci, G., Shin, H.S.: Liquidity risk and contagion. J. Eur. Econ. Assoc. **3**(2–3), 556–566 (2005)
10. Dias, A., Campos, P., Garrido, P.: An agent based propagation model of bank failures. In: Amblard, F., Miguel, F.J., Blanchet, A., Gaudou, B. (eds.) Advances in Artificial Economics. LNEMS, vol. 676, pp. 119–130. Springer, Cham (2015). doi:10.1007/978-3-319-09578-3_10
11. Hyun, S.S.: Risk and Liquidity. Oxford University Press Inc., Oxford (2010)
12. Montagna, M., Lux, T.: Hubs and resilience: towards more realistic models of the interbank markets. Some Recent Developments, Banking Integration and Financial Crisis (2015)

13. Takamasa, K., Masaaki, K., Takashi, Y., Hiroshi, T., Takao, T.: Analysis of the influences of central bank financing on operative collapses of financial institutions using agent-based simulation. In: 2016 IEEE 40th Annual Computer Software and Applications Conference (COMPSAC), pp. 95–104. IEEE (2016)
14. Maeno, Y., Morinaga, S., Matsushima, H., Amaya, K.: Risk of the collapse of a bank credit network. JSAI **27**(6), 338–345 (2012)

JURISIN 2016

Juris-Informatics

Makoto Nakamura[1], Seiichiro Sakurai[2], and Katsuhiko Toyama[1]

[1] Nagoya University, Nagoya, Japan
[2] Meiji Gakuin University, Minato, Japan

The Tenth International Workshop on Juris-Informatics (JURISIN 2016) was held with a support of the Japanese Society for Artificial Intelligence (JSAI) in association with JSAI International Symposia on AI (JSAI-isAI 2016). JURISIN was organized to discuss legal issues from the perspective of informatics. Compared with the conventional AI and law, the scope of JURISIN covers a wide range of topics, which includes model of legal reasoning, argumentation/negotiation/argumentation agent, legal term ontology, formal legal knowledge-base/intelligent management of legal knowledge-base, translation of legal documents, information retrieval of legal texts, computer-aided law education, use of Informatics and AI in law, legal issues on ubiquitous computing/multi-agent system/the Internet, social implications of use of informatics and AI in law, natural language processing for legal knowledge, verification and validation of legal knowledge systems, any theories and technologies which is not directly related with juris-informatics but has a potential to contribute to this domain.

Thus, the members of Program Committee (PC) are leading researchers in various fields: Thomas Ågotnes (University of Bergen, Norway), Floris Bex (Utrecht University, The Netherlands), Randy Goebel (University of Alberta, Canada), Guido Governatori (NICTA, Australia), Yoichi Hatsutori (IBM Japan Ltd., Japan), Tokuyasu Kakuta (Chuo University, Japan), Yoshinobu Kano (Shizuoka University, Japan), Takehiko Kasahara (Toin Yokohama University, Japan), Mi-Young Kim (University of Alberta, Canada), Beishui Liao (Zhe-jiang University, China), Makoto Nakamura (Nagoya University, Japan), Le-Minh Nguyen (Japan Advanced Institute of Science and Technology, Japan), Katumi Nitta (Tokyo Institute of Technology, Japan), Paulo Novais (University of Minho, Portugal), Julian Padget (University of Bath, UK), Ginevra Peruginelli (ITTIG-CNR, Italy), Seiichiro Sakurai (Meiji Gakuin University, Japan), Katsuhiko Sano (Hokaido University, Japan), Ken Satoh (National Institute of Informatics and Sokendai, Japan), Akira Shimazu (Japan Advanced Institute of Science and Technology, Japan), Satoshi Tojo (Japan Advanced Institute of Science and Technology, Japan), Katsuhiko Toyama (Nagoya University, Japan), and Katsumasa Yoshikawa (IBM Japan Ltd., Japan). The collaborative work of computer scientists, lawyers and philosophers is expected to contribute to the advancement of juris-informatics and it is also expected to open novel research areas.

Despite the short announcement period, eighteen papers were submitted. Each paper was reviewed by three or more members of PC. This year, we allow a double submission to JURIX 2016 and one paper was withdrawn because of acceptance to JURIX 2016 and fifteen papers were accepted in total. The collection of papers covers various topics such as legal reasoning, argumentation theory, legal compliance, dispute resolution, application of informatics and AI in law, application of natural language processing and so on.

Following the previous years, JURISIN have a session on the third Competition on Legal Information Extraction/Entailment (COLIEE 2016) which consists of a result report of the competition and eight papers for each participant.

After the workshop, eight papers were submitted for the post proceedings. They were reviewed by PC members again and five papers were finally selected. Followings are their synopses. Ryuta Arisaka and Ken Satoh present an argumentation theory which roughly follows the tradition of evidential support, but which is a meta-argumentation (or extended argumentation) where an argument can attack an attack/support arrow. They model an example of intention-to-kill in the theory. Teeradj Racharak, Satoshi Tojo, Duy Hung Nguyen, and Prachya Boonkwan show a formal and intuitive framework of analogical reasoning using an argument-based logic-programming-like language. The key idea is to use similarity information to support an inference which cannot be deductively inferred. A proof theory of the system is stated in the dialectical style, where a proof takes the form of dialogue between a proponent and an opponent of an argument. Yoichi Hatsutori, Katsumasa Yoshikawa, and Haruki Imai propose a preprocessing method to estimate document structure from unstructured legal documents. Since documents are often stored or published as unstructured documents, they need to be preprocessed to analyze them in subsequent text analytics. An experimental result showed their proposed method estimated document structure with 96.6% accuracy. The following two papers are presented at the session on COLIEE 2016. Ryosuke Taniguchi and Yoshinobu Kano developed a yes/no question answering system for legal domain. Legal yes/no question answering largely differs from other domains. The most different characteristics is that legal issues require roles and relationships of agents in sentences to be precisely analyzed. Their system performance was better than systems of previous task participants and shared first place in current year's task in Phase 2 (textual entailment). Mi-Young Kim, Ying Xu, Yao Lu, and Randy Goebel also developed an legal question answering system, which combines legal information retrieval and textual entailment, and exploits paraphrasing and sentence-level analysis of queries and legal statutes. Experimental evaluation demonstrates the value of their method, and the results show that their method outperforms previous methods. their result ranked highest in the Phase 3 (combination of information retrieval and textual entailment) in the COLIEE-2016 competition.

Finally, we wish to express our gratitude to all those who submitted papers, PC members, discussant and attentive audience.

Voluntary Manslaughter? A Case Study with Meta-Argumentation with Supports

Ryuta Arisaka$^{(\boxtimes)}$ and Ken Satoh

National Institute of Informatics, Tokyo, Japan
ryutaarisaka@gmail.com, ksatoh@nii.ac.jp

Abstract. In a criminal case, the judge's decision making often involves proving, beyond any reasonable doubt, the defendant's intention to commit a crime from material vidence. A valid decision should be supported by some material evidence, and neither the material evidence itself nor the support that it gives to the conclusion should be invalidated by any other material evidence. Luckily, this sounds a familiar topic in abstract argumentation with supports. We describe an argumentation theory, which roughly corresponds to the tradition of evidential support, but which provides a meta-argumentation (or extended argumentation) framework where an argument can attack/support other argumentation components. We model our example of intention-to-kill in the theory.

1 Introduction

The process to establish the defendant's intention to commit a crime from material evidence, an essential task in a criminal case judgement, is often non-monotonic. Suppose the judge is presented just two pieces of factual evidence by eye witnesses: (1) that the defendant threw a sharp Japanese chef knife which has a blade 13.3 cm in length; and (2) that it hit the victim's back of head to cause cerebellum stab wound and intracranial hemorrhage, then the defendant's intention to kill is most certainly inferred from them. However, we now add yet other evidence - (3) The victim is his daughter, and the defendant was very fond of her. (4) He threw the knife 3.3 m away from her. (5) He was in a habit of throwing the knife at his wife's foot once drunk and irritated. And (6) she did not die on the spot, but he did not chase her leaving the scene. Taken together with (3), (4), (5) and (6), the first two evidence seem no longer self-sufficient for proving his intention to kill [1].

Accumulation of evidence being in this manner, the reasoning process as taken by a judge tends to be complex with potentials to errors. Still, no errors are admissible. Take manslaughter cases, while voluntary manslaughter carries a sentence of death penalty, indefinite imprisonment or imprisonment for a definite term of 5 years or longer under the Criminal Justice System of Japan, involuntary manslaughter carries a much lighter sentence of imprisonment for a definite term of 3 years or longer. Hence, any error in judgement has a severe consequential implication to those involved in manslaughter cases.

© Springer International Publishing AG 2017
S. Kurahashi et al. (Eds.): JSAI-isAI 2016, LNAI 10247, pp. 241–252, 2017.
DOI: 10.1007/978-3-319-61572-1_16

In order to support the judge's decision making, in this work we investigate abstract argumentation theory [11] which is a theory for reasoning about arguments and attacks, appropriate for evidential reasoning [5,23]. As the decision in a criminal case must be a certain one, however, there is a reason to be cautious when we face the following situation: an argument a_1 is attacked by another argument a_2, which is in turn attacked by an argument a_3. Under the original theory [11], we get the conclusion that a_3 defends and accepts a_1 because a_1 counters a_2's attack on a_1. In a way, a_3 is indirectly giving a support to a_1. But real examples are not always that clean, and one argument could be attacking another without giving any support to those the latter is attacking. Consider for instance: (x) the defendant did not have any intention to kill his daughter. (1) the defendant threw a sharp Japanese chef knife which has a blade 13.3 cm in length. (y) the defendant did not have any Japanese chef knife. Here (y) attacks (1) which attacks (x), and yet it is really not the case that (y) defends (x), for the one who states (y) mentions nothing of the defendant's intention to kill (that is, he/she may be in a belief that the defendant fired a gun instead of throwing a knife). Contrast this with (x)-(1)-(z) where (z) is: the defendant was in London when his daughter was killed in Tokyo, which forms a 'clean' case where the standard notion of defence applies. In situations such as intention-to-kill inference, it is not safe to take a support for granted out of indirect supports, and such distinction as seen between (x)-(1)-(y) and (x)-(1)-(z) should be expressed. We make use of a support relation [7,9,20–22] along with the attack relation for that purpose. Also, we allow a situation where some evidence is for confirming or rejecting an attack or a support from some evidence to some evidence, which occurs naturally in situations similar to our earlier example. For instance, (1) attacks (x), and (5) is better considered attacking the attack than (1) directly. Our argumentation theory will be consequently both bipolar [7,9,20–22], i.e. containing the two relations, and meta-argumentational [7,12,17], i.e. facilitating the higher-order attacks and supports. The exact interpretation of support varies from a work to a work, as classified by Cayrol and Lagasquie-Schiex [10]. If some a_1 supports some a_2, then: deductive support [7] ensures that acceptance of a_1 licences that of a_2; necessary support [20,21] ensures that acceptance of a_2 necessitates that of a_1; and evidential support [22] guarantees: that a set of arguments, if acceptable, contains special non-attackable arguments that are always accepted; and that all the other arguments are connected by a chain of the support relation from (one or more of) the special arguments. Our interpretation will be roughly in the evidential tradition; however, our framework will be meta-argumentational and will contain no special arguments, adding to flexibility. With our frameworks, we aim to simplify verificaion of relations among evidence.

We believe that verification of relations among evidence for or against verdicts can be simplified with our frameworks. In fact, as we are to show later, it may be even possible to use this theory to study concluded past cases deeper and test their soundness.

In the rest, we will: touch upon technical background (in Sect. 2); present our meta-argumentation framework with supports (in Sect. 3); and model one actual criminal case we saw earlier (from (1) to (6)) in a little more details than given here (in Sect. 4), before drawing conclusions.

2 Technical Background

Argumentation frameworks allow one to reason about attacks and defences among arguments. In [11] an argument is an abstract entity, and an attack relation is a binary relation over a pair of the set of arguments. Suppose a_1 and a_2 are arguments, then a_1 is considered to be attacking a_2 just when the binary relation is defined for (a_1, a_2). The following definitions are taken from the original theory. An argumentation framework is a pair (A, R) where A is a finitary set of abstract entities *arguments*, and R is a binary relation defined over them. An argument a_1 is said to *attack* an argument a_2 iff $(a_1, a_2) \in$ R. A set of arguments $A_1 (\subseteq A)$ is *conflict-free* iff no $a_1 \in A_1$ is attacking any $a_2 \in A_1$. A set of arguments A_1 *defends* a_1 iff for any a_2 attacking a_1, there is some $a_3 \in A_1$ that attacks a_2. A set A_1 *accepts* a iff A_1 defends a. A set A_1 is admissible iff A_1 accepts all its members. A set A_1 is a preferred set (extension) iff A_1 is admissible and there exists no $A_1 \subset A_2$ such that A_2 is admissible. There are other notions such as complete sets (extensions), stable sets (extensions), and the grounded set (extension). More information for these semantics can be found elsewhere in the literature.

There are many extensions to the original theory. Two that are of particular importance to our technical development are: the support relation which characterises arguments supporting - instead of attacking - arguments [7,9,20–22]; and the argument-to-argument-or-attack relation [16] or even more relaxed a relation [12]. In the left drawing below, a thin arrow is an attack, while a thick arrow is a support: an argument a_1 supports a_2 and attacks a_3, or we just say $(a_1, a_2) \in$ R$_{\text{sup}}$ and $(a_1, a_3) \in$ R, treating R$_{\text{sup}}$ as the support relation, and R as the attack relation. In the right drawing below, a_1 attacks the attack of a_2 on a_3, i.e. (a_2, a_3). Any argumentation theory that allows an attack or a support to an arrow (an edge) as well as an argument (a node) is considered a meta-argumentation theory, which in the literature is also called extended argumentation.

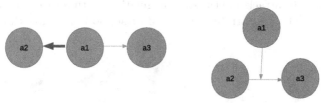

There are several proposals for the exact semantics of support, such as deductive support, necessary support and evidential support:

Deductive support [7] $(a_1, a_2) \in$ R$_{\text{sup}}$ means: (1) if a_1 is accepted, then so is a_2; and (2) if a_2 is not accepted, then a_1 is also not accepted.

Necessary support [20,21] $(a_1, a_2) \in R_{\text{sup}}$ means if a_2 is accepted, then so must a_1 be.

Evidential support [22] There are special kinds of arguments $\mathfrak{n}$ with or without a subscript. They are always unattacked, are unsupported and are accepted. Let R_{sup}^+ be such that: (1) $(\mathfrak{n}, a) \in R_{\text{sup}}$ materially implies $(\{\mathfrak{n}\}, a) \in R_{\text{sup}}^+$; and (2) $(A, a) \in R_{\text{sup}}^+$ only if, for each $a_x \in A$, $(A_x, a_x) \in R_{\text{sup}}^+$ for some $A_x \subseteq A \backslash \{a_x\}$. Then other arguments a that are accepted in A must satisfy $(A_1, a) \in R_{\text{sup}}^+$ for some $A_1 \subseteq A$. Note that in [22] an attacker is assumed to be a set of arguments in the manner similar to Nielsen-Parsons argumentation theory [19].

Among the three, our interpretation of support will be evidential, broadly construed. However, it will be a meta-argumentation framework and will contain no explicit set of $\mathfrak{n}$s. While very useful in some other applications, in this work we do not require Nielsen-Parsons group attacks or supports.

3   Our Meta-Argumentation with Supports

We assume that $\mathbb{N}$ is the class of natural numbers including 0. We assume a set of evidence made available to a judge and a set of possible decisions he/she may make. We denote the former by Evidence and the latter by Verdict. Now, let Rel^0 be Evidence $\times$ (Evidence $\cup$ Verdict) $\rightharpoonup \{-1, 1\}$, and let Rel^{k+1} be Evidence $\times$ dom(Rel^k) $\rightharpoonup \{-1, 1\}$ for any $k \geq 0$. They form our meta-argumentation framework which is: (Evidence, Verdict, Rel) for $\text{Rel} = \bigcup_{k \in \mathbb{N}} \text{Rel}^k$. The two values 1 and -1 indicate support and attack respectively.

We denote an element of Evidence by e, an element of Verdict by v, an element of Evidence $\cup$ Verdict by a, and an element of dom(Rel) by rel, all with or without a subscript. For any ordered set Γ (not the particular symbol Γ but any ordered set), $\pi(n, \Gamma)$ is: the n-th element of Γ if $0 < n \leq |\Gamma|$; or else undefined.

3.1   Acceptability Semantics

That the decision in a criminal case should be first of all a certain one imposes a requirement on the attack and the support relations with respect to acceptability of arguments. Let us illustrate our notion of support by going through a few examples.

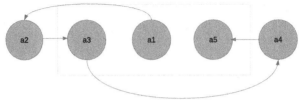

Here a_k is attacking a_{k+1} for each $1 \leq k \leq 4$. The argument a_1 is not attacked by any argument, and so it is an acceptable argument. Although a_3 is being attacked by a_2, it in turn is being attacked by a_1. In the classic acceptability semantics (Sect. 2), it means that a_1 is nullifying the allegation that a_2 raises against a_3. Since no other arguments are attacking a_3, a_1 indirectly supports a_3 which to a_1 is acceptable. In a similar manner, a_5 is also accepted by a_1.

As we mentioned in Sect. 1, however, an argument may be attacking another argument a_x without giving any support to an argument a_x is attacking. There are numerous examples of these kinds in every-day argumentation, not really limited to legal cases. Hence, there is a good reason why we should make any support explicit in our argumentation framework.

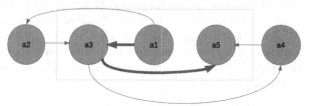

a_1 supports a_3 here, by which we more specifically mean that a_1 supports whatever attacks and whatever supports a_3 is making: a_3's attack on a_5 and a_3's support for a_5, in this example. Consequently a_1 supports a_5 through its support for a_3. This can be contrasted to the example below.

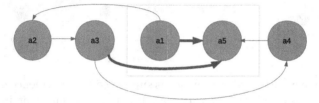

Here, a_1 cannot be considered to be supporting a_3. Finally what if we have the following:

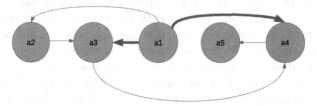

Then there is a known problem: a_1 both supports (directly) and attacks (via a_3) a_4. This we, like others, consider inconsistent.

Definition 1 (Supports and attacks). We say that a_1 supports $\pi(2, rel)$ iff $\mathsf{Rel}(\pi(1, rel), \pi(2, rel)) = 1$ **and** either a_1 is $\pi(1, rel)$ or else there exists a sequence $a_1, \ldots, a_n$, $n \geq 1$, such that $\mathsf{Rel}(a_i, a_{i+1}) = 1$ for all $1 \leq i \leq n - 1$ and $\mathsf{Rel}(a_n, \pi(1, rel)) = 1$. We say that a_1 attacks $\pi(2, rel)$ iff $\mathsf{Rel}(\pi(1, rel), \pi(2, rel)) = -1$ **and** either a_1 is $\pi(1, rel)$ or else there exists a sequence $a_1, \ldots, a_n$, $n \geq 1$, such that $\mathsf{Rel}(a_i, a_{i+1}) = 1$ for all $1 \leq i \leq n - 1$ **and** $\mathsf{Rel}(a_n, \pi(1, rel)) = 1$.

In this definition and elsewhere, we may explicitly use **and** to indicate formal 'and' with semantics of classic logic conjunction for a disambiguation purpose. **or** is its dual.

Definition 2 (Inconsistent framework). We say that (Evidence, Verdict, Rel) is inconsistent iff there exists $\pi(2, rel)$ and $e \in$ Evidence such that $\mathsf{Rel}(rel)$ is defined **and** e both attacks and supports $\pi(2, rel)$.

In view of Definition 1, if, in a given framework, a_1 supports a_x, and also a_1 supports a chain of arguments $a_2, \ldots, a_n$ and a_n attacks a_x, then the framework is inconsistent. We assume in the rest that an argumentation framework is not inconsistent. Let us now go through some examples, as shown below, to connect these supports and attacks to the judge's decisions. The idea is to decide, given a set of verdicts and a set of evidence, which verdicts are acceptable, and, moreover, which evidence actually support them.

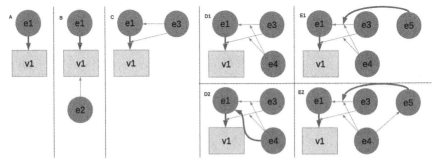

In A, we have some verdict v_1 which is supported by an evidence e_1. Intuition speaks that v_1 is acceptable in this case. But it is not sufficient just to prove existence of a supporting evidence to conclude that v_1 is acceptable in general. For example, there may be another evidence e_2 as in B attacking the same verdict. Or like in C, e_1 or the attack arrow (e_1, v_1) may be attacked which is not in turn attacked. Consequently, we need also examine all evidence attacking v_1 or supports for it, and need show that their attacks are nullified by other evidence. An example is D1 where we can conclude that v_1 be acceptable because e_4 nullifies the attacks by e_3. Similarly if e_4 attacked e_3 directly. In D2, again v_1 is acceptable, but here it is supported not just by e_1 but also by e_4. Now, a little more complicated cases could be E1 and E2 which both extend D1. In those cases like E1, we shall conclude that v_1 be *not* acceptable, the reason being that e_5 in a way, by supporting the attack (e_3, e_1), attacks e_1, and that there is no evidence attacking e_5. If, however, there is an attack arrow (e_4, e_5) as in E2, then we conclude that v_1 is acceptable, in comparison. In contrast to D2, in this case it is not that e_4 supports v_1.

We make our intuition explicit.

Definition 3 (Conflict-freeness). Let $\mathsf{Test}(rel, A)$ be a recursive predicate: $\mathsf{Test}((a_1, a_2), A)$ iff $a_1, a_2 \in A$; and $\mathsf{Test}((a_1, rel), A)$ iff $a_1 \in A$ **and** $\mathsf{Test}(rel, A)$. Let $\delta(A)$ be $A \cup \{rel \in \mathsf{dom}(\mathsf{Rel}) \mid \mathsf{Test}(rel, A)\}$. We say that $A \subseteq$ Evidence $\cup$ Verdict is conflict-free iff there is no $a \in A$ and $x \in \delta(A)$ such that a attacks x.

$\delta(A)$ contains, along with the elements of A, all the arrows (support or attack) that occur in A. Consider the following graph, $\delta(\{v_1, e_1, e_2\})$ contains all but e_3 (because it is not in the set) and the attack from e_2 to e_3 (because, while e_2 is in the set, e_3 is not in the set). $\delta(\{v_1, e_1, e_3\})$ contains v_1, e_1 and e_3 as well as the support arrow, and is conflict-free.

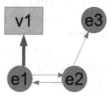

Definition 4 (Support set). For $x \in$ Evidence $\cup$ Verdict $\cup$ dom(Rel), we say that $A \subseteq$ Evidence $\cup$ Verdict is its support set iff each $a_1 \in A$ supports x **and** A is conflict-free. We say that a support set A of x is maximal iff there is no greater support set of x.

In the below diagrams, $\{e_2, e_3\}$ is a maximal support set for e_1 (left) and respectively for the support of e_1 for v_1 (right).

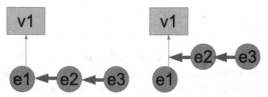

The first purpose of a maximal support set is to know all the nodes that are indirectly attacking some node through a node that directly attacks it. See E1 in the earlier figure. The second purpose is to know, for any acceptable verdict, which evidence actually support it.

Definition 5 (Defence). $A \subseteq$ Evidence $\cup$ Verdict defends $a \in A$ iff: for every $a_1 \in$ Evidence $\cup$ Verdict that attacks a, there is no $a_x \in A$ that supports a_1 (i.e. no set member supports its attacker) **and** either: a_1 and every member of every maximal support set A_1 of a_1 is attacked by some member of A; or (a_1, a) and every member of every maximal support set A_2 of (a_1, a) is attacked by some member of A.

Definition 6 (Admissible and preferred set). We say that $A \subseteq$ Evidence $\cup$ Verdict is admissible iff A is conflict-free **and** A defends every $a \in A$. We say that it is preferred iff there is no greater admissible set of A.

Lemma 1 (Fundamental Lemma). *If A is admissible and if it defends a, then $A \cup \{a\}$ is admissible.*

Proof. It is not possible that a attacks a member of A, for then A would not be conflict-free. The rest is trivial.

Definition 7 (Acceptability of verdicts). We say that v is accepted by $E \subseteq$ Evidence $\cup$ Verdict iff there is a preferred set which contains both v and E **and** E is a support set of v.

As we can see from this definition, preferred sets are important because we judge acceptance of a verdict based on them. However, it is not necessary that every member of a preferred set is connected by supports. See E2 in the earlier figure. Hence, we extract from a preferred set a support set for the verdict.

An easily understandable strategy to compute acceptable verdicts is: to obtain a preferred set for a concerned verdict v; and to extract a maximal support of v in the set. For the first part, it suffices to generate a preferred by starting with a singleton $\{v\}$ and extend it incrementally by adding all arguments that the set defends, which works thanks to Lemma 1. Specifically, it suffices to use a non-deterministic function: $f(A) := \{a \in$ Evidence $\cup$ Verdict $\mid A$ defends $a\}$, and obtain $\Gamma := \{\{A\} \mid A$ is a fixpoint of $f\}$. For the second part, it suffices to generate a maximal support set by backward-tracing support arrows from v incrementally in each $A_1 \in \Gamma$. If there exists a non-empty maximal support set, then it accepts v.

4   An Example

In Sect. 1, we illustrated our motivation with one actual criminal case. Here we show more complete picture of the case (some details are simplified) within our argumentation framework. There are two verdicts: v_1 and v_2, and several items of evidence.

v_1: Man is guilty of voluntary manslaughter.

v_2: Man is guilty of involuntary manslaughter.

e_1: Man threw a sharp knife at his daughter walking down a staircase, and the knife hit her back of head causing her to sustain cerebellum stab wound and intracranial hemorrhage.

e_2: Usually in knife-throwing, aim is poor at 3.3 m distance from a target.

e_3: Man was throwing the knife at his wife when drunk and irritated.

e_4: Man was in a habit of shouting "I kill you!" when drunk and irritated, and was throwing the knife at the floor near his wife's foot just to scare his wife and daughter.

e_5: Man shouted "I kill you!" at his daughter.

e_6: Daughter did not die on the spot and started to move away but man did not chase her.

e_7: Man loved his daughter.

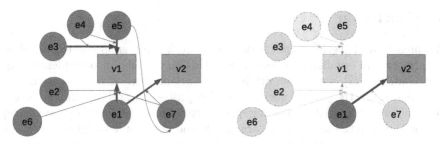

The left graph represents this criminal case. As is clear, v_1 is supported by two pieces of evidence: e_1 and e_5. Further, the support of e_5 for v_1 is supported by e_3. However, firstly, the support of e_5 for v_1 as well as that of e_3 for the support of e_5 for v_1 is attacked by e_4 which is not countered. Therefore, there is no preferred set that contains both v_1 and e_5. Similarly, e_2 and e_5 both attack the support of e_1 for v_1; it is not possible for a preferred set containing v_1 and e_1 to exist. But consider v_2 now. It is supported by e_1. While the support is attacked by e_7, we have e_5 that attacks it. Neither e_5 nor its attack is attacked by any nodes. Here, $\{e_1, e_5, v_2\}$ forms a preferred set. But among them, the only node that gives a support to v_2 is e_1, which derives the right drawing above showing that e_1 is the support for v_2.

4.1   For Analysis of the Past Cases

It may be possible to inspect the past cases with the aid of this argumentation framework, testing their soundness with additional and unconsidered evidence.

One observation which did not enter into the judge's verdict, which nonetheless might have been relevant, is the following fact: if an object is right below ourselves, it is much easier to hit it. The daughter in the above case was walking down a staircase and, according to the original case document, her head was about the man's feet hight. Therefore, there could have been e_8 : It is easier to hit an object which is down below. The modified argumentation framework is shown below, where e_2 can no longer attack the support without getting countered.

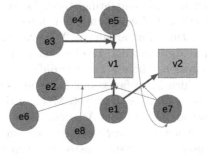

As they say, drawings help, and we believe the same for the judge's decision making.

5   Conclusion

We presented a meta-argumentation framework with supports to model an intention-to-kill criminal case judgement. Though roughly in the evidential support tradition [22], our framework is a meta-argumentation framework (as it is more natural in the judge's decision making process to presume that some evidence act for or against attack or support arrows but not directly on evidence), does not involve the set of special arguments and does not demand an acceptable argument to be supported directly or indirectly by those arguments that are by default accepted, leading to a new interpretation of support. These consideration are reasonable because we begin with a set of verdicts and start to backward-trace supports from them, unlike in [22] where the task is rather to forward-trace (in the direction of arrows) supports from special nodes. Note also in the example in Sect. 4, $\{e_1, e_5, v_2\}$ is a preferred set even though e_5's support arrow is being attacked. We believe that further research in meta-argumentation frameworks with supports will aid the judge's decision making by simplifying their reasoning process. An appeal of argumentation theory is the very intuitive character it has. While we focused on a case study involving criminal intention here, there are works that deal directly with intentionality, e.g. [13,14] in defeasible logic.

Outside abstract argumentation, there are many other works in structured argumentation [8,11,18,24] in which an argument is expressed more concretely as a pair of a set of premises and a conclusion. There is a support relation holding in each pair from the premises to the conclusion. An argument may be rebutted (the conclusion is attacked), undercut (the support is attacked) or undermined (some premise is attacked), and such greater expressiveness is an advantage. It should be noted, however, that our bipolar argumentation is not structural argumentation in disguise. Under our theory here, the relation: a_1 supports a_2, is not equivalent to a structural argument which has a_2 as the conclusion and a_1 as its premise *except for verdicts*. Let us say that a_1 supports a_2, and that a_2 supports v. In our theory, a_1 can be attacked, and v may be still acceptable. By contrast, if a_1 were the sole premise for a_2, i.e. (a_1, a_2) is a structural argument, and similarly (a_2, v), such situation would be an undermining to both arguments. In contemplating structural argumentation, we see it also true that concretisation of abstract entities can entail introduction of more assumptions about the nature of an argument. We have for instance pointed out [2] that not every natural argument may be clear-cuttingly divided into supporting premises and a conclusion. There is also the difficult issue of contrariness in the modern logic [15], which is, as far as our awareness extends, being studied [3], but which has not been fully resolved, yet. Much of these difficulties can be encapsulated in abstract argumentation. As such, there is a benefit in working within a sufficiently abstract framework if identification of some key matters around attacks and supports is an issue. Lastly, we mention that our approach, as well as argumentation-theoretic approaches, is treated as evidential reasoning. There are also studies into abductive reasoning purporting to find the best possible explanations to causes [4] or those that combine evidential and abductive reasoning [6].

References

1. Hanrei Times No. 985, p. 300 (1999)
2. Arisaka, R., Satoh, K.: Balancing rationality and utility in logic-based argumentation with classical logic sentences and belief contraction. In: Baldoni, M., Chopra, A.K., Son, T.C., Hirayama, K., Torroni, P. (eds.) PRIMA 2016. LNCS (LNAI), vol. 9862, pp. 168–180. Springer, Cham (2016). doi:10.1007/978-3-319-44832-9_10
3. Baroni, P., Giacomin, M., Liao, B.: Dealing with generic contrariness in structured argumentation. In: IJCAI, pp. 2727–2733 (2015)
4. Bex, F., Bench-Capon, T.J.M., Atkinson, K.: Did he jump or was he pushed? Abductive practical reasoning. Artif. Intell. Law **17**(2), 79–99 (2009)
5. Bex, F., Prakken, H., Reed, C., Walton, D.: Towards a formal account of reasoning about evidence: argumentation schemes and generalisations. Artif. Intell. Law **11**(2), 125–165 (2003)
6. Bex, F., van Koppen, P.J., Prakken, H., Verheij, B.: A hybrid formal theory of arguments, stories and criminal evidence. Artif. Intell. Law **18**(2), 123–152 (2010)
7. Boella, G., Gabbay, D.M., van der Torre, L., Villata, S.: Support in abstract argumentation. In: COMMA, pp. 111–122 (2010)
8. Caminada, M., Amgoud, L.: On the evaluation of argumentation formalisms. Artif. Intell. **171**(5–6), 286–310 (2007)
9. Cayrol, C., Lagasquie-Schiex, M.C.: On the acceptability of arguments in bipolar argumentation frameworks. In: Godo, L. (ed.) ECSQARU 2005. LNCS (LNAI), vol. 3571, pp. 378–389. Springer, Heidelberg (2005). doi:10.1007/11518655_33
10. Cayrol, C., Lagasquie-Schiex, M.-C.: Bipolarity in argumentation graphs: towards a better understanding. In: Benferhat, S., Grant, J. (eds.) SUM 2011. LNCS (LNAI), vol. 6929, pp. 137–148. Springer, Heidelberg (2011). doi:10.1007/978-3-642-23963-2_12
11. Dung, P.M.: On the acceptability of arguments and its fundamental role in nonmonotonic reasoning, logic programming, and n-person games. Artif. Intell. **77**(2), 321–357 (1995)
12. Gabbay, D.M.: Semantics for higher level attacks in extended argumentation frames part 1: overview. Stud. Logica. **93**(2–3), 357–381 (2009)
13. Governatori, G., Padmanabhan, V.: A defeasible logic of policy-based intention. In: Gedeon, T.T.D., Fung, L.C.C. (eds.) AI 2003. LNCS (LNAI), vol. 2903, pp. 414–426. Springer, Heidelberg (2003). doi:10.1007/978-3-540-24581-0_35
14. Governatori, G., Rotolo, A.: BIO logical agents: norms, beliefs, intentions in defeasible logic. JAAMAS **17**(1), 36–69 (2008)
15. Horn, L.R.: A Natural History of Negation, 2nd edn. The University Chicago Press, Chicago (2001)
16. Modgil, S.: Reasoning about preferences in argumentation frameworks. Artif. Intell. **173**, 901–934 (2009)
17. Modgil, S., Bench-Capon, T.J.M.: Metalevel argumentation. J. Logic Comput. **21**(6), 959–1003 (2011)
18. Modgil, S., Prakken, H.: A general account of argumentation with preferences. Artif. Intell. **195**, 361–397 (2013)
19. Nielsen, S.H., Parsons, S.: A generalization of Dung's abstract framework for argumentation. In: Argumentation in Multi-Agent Systems, pp. 54–73 (2006)
20. Nouioua, F., Risch, V.: Bipolar argumentation frameworks with specialized supports. In: ICTAI, pp. 215–218 (2010)

21. Nouioua, F., Risch, V.: Argumentation frameworks with necessities. In: Benferhat, S., Grant, J. (eds.) SUM 2011. LNCS (LNAI), vol. 6929, pp. 163–176. Springer, Heidelberg (2011). doi:10.1007/978-3-642-23963-2_14

22. Oren, N., Reed, C., Luck, M.: Moving between argumentation frameworks. In: COMMA, pp. 379–390 (2010)

23. Prakken, H.: Analysing reasoning about evidence with formal models of argumentation. Law Probab. Risk **3**(1), 33–50 (2004)

24. Prakken, H.: An abstract framework for argumentation with structured arguments. Argument Comput. **1**, 93–124 (2010)

Argument-Based Logic Programming
for Analogical Reasoning

Teeradaj Racharak[1,2(✉)], Satoshi Tojo[2], Nguyen Duy Hung[1],
and Prachya Boonkwan[3]

[1] School of Information, Computer, and Communication Technology,
Sirindhorn International Institute of Technology,
Thammasat University, Pathum Thani, Thailand
r.teeradaj@gmail.com, hung.nd.siit@gmail.com
[2] School of Information Science,
Japan Advanced Institute of Science and Technology, Ishikawa, Japan
{racharak,tojo}@jaist.ac.jp
[3] National Electronics and Computer Technology Center, Pathum Thani, Thailand
prachya.boonkwan@nectec.or.th

Abstract. Analogical reasoning can be understood as a kind of resemblance of one thing to another, thus assigning properties from one context to another. The key idea is to use similarity information to support an inference which cannot be deductively inferred. In this paper, we present a formal and intuitive framework of this phenomena using an argument-based logic-programming-like language. A proof theory of our system is stated in the dialectical style, where a proof takes the form of dialogue between a proponent and an opponent of an argument. We also discuss how the proposed framework can be fine tuned for optimistic analogical reasoning and pessimistic analogical reasoning. Finally, we discuss a design sketch of our proposed analogical reasoner called Analogist.

Keywords: Analogical argumentation · Argumentation schemes · Argument from Analogy · Argument-based logic programming

1 Introduction and Motivation

Analogical reasoning can be understood as a kind of resemblance of one thing to another, thus assigning properties from one context to another. This kind of reasoning is used quite often by human beings in real-life situations, especially when humans encounter an unseen situation. To have an intuitive understanding on the mechanism, let us take a look on the following case where attorney Gerry Spence reasons in the case of Silkwood v. Kerr-McGee Corporation (1984) [1].

Example 1 (The Silkwood case). Karen Silkwood was a technician who had the job of grinding and polishing plutonium pins used to make fuel rods for nuclear reactors. Tests in 1974 showed that she had been exposed to dangerously high levels of plutonium radiation. After she died in an automobile accident, her father

© Springer International Publishing AG 2017
S. Kurahashi et al. (Eds.): JSAI-isAI 2016, LNAI 10247, pp. 253–269, 2017.
DOI: 10.1007/978-3-319-61572-1_17

brought an action against Kerr-McGee in which the corporation was held to be at fault for her death on the basis of strict liability. In strict liability, a person can be held accountable for the harmful consequences of some dangerous activity he was engaged in, without having to prove that he intended the outcome.   □

Spence's closing argument uses the analogy of the escaping lion, which had great rhetorical effect on the jury. According to his speech (p. 129 of [1]), he emphasized the statement *If the lion got away, Kerr-McGee has to pay.*

> Some guy brought an old lion in a cage – lions are dangerous – and through no negligence of his own, the lion got away. Nobody knew how – like in the Silkwood case, *nobody knew how.* And, the lion ate up some people. And they said, you know: *Pay. It was your lion and it got away.* And, the man says: *But I did everything that I could and it isn't my fault that it got away.* They said: *You have to pay.* You have to pay because it was your lion – unless the person who was hurt let the lion out himself.

Roughly, reasoning by analogy is a form of non-deductive reasoning in which we infer a conclusion based on similarity of two situations. There is substantial work on methodology ranging from a kind of introspective folk psychology [2–4] to partial identity of Horn clause logic interpretations [5]. There are some contributions which include elements of both, e.g. [6,7]; and also, work which provides a form of analogical reasoning in terms of a system of hypothetical reasoning based on mathematical logic [8]. While there is a diversity of methodology, there is a some consensus, i.e. using similarity information to support an inference which cannot be deductively inferred.

In Waller's observation [4], the term *persuasion* is used to suggest that the analogical principle used in reasoning is defeasible. That is, the opponent is persuaded to believe the conclusion and that conclusion is inherently subject to exceptions. For example, in the case of Silkwood, by arguing analogically that plutonium processing is an ultrahazardous activity, Kerr-McGee is held strictly liable for any damages caused by the escape of its plutonium.

Walton's argumentation scheme for Argument from Analogy is an analysis tool for this form of argumentation. In this work, we investigate the nature of analogical argumentation based on the study of Walton's and Dung's theory of abstract argumentation [9]. Though many developments build on Dung's abstract theory, the study of analogical reasoning in the abstract framework is still relatively unexplored. Current frameworks (cf. Sect. 6) may treat analogical reasoning by representing similarity in some logical rules; however, most of them do not consider the interactions between analogical arguments. This is a revised and improved version of the workshop paper [10] which discusses the framework for analogical argumentation (cf. Sects. 3 and 4) and a design sketch of our proposed analogical reasoner called Analogist (cf. Sect. 5). Preliminaries and the conclusion are discussed in Sects. 2 and 7, respectively.

2   Preliminaries

In this section, we review the basics of argumentation schemes and concept similarity measure under preference profile in Description Logics (DLs).

2.1   Argumentation Schemes: Argument from Analogy

Argumentation schemes [3] are stereotypical non-deductive patterns of reasoning, consisting of a set of premises and a conclusion that is *presumed* to follow from the premises. Use of argumentation schemes is evaluated by a specific set of critical questions corresponding to each scheme. Let us illustrate this with the argumentation scheme called *Argument from Analogy* as follows:

1. A situation is described in C_1.
2. A is plausibly drawn as an acceptable conclusion in C_1.
3. Generally, C_1 is similar to C_2.
∴ A is plausibly drawn as an acceptable conclusion in C_2.

The following set of critical questions matches the scheme:

1. Are there respects in which C_1 and C_2 are different that would tend to undermine the similarity cited?;
2. Is A the right conclusion to be drawn in C_1?;
3. Is there some other situation C_3 that is also similar to C_1, but in which A is not drawn as an acceptable conclusion?

The first critical question relates to differences between the two situations that could detract from the strength of the argument from analogy. The second critical question nicely ensures the right conclusion. Lastly, the third critical question is associated with a familiar type of counter-analogy. The function of this critical question is to suggest doubt that could possibly lead to a plausible counter-argument that could be used to attack the original conclusion.

2.2   Concept Similarity Measure in Description Logics

In DLs, we assume countably infinite sets CN of concept names and RN of role names that are fixed and disjoint. The set of concept descriptions, or simply concepts, for a specific DL $\mathcal{L}$ is denoted by $\mathsf{Con}(\mathcal{L})$. The set $\mathsf{Con}(\mathcal{L})$ is inductively defined on CN and RN with the use of concept constructors. An ontology $\mathcal{O}$ is usually defined as $\mathcal{O} = \langle \mathcal{T}, \mathcal{A} \rangle$ where $\mathcal{T}$ is a terminological component or TBox and $\mathcal{A}$ is an assertional component or ABox.

An interpretation $\mathcal{I}$ is a pair $\mathcal{I} = \langle \Delta^{\mathcal{I}}, \cdot^{\mathcal{I}} \rangle$ where $\Delta^{\mathcal{I}}$ is a non-empty set representing the domain of the interpretation and $\cdot^{\mathcal{I}}$ is an interpretation function which defines on every concept name, every role name, and every concept description in the standard manner.

Let $\mathsf{CN}^{\mathsf{pri}}(\mathcal{T})$, $\mathsf{RN}^{\mathsf{pri}}(\mathcal{T})$, and $\mathsf{RN}(\mathcal{T})$ be a set of primitive concept names occurring in $\mathcal{T}$, a set of primitive role names occurring in $\mathcal{T}$, and a set of role names occurring in $\mathcal{T}$, respectively. In the following, we give formal definitions for a concept similarity measure under preference profile in DLs.

Definition 1 (Preference Profile [11]). *A preference profile (denoted by π) is a quintuple $\langle \mathfrak{i}^c, \mathfrak{i}^r, \mathfrak{s}^c, \mathfrak{s}^r, \mathfrak{d} \rangle^1$ where*

- *$\mathfrak{i}^c : \mathsf{CN} \to [0,2]$ where $\mathsf{CN} \subseteq \mathsf{CN}^{\mathsf{pri}}(\mathcal{T})$ is primitive concept importance;*
- *$\mathfrak{i}^r : \mathsf{RN} \to [0,2]$ where $\mathsf{RN} \subseteq \mathsf{RN}(\mathcal{T})$ is role importance;*
- *$\mathfrak{s}^c : \mathsf{CN} \times \mathsf{CN} \to [0,1]$ where $\mathsf{CN} \subseteq \mathsf{CN}^{\mathsf{pri}}(\mathcal{T})$ is primitive concepts similarity;*
- *$\mathfrak{s}^r : \mathsf{RN} \times \mathsf{RN} \to [0,1]$ where $\mathsf{RN} \subseteq \mathsf{RN}^{\mathsf{pri}}(\mathcal{T})$ is primitive roles similarity; and*
- *$\mathfrak{d} : \mathsf{RN} \to [0,1]$ where $\mathsf{RN} \subseteq \mathsf{RN}(\mathcal{T})$ is role discount factor.*

We discuss the interpretation of each above function in order. Firstly, for any $A \in \mathsf{CN}^{\mathsf{pri}}(\mathcal{T})$, $\mathfrak{i}^c(A) = 1$ captures an expression of normal importance on A, $\mathfrak{i}^c(A) > 1$ ($\mathfrak{i}^c(A) < 1$) indicates that A has higher (and lower, respectively) importance, and $\mathfrak{i}^c(A) = 0$ indicates that A is of no importance to the similarity identification. Secondly, we define the interpretation of $\mathfrak{i}^r$ in the similar fashion as $\mathfrak{i}^c$ for any $r \in \mathsf{RN}(\mathcal{T})$. Thirdly, for any $a, b \in \mathsf{CN}^{\mathsf{pri}}(\mathcal{T})$, $\mathfrak{s}^c(A, B) = 1$ captures an expression of total similarity between A and B and $\mathfrak{s}^c(A, B) = 0$ captures an expression of total dissimilarity between A and B. Fourthly, the interpretation of $\mathfrak{s}^r$ is defined in the similar fashion as $\mathfrak{s}^c$ for any $r, s \in \mathsf{RN}^{pri}(\mathcal{T})$. Lastly, for any $r \in \mathsf{RN}(\mathcal{T})$, $\mathfrak{d}(r) = 1$ captures an expression of total importance on a role (over a corresponding nested concept) and $\mathfrak{d}(r) = 0$ captures an expression of total importance on a nested concept (over a corresponding role).

Definition 2 ([12]). *Given a preference profile π, two concepts $C, D \in \mathsf{Con}(\mathcal{L})$, and a TBox $\mathcal{T}$, a concept similarity measure under preference profile w.r.t. a TBox $\mathcal{T}$ is a function $\sim^{\pi}_{\mathcal{T}} : \mathsf{Con}(\mathcal{L}) \times \mathsf{Con}(\mathcal{L}) \to [0,1]$. A function $\sim^{\pi}_{\mathcal{T}}$ is called preference invariance w.r.t equivalence if $C \equiv D \Leftrightarrow (C \sim^{\pi}_{\mathcal{T}} D = 1 \text{ for any } \pi)$.*

Several concept similarity measures abound. Unfortunately, they may not consider preferential aspects for identifying similarity, i.e. $\sim_{\mathcal{T}}$. Though these measures are also applicable for our proposed framework (cf. Sect. 3), we recommend to use measures exhibiting aspects of preference profile, e.g. sim^{π} [12], when preferences are to be used for identifying the degree of relevant similarity.

3   The Framework of Analogical Reasoning

3.1   The Language

Our object language conforms to the familiar logic-programming-like style. That is, let $\Sigma = \langle \mathcal{C}, \mathcal{V}, \mathcal{P} \rangle$ be a signature with a finite set of constants $\mathcal{C}$, an infinite set of variables $\mathcal{V}$, and a finite set of predicate symbols $\mathcal{P}$. Let L_Σ be the first-order language constructed over Σ. There are two types of literals. A *strong literal* is an atomic first-order formula A (of L_Σ) or such a formula preceded by the classical negation, i.e. $\neg A$. A *weak literal* is a literal of the form *not A*, where A is a strong literal and *not* denotes *negation-as-failure* (or default

---

[1] In the original definition of preference profile [11,12], both $\mathfrak{i}^c$ and $\mathfrak{i}^r$ are mapped to $\mathbb{R}_{\geq 0}$ which is a minor error.

negation). Informally, *not A* reads as *there is no evidence that A is the case* whereas ¬*A* reads as *A is definitely not the case*. In what follows, we use the standard typographic conventions of Logic Programming.

Definition 3 (Strict Rule). *A strict rule is an expression of the form* $L_0 \leftarrow L_1, \ldots, L_n$ *where* $n \geq 0$ *and* L_i *(*$0 \leq i \leq n$*) is a strong literal. If* $n = 0$*, it is referred to as a* fact.

Definition 4 (Defeasible Rule). *A defeasible rule is an expression of the form* $L_0 \Leftarrow L_1, \ldots, L_n$ *where* $n \geq 0$*,* L_0 *is a strong literal, and* L_i *(*$1 \leq i \leq n$*) is a literal. If* $n = 0$*, it is referred to as a* presumption.

Defeasible rules are used to represent tentative information which will be used if no one could disprove it whereas strict rules are used to represent strict information. For example, $fly(X) \Leftarrow bird(X)$ expresses *usually, a bird can fly* whereas $bird(X) \leftarrow penguin(X)$ expresses *all penguins are birds*.

It is worth noting that in general account of defeasible logic, particularly Nute's d-Prolog [13], it contains a facility to define *defeater rules*, e.g. sick birds do not fly. The purpose of defeater rules is to express exceptions to defeasible rules. However, [14] shows that defeater rules can be simulated by means of strict and defeasible rules (in Nute's sense). As we will also show soon, our system does not need to supply with defeater rules. The system will find counter-arguments, including counter-analogies, among arguments it is able to build.

Definition 5 (Similarity Rule). *A similarity rule is an expression of the form* $L_0 \overset{x}{\Leftarrow} L_1$[2] *where* L_0, L_1 *are first-order predicates and* $0 < x \leq 1$ *for any real number* x*, such that* $L_0 \overset{x}{\Leftarrow} L_1$ *means* L_1 *is similar to* L_0 *at* x *degree (but, not vice versa) and* $L_0 \overset{1}{\Leftarrow} L_1$ *indicates* L_1 *is totally similar to* L_0*, i.e.* $L_1 \equiv L_0$*.*

Similarity rules are used to form similarity premises, e.g. $plutonium\_plant \overset{1}{\Leftarrow} lion$ expresses *lions are totally similar to plutonium plants*. One methodology for building similarity rules is to query similarity information of corresponding concepts defined in DL-based ontologies (Definition 2). To the best of our knowledge, $\overset{x}{\Leftarrow}$ is first introduced for analogical reasoning in this work.

Definition 6 (Logic Program). *A logic program* $\mathcal{P}$ *is a triple* $\langle SR, DR, SIM \rangle$ *where* SR *is a finite set of strict rules,* DR *is a finite set of defeasible rules, and* SIM *is a finite set of similarity rules.*

We assume that every rule of $\mathcal{P}$ is grounded. Nevertheless, our examples may use the usual convention, i.e. *schematic rules with variables*, in logic programs.

Definition 7 (Derivation). *Let* $\mathcal{P}$ *be a logic program and* L *be a ground literal. A derivation for* L *from* $\mathcal{P}$ *with an* analogy degree w*, in symbols* $\mathcal{P} \overset{w}{\vdash} L$[3]*, is a*

---

[2] When $x = 1$, we may remove it.
[3] When $w = 1$, we may remove it.

finite sequence $L_1, \ldots, L_n$ of ground literals such that $L_n = L$ and each literal L_i $(1 \leq i \leq n)$ satisfies the following conditions:

1. L_i *is a fact or a presumption;*
2. *There exists a rule R_i of $\mathcal{P}$ (strict or defeasible) with head L_i and body $L_1, \ldots, L_j$ such that every literal of the body, except ones preceded by negation-as-failure, is an element of the sequence appearing before L_i $(j < i)$;*
3. *There exist similarity rules $A_1' \overset{x_1}{\rightleftharpoons} L_1', \ldots, A_k' \overset{x_k}{\rightleftharpoons} L_k'$ of $\mathcal{P}$ and another rule R_i of $\mathcal{P}$ (strict or defeasible) with head L_i and body $L_1, \ldots, L_k, \ldots, L_j$ such that the predicate of $L_1, \ldots, L_k$ are $L_1', \ldots, L_k'$, respectively. The substitution on each predicate of $L_1, \ldots, L_k$ with $A_1', \ldots, A_k'$, respectively, and other non-substituted literals, except ones preceded by negation-as-failure, is an element of the sequence appearing before L_i $(k \leq j < i)$.*
4. $w = \bigotimes\limits_{A_i' \overset{}{\rightleftharpoons} L_i'} \{x_i\}$, *where $\bigotimes$ is a triangular norm (t-norm)[4] and $A_i' \overset{x_i}{\rightleftharpoons} L_i'$ is used to derive L. Otherwise, we set $w = 1$ as the default value.*

Basically, $\mathcal{P} \overset{1}{\vdash} L$ means that L may be derived without any use of analogies. Condition 2 to 3 assume two implicit inference rules, viz. a rule of defeasible modus ponens (MP) and a rule of defeasible analogical rule (AR) as follows:

$$\frac{L_1, \ldots, L_j \qquad L_i \Leftarrow L_1, \ldots, L_j, not\ L_m, \ldots, not\ L_n}{L_i} \ \text{MP}$$

$$\frac{A_1' \overset{x_1}{\rightleftharpoons} L_1' \ \ldots \ A_k' \overset{x_k}{\rightleftharpoons} L_k' \qquad L_i \Leftarrow L_1, \ldots, L_k, \ldots, L_j, not\ L_m, \ldots, not\ L_n}{L_i \Leftarrow A_1, \ldots, A_k, \ldots, L_j, not\ L_m, \ldots, not\ L_n} \ \text{AR}$$

Since this work employs the notion of t-norm, we include its definition here for self-containment. A function $\otimes : [0,1]^2 \to [0,1]$ is called a t-norm iff it fulfills the following properties for all $x, y, z, w \in [0,1]$: **(1)** $x \otimes y = y \otimes x$ (commutativity); **(2)** $x \leq z$ and $y \leq w \Rightarrow x \otimes y \leq z \otimes w$ (monotonicity); **(3)** $(x \otimes y) \otimes z = x \otimes (y \otimes z)$ (associativity); **(4)** $x \otimes 1 = x$ (identity). A t-norm is called *bounded* iff $x \otimes y = 0 \Rightarrow x = 0$ or $y = 0$. There are several reasons for the use of a t-norm. Firstly, it is the generalization of the conjunction in propositional logic. Secondly, the operator *min* (i.e. $x \otimes y = min\{x, y\}$) is an instance of a bounded t-norm. This reflects an intuition on the use of analogical reasoning that the strength of a consequence depends upon the use of similarities. Lastly, 1 acts as the neutral element for t-norms. Table 1 shows other instances of $\otimes$.

Example 2 (Continuation of Example 1). We translate the Silkwood case into our logic program $\mathcal{P}$ as follows. The literal *exception(X)* means *X is an exception to inactivate the goal.* To avoid confusion, we separate each rule by a semicolon.
    SR = {*defendant(X)* ← *owner(X, Y), danger(Y), killer(Y, Z); lion(l_1); owner (guy, l_1); plutonium_plant(p_1); owner(kerr_mcgee, p_1); person(man); killer(l_1, man); person(silkwood); killer(p_1, silkwood)*};

---

[4] The precise definition of t-norm is given later.

Table 1. Some instances of the operator $\otimes$

Name	Notation	$x_1 \otimes x_2 =$
Minimum	$\otimes_{\min}$	$\min\{x_1, x_2\}$
Product	$\otimes_{\text{prod}}$	$x_1 \cdot x_2$
Hamacher product	$\otimes_{H_0}$	0 if $x_1 = x_2 = 0$ or $\frac{x_1 \cdot x_2}{x_1 + x_2 - x_1 \cdot x_2}$ if otherwise

$$\mathsf{DR} = \{liable(X) \Leftarrow defendant(X), not\ exception(X);\ danger(X) \Leftarrow lion(X)\};$$
$$\mathsf{SIM} = \{plutonium\_plant \stackrel{0.8}{\Leftarrow} lion\}^5.$$

Let $\bigotimes$ be the min operator, we have $\mathcal{P} \stackrel{0.8}{\vdash} liable(kerr\_mcgee)$. $\square$

3.2 Structured Argument

We are ready to extend the notion of derivation (Definition 7) for constructing arguments. Our notion of building arguments is adapted from [15–17] for
discovering inconsistency in the strict knowledge base (cf. Sect. 6).

Definition 8 (Argument). *Let* $\mathcal{P} = \langle \mathsf{SR}, \mathsf{DR}, \mathsf{SIM} \rangle$ *be a logic program and*
Q *be a ground literal. A structure of an argument* $\mathcal{A}$ *for* Q *is a quadruple*
$\langle \mathsf{D}, \mathsf{S}, Q, w \rangle$, *where* $\mathsf{D} \subseteq \mathsf{DR}$ *and* $\mathsf{S} \subseteq \mathsf{SIM}$ *such that:*

1. *There exists a derivation for* Q *from* $\langle \mathsf{SR}, \mathsf{D}, \mathsf{S} \rangle$, *i.e.* $\langle \mathsf{SR}, \mathsf{D}, \mathsf{S} \rangle \stackrel{w}{\vdash} Q$;
2. *If* L *is a literal in the derivation for* Q, *then there is no defeasible rule in* D *containing* $not\ L$ *in the body;*
3. $\langle \mathsf{SR}, \mathsf{D}, \mathsf{S} \rangle$ *is consistent, i.e.* $\langle \mathsf{SR}, \mathsf{D}, \mathsf{S} \rangle \stackrel{w}{\nvdash} A, \overline{A}$ *for some ground literal* A;
4. $\mathcal{A}$ *is minimal, i.e. there is no* $\mathsf{D}' \cup \mathsf{S}' \subseteq \mathsf{D} \cup \mathsf{S}$ *such that* $\langle \mathsf{D}', \mathsf{S}', Q, w \rangle$.

It is worth noting that Condition 2 helps avoiding an introduction of a self-
conflicting argument. For example, let $\mathcal{P} = \langle \mathsf{SR}, \mathsf{DR}, \mathsf{SIM} \rangle$ where $\mathsf{SR} = \emptyset$, $\mathsf{DR} =$
$\{a \Leftarrow b; b \Leftarrow not\ a\}$, and $\mathsf{SIM} = \emptyset$. Without Condition 2, it is possible to derive
a from $\mathcal{P}$, i.e. assuming $not\ a$ obtains a.

By distinguishing a set of defeasible rules of DR and a set of similarity rules
of SIM in an argument's structure, we can clearly identify analogical arguments
(or arguments from analogy) apart from standard arguments.

Definition 9. *An argument* $\mathcal{A} = \langle \mathsf{D}, \mathsf{S}, Q, w \rangle$ *is called an* analogical argument
if $\mathsf{S} \neq \emptyset$ *and is called a* standard argument *if otherwise. Alternatively,* $\mathcal{A}$ *is called*
a strict argument *if* $\mathsf{D} = \mathsf{S} = \emptyset$ *and is called a* defeasible argument *if otherwise.*

From the definition, $\mathcal{A}_1 = \langle \emptyset, \mathsf{S}_1, Q_1, 0.8 \rangle$, where $\mathsf{S}_1 = \{plutonium\_plant \stackrel{0.8}{\Leftarrow}$
$lion\}$ and $Q_1 = danger(kerr\_mcgee)$, is an analogical argument. Observe that
an analogical argument is also a defeasible argument (such as $\mathcal{A}_1$).

---

[5] We may employ the notion of $\stackrel{\pi}{\sim}_T$ to obtain 0.8 from realistic ontologies.

Definition 10. *Let* $\mathcal{P} = \langle \mathsf{SR}, \mathsf{DR}, \mathsf{SIM} \rangle$ *be a logic program and* $\mathcal{A}_1 = \langle D_1, S_1, Q_1, w_1 \rangle$, $\mathcal{A}_2 = \langle D_2, S_2, Q_2, w_2 \rangle$ *be arguments. Then, we say that* $\mathcal{A}_1$ *attacks* $\mathcal{A}_2$ *iff one of the following conditions hold:*

1. $\langle \mathsf{SR}, D_1, S_1 \rangle \overset{w_1'}{\vdash} A$ *and* $\langle \mathsf{SR}, D_2, S_2 \rangle \overset{w_2'}{\vdash} \overline{A}$ *for some ground literal A;*

2. $\langle \mathsf{SR}, D_1, S_1 \rangle \overset{w_1'}{\vdash} A$ *and there exists* $r \in D_2$ *which contains* not A *in the body;*

Definition 10 does not include ways of comparing which arguments are better. It only says which arguments are in conflict. We illustrate this in Example 3.

Example 3 (Continuation of Example 2). Let us enrich the example that
$\mathsf{SR} = \{ defendant(X) \leftarrow owner(X,Y), danger(Y), killer(Y,Z); lion(l_1); owner (guy, l_1); plutonium\_plant(p_1); owner(kerr\_mcgee, p_1); person(man); killer(l_1, man); person(silkwood); killer(p_1, silkwood); \neg danger(X) \leftarrow green\_environment (X) \};$
$\mathsf{DR} = \{ liable(X) \Leftarrow defendant(X), not\ exception(X); danger(X) \Leftarrow lion(X); green\_environment(X) \Leftarrow wind\_turbine(X); federal\_law(X) \Leftarrow owner(X,Y), plutonium\_plant(Y); \neg liable(X) \Leftarrow federal\_law(X) \};$
$\mathsf{SIM} = \{ plutonium\_plant \overset{0.8}{\Leftarrow} lion; plutonium\_plant \overset{0.4}{\Leftarrow} wind\_turbine \}.$

We note that $federal\_law(X)$ refers to the federal preemption of state regulation of the safety aspects of nuclear energy. Hence, we can find that:
$\mathcal{A}_1 = \langle \{ danger(X) \Leftarrow lion(X); liable(X) \Leftarrow defendant(X), not\ exception(X) \}, \{ plutonium\_plant \overset{0.8}{\Leftarrow} lion \}, liable(kerr\_mcgee), 0.8 \rangle$ as an analogical argument, $\mathcal{A}_2 = \langle \{ federal\_law(X) \Leftarrow owner(X,Y), plutonium\_plant(Y); \neg liable(X) \Leftarrow federal\_law(X) \}, \emptyset, \neg liable(kerr\_mcgee), 1 \rangle$ as a defeasible argument, and $\mathcal{A}_3 = \langle \{ green\_environment(X) \Leftarrow wind\_turbine(X) \}, \{ plutonium\_plant \overset{0.4}{\Leftarrow} wind\_turbine \}, \neg danger(p_1), 0.4 \rangle$ as an analogical argument. Thus, we have $\mathcal{A}_1$ attacks $\mathcal{A}_2$ and $\mathcal{A}_2$ attacks $\mathcal{A}_1$ by Condition (1). In addition, $\mathcal{A}_1$ attacks $\mathcal{A}_3$ and $\mathcal{A}_3$ also attacks $\mathcal{A}_1$ by Condition (1). □

The argument $\mathcal{A}_1$ attacks $\mathcal{A}_2$ since it satisfies the first condition, i.e. $\langle \mathsf{SR}, D_1, S_1 \rangle \overset{w_1'}{\vdash}$ $liable(kerr\_mcgee)$ and $\langle \mathsf{SR}, D_2, S_2 \rangle \overset{w_2'}{\vdash} \neg liable(kerr\_mcgee)$. This way of attack is called *rebuttal*. Basically, a rebuttal attacks an argument by drawing the complement of a derived literal. On the other hand, the second condition is called *undercut*. An undercut attacks by showing an exceptional situation without drawing the complement of a literal.

Formalizing the scheme makes a special way of comparing arguments. This is because use of the scheme imposes some specialties for adjudicating conflicting arguments. There are three kinds of the comparison as discussed following.

Firstly, we consider the case of a standard argument $\mathcal{A}_1$ attacking another standard argument $\mathcal{A}_2$, i.e. $\mathcal{A}_1 = \langle D_1, S_1, Q_1, w_1 \rangle$, $\mathcal{A}_2 = \langle D_2, S_2, Q_2, w_2 \rangle$, and $S_1, S_2 = \emptyset$. Comparing arguments is treated in usual ways, e.g. some preference criteria are required to compare which argument is better for the rebuttal case. Such criteria can be defined as a relation $> \subseteq \mathsf{DR} \times \mathsf{DR}$ in a standard way, i.e. $r_1 > r_2$ means r_1 is preferred over r_2 for any $r_1, r_2 \in \mathsf{DR}$. Then, we say an argument $\mathcal{A}_1 = \langle \mathsf{S}_1, \mathsf{D}_1, Q_1, w_1 \rangle$ is better than $\mathcal{A}_2 = \langle \mathsf{S}_2, \mathsf{D}_2, Q_2, w_2 \rangle$ (denoted by $\mathcal{A}_1 \succ \mathcal{A}_2$[6]) if (1) $\exists r_1 \in \mathsf{D}_1. \exists r_2 \in \mathsf{D}_2 : r_1 > r_2$; and (2) $\forall r_1 \in \mathsf{D}_1. \forall r_2 \in \mathsf{D}_2 : r_2 \not> r_1$. We also establish that an argument structure based on facts is preferable to an argument structure based on presumptions.

Secondly, we consider the case of an analogical argument $\mathcal{A}_1$ attacking another analogical argument $\mathcal{A}_2$, i.e. $\mathcal{A}_1 = \langle \mathsf{D}_1, \mathsf{S}_1, Q_1, w_1 \rangle$, $\mathcal{A}_2 = \langle \mathsf{D}_2, \mathsf{S}_2, Q_2, w_2 \rangle$, and $\mathsf{S}_1, \mathsf{S}_2 \neq \emptyset$. In these cases, we compare the analogy degrees of both arguments. To defeat another, the degree must be at least equal to one's another.

Lastly, we consider the case of an analogical argument $\mathcal{A}_1$ attacking another standard argument $\mathcal{A}_2$, i.e. $\mathcal{A}_1 = \langle \mathsf{D}_1, \mathsf{S}_1, Q_1, w_1 \rangle$, $\mathcal{A}_2 = \langle \mathsf{D}_2, \mathsf{S}_2, Q_2, w_2 \rangle$, $\mathsf{S}_1 \neq \emptyset$, and $\mathsf{S}_2 = \emptyset$. In these cases, we base our reasoning on the use of argumentation schemes. Uttering an instance of Argument from Analogy like this in a dialogue creates a presumption in favor of the conclusion from analogy and a corresponding proof for the other side in the dialogue to defeat the conclusion by asking critical questions. This also conforms to Waller's use of analogical arguments [4] in the way of persuasion. Thus, an analogical argument is preferable.

In the following, conditions 1 to 3 capture these three kinds of comparing for rebuttal. Conditions 4 to 6 capture these three kinds of comparing for undercut.

Definition 11. *Let $\mathcal{A}_1 = \langle \mathsf{D}_1, \mathsf{S}_1, Q_1, w_1 \rangle$ and $\mathcal{A}_2 = \langle \mathsf{D}_2, \mathsf{S}_2, Q_2, w_2 \rangle$ be two arguments. Then, $\mathcal{A}_1$ defeats $\mathcal{A}_2$ iff one of the following holds:*

1. *$\mathcal{A}_1$ attacks $\mathcal{A}_2$ under Condition (1) of Definition 10, $(\mathsf{S}_1, \mathsf{S}_2 = \emptyset)$, $\mathcal{A}_2 \not\succ \mathcal{A}_1$, and $\mathcal{A}_2$ does not attack $\mathcal{A}_1$ under Condition (2) of Definition 10;*
2. *$\mathcal{A}_1$ attacks $\mathcal{A}_2$ under Condition (1) of Definition 10, $(\mathsf{S}_1, \mathsf{S}_2 \neq \emptyset)$, $w_1 \geq w_2$, and $\mathcal{A}_2$ does not attack $\mathcal{A}_1$ under Condition (2) of Definition 10;*
3. *$\mathcal{A}_1$ attacks $\mathcal{A}_2$ under Condition (1) of Definition 10, $\mathsf{S}_1 \neq \emptyset$, $\mathsf{S}_2 = \emptyset$, and $\mathcal{A}_2$ does not attack $\mathcal{A}_1$ under Condition (2) of Definition 10;*
4. *$\mathcal{A}_1$ attacks $\mathcal{A}_2$ under Condition (2) of Definition 10 and $(\mathsf{S}_1, \mathsf{S}_2 = \emptyset)$;*
5. *$\mathcal{A}_1$ attacks $\mathcal{A}_2$ under Condition (2) of Definition 10, $(\mathsf{S}_1, \mathsf{S}_2 \neq \emptyset)$, and $w_1 \geq w_2$; and*
6. *$\mathcal{A}_1$ attacks $\mathcal{A}_2$ under Condition (2) of Definition 10, $\mathsf{S}_1 \neq \emptyset$, and $\mathsf{S}_2 = \emptyset$.*

We say that $\mathcal{A}_1$ strictly defeats $\mathcal{A}_2$ iff $\mathcal{A}_1$ defeats $\mathcal{A}_2$ and $\mathcal{A}_2$ does not defeat $\mathcal{A}_1$.

We note that many researchers in the area of argumentation with priority have defined many ways to compare arguments. Addressing this issue is outside the scope and is irrelevant to the Condition 1. We leave this as a future task.

Example 4 (Continuation of Example 3). We have that $\mathcal{A}_1$ *strictly* defeats $\mathcal{A}_2$ and also *strictly* defeats $\mathcal{A}_3$. ☐

Theorem 1. *There does not exist an argument which attacks an argument $\mathcal{A} = \langle \emptyset, \emptyset, Q, w \rangle$, where Q is a ground literal and $w \in (0, 1]$.*

---

[6] Later, this definition is used by Definition 11 for comparing between rebuttal attacks.

Proof. *Let* $\mathcal{P} = \langle \mathsf{SR}, \mathsf{DR}, \mathsf{SIM} \rangle$ *be a logic program and suppose that there exists an argument* $\mathcal{B} = \langle \mathsf{D}_b, \mathsf{S}_b, Q_b, w_b \rangle$ *which attacks* $\mathcal{A}$. *By Definition 10, we show contradiction by cases:*

- *(Rebuttal) We have* $\langle \mathsf{SR}, \mathsf{D}_b, \mathsf{S}_b \rangle \overset{w_b}{\vdash} A$ *and* $\langle \mathsf{SR}, \emptyset, \emptyset \rangle \overset{w_a}{\vdash} \overline{A}$ *for some ground literal* A. *This means* $\langle \mathsf{SR}, \mathsf{D}_b, \mathsf{S}_b \rangle$ *is inconsistent and* $\mathcal{B}$ *is not an argument.*
- *(Undercut) We have* $\langle \mathsf{SR}, \mathsf{D}_b, \mathsf{S}_b \rangle \overset{w_b}{\vdash} A$ *and there exists* $r \in \emptyset$ *which contains* not A *in the body. This case is trivial.* □

Corollary 1. *Strict arguments always* strictly defeat *defeasible arguments, including analogical arguments.*

Proof. *This immediately follows from Theorem 1 and Definition 11.* □

Corollary 1 exhibits that any conclusions drawn from analogy cannot strictly defeat conclusions drawn from the use of only strict rules. This shows that our system conforms to how the legal (and similar) uses analogy in reasoning.

3.3 Justification Through Dialectical Analysis

In this work, we base our semantics on the grounded semantics of Dung's theory [9]. A sound and complete calculus under the grounded semantics has a form of the dialectical style between a proponent (P) and an opponent (O) of an argument. A proponent starts with an argument to be justified and then the players take turn. An opponent must defeat or strictly defeat a proponent's last argument while a proponent must strictly defeat opponent's last argument. Moreover, a proponent is not allowed to repeat his arguments.

Definition 12 ([18]). *A* dialogue *is a finite nonempty sequence of moves* $move_i = \langle Player_i, \mathcal{A}_i \rangle$ *where* $i > 0$ *such that:*

1. $(Player_i = P \Leftrightarrow i$ *is odd) and* $(Player_i = O \Leftrightarrow i$ *is even);*
2. $Player_i = Player_j = P$ *and* $i \neq j \Rightarrow \mathcal{A}_i \neq \mathcal{A}_j$;
3. $Player_i = P$ $(i > 1) \Rightarrow \mathcal{A}_i$ *strictly defeats* $\mathcal{A}_{i-1}$;
4. $Player_i = O \Rightarrow \mathcal{A}_i$ *defeats* $\mathcal{A}_{i-1}$.

Definition 13 ([18]). *A* dialogue tree *is a finite tree of moves such that:*

1. *Each path of the tree is a dialogue;*
2. *If* $Player_i = P$, *then children of* $move_i$ *are all defeaters of* $\mathcal{A}_i$.

A player wins a dialogue if there are no moves for another player. Furthermore, a player wins a dialogue tree iff that player wins all paths of the tree.

Definition 14. *An argument* $\mathcal{A} = \langle \mathsf{D}, \mathsf{S}, Q, w \rangle$ *is a* justified argument *from a logic program* $\mathcal{P}$ *(in symbols,* $\mathcal{P} \vdash \mathcal{A}$*) iff there exists a dialogue tree which* $\mathcal{A}$ *appears at the root and is won by the proponent. If* $\mathcal{A}$ *is justified, then* Q *is* *called a* justified conclusion *of* $\mathcal{A}$. *The justification degree, denoted by* $|\mathcal{A}|$, *is defined as follows:*

$$|\mathcal{A}| = \bigoplus_{\mathcal{A}_i = \langle D_i, S_i, Q_i, w_i \rangle} \{w_i\} \text{ , for all } \mathcal{A}_i \text{ of each } P \text{ along the tree and}$$

an accumulator $\bigoplus : [0,1]^n \to [0,1]$ *holds the following properties where* n *is the cardinality of the set* $\{w_i\}$:

- *Identity closed:* $\forall a, \ldots, z \in \{w_i\} : \bigoplus\{a, \ldots, z\} = 1 \Leftrightarrow a = \cdots = z = 1$;
- *Monotonicity:* $\forall a, \ldots, z, a', \ldots, z' \in \{w_i\} : a \le a' \wedge \cdots \wedge z \le z' \Rightarrow \bigoplus\{a, \ldots, z\} \le \bigoplus\{a', \ldots, z'\}$.

Theorem 2. *If an argument* $\mathcal{A} = \langle D, S, Q, w \rangle$ *is justified and its dialogue tree does not use analogical arguments, then* $|\mathcal{A}| = 1$.

Let us illustrate this based on our motivative example. In the following, we define $\otimes$ as the *min* operator and $\bigoplus$ as the *arithmetic mean*.

Example 5 (Continuation of Example 4). There are 10 arguments from the logic program, i.e. $\mathcal{A}_1 = \langle\{danger(X) \Leftarrow lion(X)\}, \{plutonium\_plant \overset{0.8}{\Leftarrow} lion\}, danger(p_1), 0.8\rangle$, $\mathcal{A}_2 = \langle\{green\_environment(X) \Leftarrow wind\_turbine(X)\}, \{plutonium\_plant \overset{0.4}{\Leftarrow} wind\_turbine\}, green\_environment(p_1), 0.4\rangle$, $\mathcal{A}_3 = \langle\{danger(X) \Leftarrow lion(X)\}, \emptyset, danger(l_1), 1\rangle$, $\mathcal{A}_4 = \langle \mathcal{A}_3^D, \emptyset, defendant(guy), 1\rangle$[7], $\mathcal{A}_5 = \langle \mathcal{A}_1^D, \mathcal{A}_1^S, defendant(kerr\_mcgee), 0.8\rangle$, $\mathcal{A}_6 = \langle \mathcal{A}_4^D \cup \{liable(X) \Leftarrow defendant(X), not\ exception(X)\}, \emptyset, liable(guy), 1\rangle$, $\mathcal{A}_7 = \langle \mathcal{A}_1^D \cup \{liable(X) \Leftarrow defendant(X), not\ exception(X)\}, \mathcal{A}_1^S, liable(kerr\_mcgee), 0.8\rangle$, $\mathcal{A}_8 = \langle\{federal\_law(X) \Leftarrow owner(X,Y), plutonium\_plant(Y)\}, \emptyset, federal\_law(kerr\_mcgee), 1\rangle$, $\mathcal{A}_9 = \langle \mathcal{A}_8^D \cup \{\neg liable(X) \Leftarrow federal\_law(X)\}, \emptyset, \neg liable(kerr\_mcgee), 1\rangle$, $\mathcal{A}_{10} = \langle \mathcal{A}_2^D, \mathcal{A}_2^S, \neg danger(p_1), 0.4\rangle$.

We can show that $\mathcal{A}_7$ is a justified argument through the dialectical analysis by starting a dispute with $\mathcal{A}_7$. Now, the opponent has to defeat this argument. However, there are no arguments defeating $\mathcal{A}_7$. Since the opponent runs out of moves, $\mathcal{A}_7$ is a justified argument with the degree $|\mathcal{A}_7| = \frac{0.8}{1} = 0.8$. $\square$

The following definition identifies components for the implementation of an analogical reasoner. Intuitively, we would like to make the reasoner possible to also fine tune with a user-defined similarity threshold. Hence, we include a *relevant degree* s where $s \in (0,1]$ apart from the logic program.

Definition 15 (Analogical Reasoner). *An analogical reasoner* $\mathcal{R}$ *is a pair* $\langle \mathcal{P}, s \rangle$ *where* $\mathcal{P} = \langle SR, DR, SIM \rangle$ *is a logic program and* $s \in (0,1]$ *is a relevant degree. An argument* $\mathcal{A}$ *is justified by* $\mathcal{P}$ *under* s *iff* $\mathcal{A}$ *is a justified argument of* $\mathcal{P}'$ *where* $\mathcal{P}' = \langle SR, DR, SIM' \rangle$ *and* $SIM' = \{L_0 \overset{x}{\Leftarrow} L_1 \subseteq SIM : x \ge s\}$.

---

[7] For the sake of succinctness, $\cdot^D$ (and $\cdot^S$) indicates a duplicated set of defeasible rules (and similarity rules, respectively) from a specified argument structure.

4 Guideline of Choosing Operator $\otimes$ and $\oplus$

The following theorems are aids to help deciding which operator $\otimes$ and $\oplus$ to choose. We remind that $\otimes$ determines the analogy degree of an argument and $\oplus$ determines the overall degree of a justified conclusion.

Theorem 3. *From Table 1 and let* $x_1, x_2 \in (0, 1]$. *Then,* $\otimes_{prod} \leq \otimes_{H_0} \leq \otimes_{min}$.

Proof *(Sketch).* We show the following inequality:

$$x_1 \cdot x_2 \leq \frac{x_1 \cdot x_2}{x_1 + x_2 - x_1 \cdot x_2} \leq \min\{x_1, x_2\}$$

That is, we show $x_1 \cdot x_2 \leq \frac{x_1 \cdot x_2}{x_1 + x_2 - x_1 \cdot x_2}$ *as follows:*

$$x_1 \cdot x_2 \leq \frac{x_1 \cdot x_2}{x_1 + x_2 - x_1 \cdot x_2} \Leftrightarrow 1 \leq \frac{1}{x_1 + x_2 - x_1 \cdot x_2} \Leftrightarrow x_1 + x_2 - x_1 \cdot x_2 \leq 1$$

$$\Leftrightarrow x_2 - x_1 \cdot x_2 \leq 1 - x_1 \Leftrightarrow (1 - x_1) \cdot x_2 \leq 1 - x_1 \Leftrightarrow x_2 \leq 1 \, (by \, assumption)$$

Lastly, we show $\frac{x_1 \cdot x_2}{x_1 + x_2 - x_1 \cdot x_2} \leq \min\{x_1, x_2\}$ *in the similar fashion.* $\square$

Table 2. Some instances of the operator $\oplus$

Name	Notation	$x_1 \oplus \cdots \oplus x_n =$
Minimum	$\oplus_{min}$	$\min\{x_1, \ldots, x_n\}$
Product	$\oplus_{prod}$	$x_1 \cdot x_2 \cdots \cdots x_n$
Average	$\oplus_{avg}$	$\frac{x_1 + \cdots + x_n}{n}$

Theorem 4 *From Table 2 and let* $x_1, \ldots, x_n \in (0, 1]$. *Then,* $\oplus_{prod} \leq \oplus_{min} \leq \oplus_{avg}$.

Proof *(Sketch).* We show $x_1 \oplus_{prod} \cdots \oplus_{prod} x_n \leq x_1 \oplus_{min} \cdots \oplus_{min} x_n \leq x_1 \oplus_{avg} \cdots \oplus_{avg} x_n$ *in the similar fashion of Theorem 3 by induction on* n. $\square$

Theorems 3 and 4 shows ordering of those instances of $\otimes$ and $\oplus$. This suggests that we can choose specific operators based on an application domain. For an analogical reasoner which strongly recognizes analogical principles, we may choose the weakest operators for both (i.e. $\otimes_{min}$ and $\oplus_{avg}$). On the other hand, we may choose the strongest ones (i.e. $\otimes_{prod}$ and $\oplus_{prod}$) for an analogical reasoner which weakly recognizes analogical principles. We generalize this observation toward the nature of pessimistic in analogical reasoning. For the sake of succinctness, we may simply denote the chosen operators with superscripts, e.g. $\mathcal{P}^{\otimes_{min}, \oplus_{avg}}$ and $|\cdot|^{\otimes_{min}, \oplus_{avg}}$.

Definition 16. *Let $\mathcal{P}$ be a logic program and $\otimes_1, \otimes_2, \oplus_1, \oplus_2$ be concrete operators. Let $\mathcal{A}^*$ be the set of all arguments from $\mathcal{P}$. Then, $\mathcal{P}^{\otimes_1,\oplus_1}$ is more* pessimistic *than $\mathcal{P}^{\otimes_2,\oplus_2}$ if $\forall a \in \mathcal{A}^* : (\mathcal{P}^{\otimes_1,\oplus_1} \vdash a$ and $\mathcal{P}^{\otimes_2,\oplus_2} \vdash a \Rightarrow |a|^{\otimes_1,\oplus_1} < |a|^{\otimes_2,\oplus_2})$.*

Dually, the nature of optimistic analogical reasoning is defined in the opposite direction, i.e. $\mathcal{P}^{\otimes_1,\oplus_1}$ is more *optimistic* than $\mathcal{P}^{\otimes_2,\oplus_2}$ if $\forall a \in \mathcal{A}^* : (\mathcal{P}^{\otimes_1,\oplus_1} \vdash a$ and $\mathcal{P}^{\otimes_2,\oplus_2} \vdash a \Rightarrow |a|^{\otimes_1,\oplus_1} > |a|^{\otimes_2,\oplus_2})$.

5 Sketch of Analogist: A Hybrid Analogical Reasoner

Analogist is an inference system which reasons from analogy. Basically, it is a hybrid reasoner of DL-based similarity measures and the argument-based logic program. We discusses a design sketch of the system in this section. It is also worth to mention that the argument from analogy has a great importance in legal practice, both in common law (because of the *stare decisis* principle, which implies that a court should use precedents to guide new decisions) and in civil law [19]. Unfortunately, this canon is prohibited in criminal law as it is the shared idea that judges shall not create new law in criminal matters.

The framework presented in Sect. 3 supplies the mechanism to define knowledge base in declarative ways. Taking this as advantage, we employ concept similarity measure techniques in DLs (cf. Subsect. 2.2) to induce similarity rules used in Analogist. Figure 1 depicts this conceptual idea.

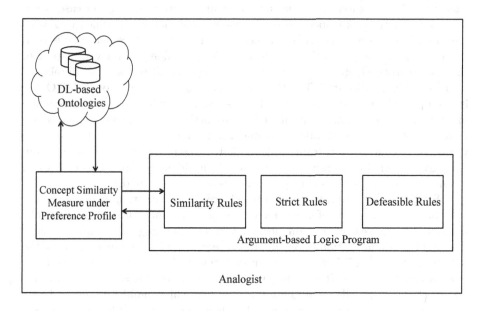

Fig. 1. The design of Analogist

With the use of DLs, the semantics of each axiomatic information is formally defined. Traditional work on analogical reasoning indicates that, to measure similarity between two situations, both mapping of relation among objects and mapping of individual objects should take into account. Choosing appropriate measures can remedy this difficulty. For example, the measure sim^π [12] determines the similarity of $\mathcal{ELH}$-concepts based on axiomatic information in TBox.Preference profile (Definition 1) may be also used to define the preferred aspects for similarity identification at stake. For instance, Spence may define a highly importance on Danger and omit to consider other aspects, i.e. $i^c(\mathsf{Danger}) = 2$ and $i^c(A) = 0$ for $A \in \mathsf{CN}^{\mathsf{pri}}(\mathcal{T}) \setminus \{\mathsf{Danger}\}$, and thereby Lion is totally similar to PowerPlant, i.e. Lion $\sim^\pi_{\mathcal{T}}$ PowerPlant $= 1$. This adds a similarity rule $plutonium\_plant \overset{1}{\Leftarrow} lion$ (and vice versa) into the program.

6   Related Work

After surveying the literature on Argument from Analogy in many fields, such as logic, law, philosophy of science, and computer science, it appears to us that there are two different forms of Argument from Analogy. The first form (cf. Subsect. 2.1), on which our work is based, is the most widely accepted version whereas the second one compares factors of two cases (e.g. [20,21]), which may be regarded as an instance of the first form. As the second one makes no reference to the notion of similarity, it becomes simpler to use, such as in standard case-based reasoning. The method of evaluating an argument from analogy in case-based reasoning (CBR) uses respects (i.e. dimensions and factors) in which two cases are similar or different. In CBR, the decision in the best precedent case is then taken as the decision into the current case. A dimension is a relevant aspect of the case whereas a factor is an argument favoring one side or the other in relation to the issue being disputed. The HYPO system [20] uses dimensions. CATO [21] is a simpler CBR system that uses factors. Systems which employ factors use pro factors to represent similarities for supporting an argument whereas con factors representing dissimilarity to undermine the argument. Factors may be weighted. In contrast, our framework formalizes the first form of Argument from Analogy and Analogist employs $\sim^\pi_{\mathcal{T}}$ for similarity identification. This creates an advantage, i.e. many aspects of preference profile can be used.

ASPIC+ [22] is also a framework which constructs an argumentation framework based on the notion of proof tree with strict rules and defeasible rules. In our literature, there are two common methodologies for building arguments, i.e. proof tree and the similar notion to our approach (e.g. [15–17]). The motivation of the notion used in [15–17] is to discover inconsistency in the strict knowledge base. For instance, let $\mathcal{P} = \langle \mathsf{SR}, \mathsf{DR}, \mathsf{SIM} \rangle$ be a program where $\mathsf{SR} = \{a \leftarrow b; \neg a \leftarrow b\}$, $\mathsf{DR} = \{b \Leftarrow\}$, and $\mathsf{SIM} = \emptyset$. Those approaches which build arguments based on proof tree will accept an argument $\{b\}$; however, there will be no arguments for those systems similar to us since accepting b will derive inconsistency in the strict knowledge base through $a \leftarrow b$ and $\neg a \leftarrow b$.

There are also substantial efforts of linking analogical reasoning to existing logical models of non-monotonic reasoning. For example, [8] proposes a form of analogical reasoning based on hypothetical reasoning. In that approach, similarity is expressed as an equality hypothesis and a goal-directed theorem prover is used to search relevant hypotheses. In [6,7], an analogical reasoning is considered as deductive reasoning by inserting the rule (in our language): $has\_property(t, p) \leftarrow has\_property(s, p), similar(s, t)$ as a strict rule to be used in deriving analogical conclusions. Definition 7 also uses a rule which functions similar to the above. However, our approach is different to those on constraint checking and similarity identification. On the one hand, existing logical approaches require consistency on logic programs and uses the number of common properties (in the model theory) for similarity identification. On the other hand, our approach relies on the notion of counter-analogy and uses similarity rules to define similarity of two predicates. Using only consistency for constraint checking is not enough. We exemplify this case in our system language. For instance, let $\mathcal{P} = \langle \mathsf{SR}, \mathsf{DR}, \mathsf{SIM} \rangle$ be a program where $\mathsf{SR} = \{plutonium\_plant(p)\}$, $\mathsf{DR} = \{danger(X) \Leftarrow lion(X); \neg danger(X) \Leftarrow solar\_plant(X)\}$, and $\mathsf{SIM} = \{plutonium\_plant \Leftarrow lion; plutonium\_plant \Leftarrow solar\_plant\}$. Most existing approaches conclude either $danger(p)$ or $\neg danger(p)$ from logic programs. However, our approach concludes nothing as a counter-analogy is discovered. Using similarity rules is also a pro since it enables integration with external systems for building similarity rules, as designed in Analogist. This invokes more dimensions of configurable aspects for evaluating the similarity of situations.

7   Conclusion

In this paper, we make contributions to the logical study of analogical reasoning under the lens of structured argumentation. The key idea behind our approach lies in the scheme Argument from Analogy proposed by Walton [3]. As a result, this work provides the possibility of representing information in the form of strict, defeasible, and similarity rules in a declarative manner (cf. Sect. 3). Critical questions used in the argumentation scheme are reconfigured as defeating relation; thereby the acceptability of an analogical argument is evaluated. The proposed framework also makes it possible to integrate with external systems for forming similarity rules and is also flexible to be tuned by users. That is, the chosen operators for $\otimes$ and $\oplus$ are dependent on an application, e.g. optimistic analogical reasoning and pessimistic analogical reasoning (cf. Sect. 4).

We also present a design sketch of Analogist (cf. Sect. 5) which is a hybrid analogical reasoner based on our proposed framework. The argument from analogy has a great potential in common law and civil law [19]. Thus, it would be interesting to capture real legal cases and apply Analogist as for practical evaluation. Using $\overset{\pi}{\sim}_{\mathcal{T}}$ to create similarity rules also has the advantage, i.e. aspects of preference profile are used for similarity identification.

Finally, we note that the paper focuses on the grounded semantics of Dung's abstract theory. Therefore, it appears to be a natural step to investigate other semantics of Dung's theory in the context of analogical argumentation. Also, similarity rules introduced in the framework are attached with numerical values. Hence, it would be interesting to link the proposed framework with value-based argumentation. We leave these for our theoretical future research directions.

Acknowledgments. This research is supported by JAIST, NECTEC, and SIIT under the dual doctoral degree program; and is partly supported by CILS of Thammasat University and the NRU project of Thailand Office of Higher Education Commission.

References

1. Lief, M.S., Caldwell, H.M., Bycel, B.: Ladies and Gentlemen of the Jury: Greatest Closing Arguments in Modern Law. Scribner, New York (2000)
2. Hofstadter, D., Mitchell, M.: Concepts, analogies, and creativity. In: Proceedings of CSCSI 1988, pp. 94–101, June 1988
3. Walton, D., Reed, C., Macagno, F.: Argumentation Schemes. Cambridge University Press, Cambridge (2008)
4. Waller, B.N.: Classifying and analyzing analogies. Informal Logic **21**(3), 199–218 (2001)
5. Haraguchi, M., Arikawa, S.: Reasoning by Analogy as a Partial Identity between Models. Springer, Heidelberg (1987). pp. 61–87
6. Greiner, R.: Learning by Understanding Analogies. Springer, Boston (1986). pp. 81–84
7. Winston, P.H.: Learning and reasoning by analogy. Commun. ACM **23**(12), 689–703 (1980)
8. Goebel, R.: A Sketch of Analogy as Reasoning with Equality Hypotheses. Springer, Heidelberg (1989). pp. 243–253
9. Dung, P.M.: On the acceptability of arguments and its fundamental role in non-monotonic reasoning, logic programming and n-person games. Artif. Intell. **77**(2), 321–358 (1995)
10. Racharak, T., Tojo, S., Hung, N.D., Boonkwan, P.: Argument-based logic programming for analogical reasoning. In: Proceedings of the 10th International Workshop on Juris-Informatics (JURISIN), pp. 73–86, November 2016
11. Racharak, T., Suntisrivaraporn, B., Tojo, S.: Identifying an Agent's Preferences Toward Similarity Measures in Description Logics. Springer International Publishing, Cham (2016). pp. 201–208
12. Racharak, T., Suntisrivaraporn, B., Tojo, S.: sim^{π}: a concept similarity measure under an agent's preferences in description logic $\mathcal{ELH}$. In: Proceedings of the 8th International Conference on Agents and Artificial Intelligence, pp. 480–487 (2016)
13. Nute, D.: Handbook of Logic in Artificial Intelligence and Logic Programming, vol. 3, pp. 353–395. Oxford University Press Inc, New York (1994)
14. Antoniou, G., Billington, D., Governatori, G., Maher, M.J.: Representation results for defeasible logic. ACM Trans. Comput. Logic **2**(2), 255–287 (2001)
15. Amgoud, L., Cayrol, C.: On the acceptability of arguments in preference-based argumentation. In: Proceedings of the 14th Conference on Uncertainty in Artificial Intelligence, UAI 1998, San Francisco, CA, USA, pp. 1–7. Morgan Kaufmann Publishers Inc. (1998)
16. Elvang-Gøransson, M., Krause, P.J., Fox, J.: Acceptability of Arguments as 'Logical Uncertainty'. Springer, Heidelberg (1993). pp. 85–90

17. García, A.J., Simari, G.R.: Defeasible logic programming: an argumentative approach. Theory Pract. Log. Program. 4(2), 95–138 (2004)
18. Prakken, H., Sartor, G.: Argument-based extended logic programming with defeasible priorities. J. Appl. Non-classical Logics 7, 25–75 (1997)
19. Damele, G.: Analogia Legis and Analogia Luris: An Overview from a Rhetorical Perspective. Springer International Publishing, Cham (2014). pp. 243–256
20. Ashley, K.: Case-Based Reasoning. Springer, Netherlands (2006). pp. 23–60
21. Aleven, V.: Teaching case-based argumentation through a model and examples (1997)
22. Prakken, H., Wyner, A.Z., Bench-Capon, T.J.M., Atkinson, K.: A formalization of argumentation schemes for legal case-based reasoning in ASPIC+. J. Log. Comput. 25(5), 1141–1166 (2015)

Estimating Legal Document Structure by Considering Style Information and Table of Contents

Yoichi Hatsutori[✉], Katsumasa Yoshikawa, and Haruki Imai

IBM Research, Tokyo, IBM Japan Ltd.,
19-21 Nihonbashi Hakozaki-cho Chuo-ku, Tokyo 103-8510, Japan
{yoichih,katsuy,imaihal}@jp.ibm.com
http://www.research.ibm.com/labs/tokyo/index_j.shtml

Abstract. Text analytics is used to analyze diverse documents. For example, legal documents (such as contracts, ordinances, regulations, and global standards) must be analyzed for corporations to manage their business risk and meet compliance requirements. However, since documents are often stored or published as documents without a common structure, they need to be preprocessed to analyze them in subsequent text analytics. In particular, the following two forms of preprocessing are useful for text analytics: (1) extracting text, and (2) estimating document structure (such as chapters, sections, and subsections), which is used to define the range of topics or articles in a document. This paper presents a preprocessing method to estimate document structure from documents without a common structure. The proposed method follows rule-based approach, and consists of three algorithms: (1) one is based on style information, such as bold font; (2) another is based on numbered objects, such as sections; and (3) the other is based on a document's Table of Contents, which summarizes the document's structure. The accuracy of the proposed method is also evaluated by using 102 documents. The proposed method was found to be able to estimate document structure with 96.6% accuracy.

Keywords: Document structure · Article extraction · Article comparison · Text analytics · Law articles

1 Introduction

Electronic text documents (e.g. office documents, web contents, articles, and technical papers) are widely read and available online. The growth in available electronic text documents has been accompanied by active research on text analytics, like natural language processing (NLP) and text mining. Text analytics is important to analyze legal documents (such as contracts, ordinances, regulations, or global standards), for example, so that corporations can identify, evaluate, and manage their business risk and meet compliance requirements. However, since

© Springer International Publishing AG 2017
S. Kurahashi et al. (Eds.): JSAI-isAI 2016, LNAI 10247, pp. 270–283, 2017.
DOI: 10.1007/978-3-319-61572-1_18

most documents are stored as documents without a common structure, they need to be converted into a structured format for text analytics. For example, this paper has user-defined document structure such as "1 Introduction", "2 Definitions of Document Structure", and "3 Related Work". However, there are no links to document structure in the PDF files. Thus, processing diverse documents without a common structure raises challenges in the preprocessing phase of text analytics.

Key tasks in the preprocessing phase of text analytics are; (1) extracting text from documents, and (2) estimating document structure (chapters, sections, subsections, and paragraphs), which is used to define the range of topics in a document. For example, Mao et al. [6] mentioned that structural information can be very useful in indexing and retrieving information contained in a document. Pasetto et al. [7] pointed out that document structure needs to be obtained to be able to compare semantics in legal document analysis.

For extracting text, open source libraries can be used to extract text descriptions and style information from diverse documents. For example, Apache Tika[1] can extract text descriptions and style information from over 1,000 different file types. Therefore, we focus on the other task: estimating document structure.

To estimate document structure, estimation algorithms need to consider characteristics of diverse documents without a common structure. In addition, the definition of document structure can differ even if documents are stored in the same file format because ways to define document structure vary in accordance with the document's objective, its writers, and so on. Thus, document structure should be estimated from contents of a document, e.g. text description or style information.

We have already proposed a method to estimate document structure from text description, which is the unique common information among diverse documents. [2] The method is rule-based and estimates document structure by searching for numbered objects like "Sect. 1." and "Sect. 2." or "(1)," "(2)," and "(3)" as shown in Fig. 1. This method is applicable when document structure is defined by numbered objects. However, while some documents use numbered objects to define document structure, others use style information like **bold** or *italic* font as shown in Fig. 2. When document structure is defined by style information, our previous approach cannot estimate document structure. In addition, we also consider the "Table of Contents," which outlines a document structure, as shown in Fig. 3, since it is another clue for estimating document structure.

The contribution of this paper is to expand the method described in the previous study [2] and present a novel preprocessing method considering style information, numbering information, and the Table of Contents. The proposed method is designed to preprocess diverse documents without a common structure and enables us to add structure information to them.

The rest of this paper is as follows. Section 2 introduces types of document structure definition. Section 3 summarizes related work on estimating document structure. Section 4 presents our new preprocessing method. Section 5 evaluates

---

[1] https://tika.apache.org/.

and discusses the accuracy of the proposed method. In Sect. 6, the proposed method is used to analyze some Terms of Use published on websites of web-service companies. Finally, Sect. 7 concludes this paper.

2   Definitions of Document Structure

There are two ways to define document structure. One is to insert an additional text which describes structural information. The other is to change the style of text.

In the first way, serial numbers are used for defining document structure. For example, numbered objects such as "1." and "2." or numbered objects with a prefix such as "Chap. 1" and "Chap. 2" are added before each chapter, section, or subsection. Figure 1 shows an example in which additional text (i.e. "Sect. 1" and "Sect. 2") is used for structural definition.

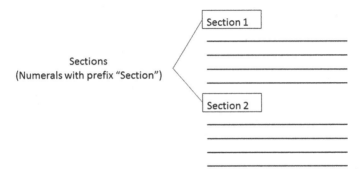

Fig. 1. Example of structural definition by additional text

Changing style is another way to define structure. For example, font size, font style (e.g. bold, and italics), alignment, indent, and so on can be used. Figure 2 shows an example in which bold font is used to define the document structure.

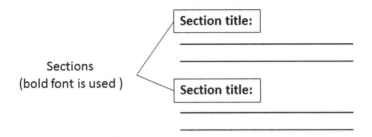

Fig. 2. Example of structural definition by style information

In addition, some documents have a Table of Contents, which outlines document structure. Since the structure shown in a Table of Contents is the same as that of the body text, this information is useful for estimating document structure. Figure 3 shows an example in which "Article 1" and "Article 2" are defined in body text and "Table of Contents" shows the same structure.

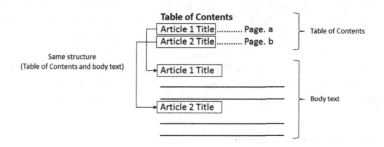

Fig. 3. Example of Table of Contents and body text

In this paper, both additional text and style information are considered to estimate document structure. Moreover, "Table of Contents" are also used for improve the results of the estimation. Details are described in Sect. 4.

3   Related Work

As described in Sect. 2, there are many ways to define document structures and algorithms to estimate them. In this section, related work is summarized and the originality of the paper over the related work is described. The first approach is based on text description. The second approach is based on additional information attached to text, e.g. font information or tag information describing structural information. The third approach is based on syntactic rules.

3.1   Text-Based Approach

Iwai et al. [4] proposed a text-description-based approach to extract document structure. In their research, pattern match is used for each line to extract document structure. Each line is assumed to conform to their knowledge-based model called the "Intra-line Structure Model," which consists of heading, delimiter, and content parts. They prepared about 300 patterns for the heading part, which consists of a numeric part, punctuation, and a reserved heading word. If a line has a heading part that is matched to a predefined pattern, then the line is extracted as a start point of document structure. The average accuracy of their structure extraction is 89.0%. However, their intra-line structure model assumes that only the heading part can be the head of a document structure and that reserved heading words are required for detecting structure.

Dejean [1] proposed a robust method to detect long-range numbered sequences in different types of documents. He specifically focused on scanned, optical character recognition (OCR), and PDF documents. Only text description is used for estimating document structure. In Dejean's research, incremental types (e.g. digits, upper- and lower-case letters, Roman letters, Chinese characters, or "Bis" sequences) are detected and extracted by predefined regular expressions. The document is scanned line by line, and numbered objects are extracted by pattern match. For every match, a pattern is associated with the line. A numbered object is recognized as a header part of the line, and the remaining part is recognized as the body part of the line. Document structure is estimated from extracted objects. The originality of this research is the notation of missing items (often due to OCR errors), and the notation of multi-increment items. This research also addresses intra-line structure, like that of Iwai et al. [4].

3.2 Markup-Based Approach

Some documents have useful information for estimating document structure. For example, "heading" tags are available in Hyper Text Markup Language (HTML). If a document has meaningful tag information describing structural information, document structure can be extracted from the document. Some open source libraries are available for extracting document structure from markups. For example, Apache Tika can extract metadata and text. However, users can use not only heading tags but also other tags such as style information, i.e. bold font as shown in Fig. 2, for defining document structure. Therefore, even if a document contains markups, an algorithm to estimate document structure from style information is required for analyzing diverse documents.

3.3 Syntactic Approach

Igari et al. [3] proposed an approach based on syntactic rules. They claim that legal judgments are described in a common format, often have structures that are highly similar, and often include the same type of information in similar parts. [3] When documents are described in a common format, the document structure can be expressed by syntax rules for the document text. Because Igari et al. focus on analyzing and parsing legal judgments, which have a well-defined common format, their method has generality, extensibility, and high accuracy. This approach is very useful and highly accurate for documents that have a common format. However, we assume that diverse documents do not share a common format.

3.4 Scope of This Paper

As described in Subsect. 3.3, we focus on analyzing diverse documents. This means that there are no common structural formats in documents, so the syntactic approach is not considered but the text-based and markup-based approaches are. In particular, style information is considered in the markup-based approach.

In addition, there are two ways for extracting document structure: a machine learning approach and a rule-based approach. In the machine learning approach, a large amount of training data is required to train a model for estimating document structure. However, preparing training data is an obstacle for practical use. On the other hand, the rule-based approach does not need training data to be prepared. Therefore, the rule-based approach is used for extracting document structure. Since this paper focuses on a method to estimate document structure, we assume that text is extracted from diverse documents preliminarily.

4   Proposed Method

This section describes a three-algorithms method to estimate document structure from documents without a common structure. As mentioned in Sect. 2, there are two types of definitions of document structure: document structure is defined by additional text or style information. Since the proposed method considers these two definitions, the method includes numbering-based and style-based algorithms. In addition, since some documents have a Table of Contents describing document structure, the proposed method also utilizes this additional information to estimate the document structure. Subsect. 4.1 details the style-based algorithm, Subsect. 4.2 the numbering-based algorithm, and Subsect. 4.3 the Table-of-Contents-based algorithm. Then the whole processing flow and some configuration parameters are described in Subsect. 4.4.

4.1   Style-Based Algorithm

This subsection describes the style-based algorithm to estimate document structure. This algorithm can be used when a given document structure is defined in terms of style information.

First, users define style information for each document structure. For example, correspondence such as bold for sections or italics for subsections are defined by users.

Second, text having the same style information is searched for in a document. When a predefined style is detected in the body text, the text is extracted as a head of each document structure. Note that searching text is processed line-by-line. When style information matches only a part of a line, the line is skipped.

Third, text belonging to each section is estimated. Basically, text at the head of a section is assumed to be text of that section. Similarly, text at the end of a document is assumed to be text of the last section. For subsections, a similar algorithm is used to estimate the range of subsections. Text at the head of a subsection is assumed to be a text of that subsection. Text at the end of the section is assumed to be text of the last subsection.

Figures 4 and 5 show examples of style-based estimation. In these figures, bold is used for sections, and italics for subsections. These relationships are predefined by users. In Fig. 4, two instances of "Section title" in bold are extracted as a section-level structure. Next, text from the first head to the second is extracted

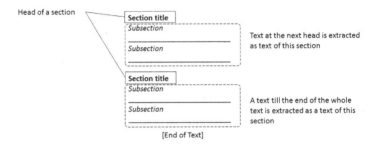

Fig. 4. Example of style-based estimation (section level)

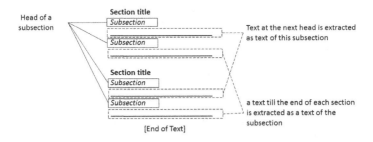

Fig. 5. Example of style-based estimation (subsection level)

as text of the first section. Text from the second head to the end of a document is extracted as text of the second section. In Fig. 5, four instances of "Subsection" in italics are extracted as a subsection-level structure. Text at the head or the end of the subsection is extracted as a text of the subsection.

4.2 Numbering-Based Algorithm [3]

This subsection briefly introduces a numbering-based algorithm to estimate document structure [2] that can be used when document structure is defined by numbered objects. First, regular expressions to search for numbered objects are predefined by users. Next, numbered objects matching regular expressions are searched for in a document. Tree structures are built by all candidates of the searched objects, and a pruning technique is applied to them to remove unnecessary leaves.

An example is shown in Fig. 6. In the left image, numbered objects are extracted by predefined regular expressions: "$\backslash d \backslash.$" and "$\backslash(\text{[a-z]}\backslash).$" In the center image, extracted objects are classified and grouped by the left-hand-side characters. Candidates of tree structures are built in each group and leaves of the tree are pruned. In this example, since all candidates are unbranched trees, the pruning technique is not needed. Then, a super-sub relationships among unbranched trees are estimated in the right image. Since (a) and (b) are covered by 1. and 2., we can estimate (a) and (b) to be a lower level structure than 1. and 2.

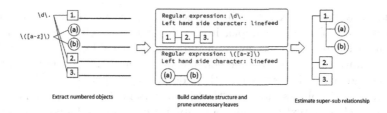

Fig. 6. Example of numbering-based estimation

4.3 Table-of-Contents-Based Algorithm

As mentioned in Sect. 2, some documents have a Table of Contents to outline document structure. Since the structure in a Table of Contents is the same as that of body text, document structure of body text can be estimated from the Table of Contents. The merits of using a Table of Contents are (1) document structure can be estimated even if document structure is not defined by either style information or numbered objects and (2) the results estimated by style- and numbering-based algorithms can be verified. Therefore, this subsection explains an algorithm that uses a Table of Contents to estimate document structure. This algorithm consists of four steps.

In the first step, the following parameters are predefined to estimate a Table of Contents and its range: (1) key-words representing the head of a Table of Contents are defined, such as Table of Contents, Contents, and Summary; (2) regular expressions to detect each line contained in a Table of Contents, for example, "(Article\s\d)([a-zA-Z0-9\s]+)(\s\.+\sPage\.\s[ab])," are defined as in the case in Fig. 3; (3) style information described in Subsect. 4.1 is defined to detect document structure; and (4) regular expressions described in Subsect. 4.2 are defined to analyze numbered objects in a Table of Contents. Note that regular expressions defined in (2) consist of three parts. The first is the prefix text part. The text in this part can be used in the body text. In Fig. 3, "Article 1" and "Article 2" are used in the body text and the regular expression "(Article \s\d)" is the first part of the regular expression. The second is the main text part. This is mainly used as a title of each section. The third is the additional text part. The text in this part is mainly used in the Table of Contents only. For example, in Fig. 3, "........ Page. a" and "........ Page. b" are the additional text part describing a page number and used only in the Table of Contents.

In the second step, the scope of the Table of Contents is estimated by predefined parameters: (1) key-words and (2) regular expressions. The text matching the prefix and main text parts of regular expressions is extracted from the Table of Contents. In Fig. 3, for example, "Article 1 Title" and "Article 2 Title" are extracted from the Table of Contents.

In the third step, text in the Table of Contents is analyzed by predefined parameters: (3) style information for document structure and (4) regular expressions for numbered objects. As a result, document structure in the Table of Contents is estimated.

Finally, by the text extracted in the second step, corresponding text in body text is searched for. For example, in Fig. 3, "Article 1 Title" and "Article 2 Title" are extracted in the second step and then, searched for in body text. Note that we empirically find that some documents use the same text in the Table of Contents and the body text while others do not, e.g. "Article 1 Title" appears in the Table of Contents but only "Title" appears in the body text. Therefore, first, we search for a text containing a prefix. If we cannot find any such sections, the text without a prefix is searching to find corresponding text.

4.4   Processing Flow and Configuration Parameters

This subsection explains processing flow of the whole process and then describes necessary configuration parameters.

Since a Table of Contents is an outline of document structure, the Table-of-Contents-based algorithm is applied to estimate the document structure first. If we cannot extract any document structure from a Table of Contents, e.g. a document does not have a Table of Contents, the style- and numbering-based are applied to estimate the document structure. Here, to select the style- or numbering-based algorithm, a configuration parameter is used.

Configuration parameters required to estimate document structure are listed in Table 1. These parameters should be predefined by users. In our implementation, a user can set configuration parameters for each document. In this study, we predefined about 70 regular expressions, e.g. "$\backslash d\backslash.$", "$\backslash[a-z]\backslash)$", "$[a-z]\backslash.$", and "$[Se][Ee][Cc][Tt][Ii][Oo][Nn]\ \backslash s[0-9]+\backslash.?\backslash s$", to detect numbering objects used in diverse documents.

Table 1. Configuration parameters required to estimate document structure

Parameters	Description
Algorithm selection	Select style-based or numbering-based
Key words for "Table of Contents"	Search for the starting point of Table of Contents
Regular expressions	Search for lines in Table of Contents
Style information	Estimate document structure in Table of Contents
Regular expressions	Estimate document structure in Table of Contents
Style information	Estimate document structure in body text
Regular expressions	Estimate document structure in body text

5   Evaluation

In this section, the proposed three-algorithms method is evaluated using legal documents available online. First, a correct dataset, i.e. ground truth, is created manually. Next, a document structure is estimated by the proposed method. Third, the result of document structure estimation is compared with the correct data set. We used some publications of US Federal Register, IT Booklet of

Table 2. Statistics of documents used in the evaluation

Definition of document structure	Number of documents	Number of sections
Style information	12	292
Numbering	90	2430
Total	102	2722

Table 3. Confusion matrix [5] and meaning of each cell

Actual	Predicted	
	Negative	Positive
Negative	True Negative (TN)	False Positive (FP)
Positive	False Negative (FN)	True Positive (TP)

FFIEC (The Federal Financial Institution Examination Council), and European Central Bank etc. Types of input legal documents, the number of documents, and the number of sections are listed in Table 2. As shown in Table 2, we use 102 documents for the evaluation. Here, the accuracy and the true positive rate are evaluated using legal documents. The confusion matrix [5], showing the predicted and actual classifications, is considered in the evaluation. A confusion matrix and the meaning of each cell are shown in Table 3.

Kohavi and Provost [5] define the following terms for a 2×2 confusion matrix;

- Accuracy: $(TN + TP)/(TN + FP + FN + TP)$
- True positive rate (recall, sensitivity): $TP/(FN + TP)$

In this study, "true positive" means that the proposed method estimates a correct section. "False positive" means that the proposed method estimates an incorrect section. "False negative" means that the proposed method does not estimate a correct section. "True negative" means that the proposed method does not estimate an incorrect section. In accordance with the above definitions, the accuracy and the true positive rate are calculated to evaluate the proposed method. The confusion matrix generated from the results of document structure detection is shown in Table 4. As a result of evaluation with 102 documents, the proposed method estimated 2,669 sections including 20 false positives. However, it failed to estimate 73 sections that should have been estimated.

Table 4. Results of document structure detection

Actual	Predicted	
	Negative	Positive
Negative	0	20
Positive	73	2649

As a result, the accuracy and the true positive rate are calculated as follows:

- Accuracy: $(0 + 2649)/(0 + 20 + 73 + 2649) = 96.6\%$
- True positive rate (recall, sensitivity): $2649/(73 + 2649) = 97.3\%$

False positives were mainly occurred at style-based algorithm. Some captions of figures and tables, which have only style information, such as bold style, and do not have numbered object, such as "Fig." or "Table", are recognized as a title. On the other hand, false negatives are caused by skipped numbers. Some documents have irregular numbering, e.g. "Sect. 1", "Sect. 2" and "Sect. 4". "Section 4" is not recognized as an element of document structure.

6   Application

In this section, the proposed method is applied to analyze legal documents. We focused on analyzing and comparing some Terms of Use of web services. Terms of Use is a legal document, so comparing some Terms of Use among web services is useful for obtaining characteristics of each service, comparing each service, and identifying risks of the service. However, Terms of Use published online have two types of document structure definitions: some use numbering-based format while others use style-based format. We can assume these Terms of Use to be examples of diverse documents without a common structure. The proposed method is used to analyze both types of definitions.

First, in Subsect. 6.1, sample Terms of Use are analyzed by the proposed method and articles defined in them are extracted. Second, in Subsect. 6.2, extracted articles are compared among web services. Through this application, the effectiveness of proposed method is discussed.

6.1   Sample Terms of Use and Results of Document Structure Estimation

In this section, the Terms of Use of five web services are targeted. Two of them use style-based format while others use numbering-based format. Parts of Terms of Use are shown in Fig. 7. In Fig. 7, one uses style-based format while another uses numbering-based format. In both, each article has a title and its description.

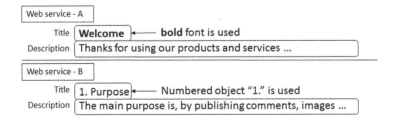

Fig. 7. Part of terms of use of two web-services

Table 5. Results of document structure estimation

	Truth	Extracted (TP)
Web service - A	12	12
Web service - B	13	13
Web service - C	16	16
Web service - D	16	16
Web service - E	22	22

First, heads of each section are detected by predefined style or regular expressions for numbering objects. For example, the regular expression "$([\s\n\r]+|^)([0-9]+)(\.\s+)$" is used for numbering-based format. Bold fonts are searched in Terms of Use written in style-base format. The results of document structure estimation are shown in Table 5. We can see that both style- and numbering-based algorithms can estimate document structure correctly.

6.2 Comparison of Articles

In this subsection, articles extracted in the previous subsection are compared. First, each article extracted by the proposed method is represented by a Vector Space Model with TF-IDF term weighting [8,10]. Second, an article is selected

Table 6. A part of comparison table

		Query article	Similar article
Result 1	Company	Web service - E	Web service - C
	Title	Disputes	Disputes between You and Other Users
	Description	Any dispute or claim relating in any way to your use of any our services will be resolved by binding arbitration … Conflicts that arise from the Service will be governed primarily under the exclusive jurisdiction of the District Court of Tokyo or the Tokyo Summary Court	You are solely responsible for your interaction with other users of the Service and other parties that you come in contact with through your use of the Service …
Result 2	Company	Web service - E	Web service - D
	Title	Disputes	Governing Law and Jurisdiction
	Description	Any dispute or claim relating in any way to your use of any our services will be resolved by binding arbitration … Conflicts that arise from the Service will be governed primarily under the exclusive jurisdiction of the District Court of Tokyo or the Tokyo Summary Court	These Terms and Conditions will be governed ny the laws of Japan. Conflicts that arise from the Service or conflicts between Users and the Company related to the Service will be governed primarily under the exclusive jurisdiction of the District Court of Tokyo or the Tokyo Summary Court

as a query and similar articles are searched for by calculating cosine similarity [9] between vectorized articles in other services. Here, the range of similarity scores is 0.0–1.0, and higher scores are better. Since there is no common format of Terms of Use, contents of Terms of Use or granularity of articles can differ. Therefore, we use two types of clues to obtain similar articles: a title and its text description. This means that similar articles are searched for by title-title similarity and text-text similarity. Finally, similar articles are aligned to the query article, and a comparison table is created.

A part of the comparison table is shown in Table 6, which lists two results. Note that a part of description is shown to sanitize in Table 6. The query article is the same. However, a similar article searched for by its title is aligned in the "result 1" row, and a similar article searched for by its description is aligned in the "result 2" row. In "result 1", although both articles have similar titles, descriptions are different from each other. For example, "Web service - E" mentions its jurisdiction, "Web service - C" claims disputes between users. On the other hand, in "result 2", though two articles have different titles, both descriptions claim their jurisdiction. Actually, "Web service - E" is related to online shopping, while "Web service - C" is related to social network service. Then, the characteristics of web services are appeared in comparison table.

7 Conclusion

Active research on text analytics is accompanying the growth in available electronic documents. One critical challenge for obtaining meaningful insights from these document in text analytics is preprocessing of diverse documents without a common structure. In this paper, we presented a novel preprocessing method to estimate document structure that uses numbering-based and style-based algorithms. In addition, a new Table-of-Contents-based algorithm was also used. Proposed method refers style information, numbering object, and Table-of-Contents. These clues are commonly used for documents, and do not depend on domains. It means that our method can be applied to not only legal domain, but also other domains.

The proposed three-algorithms method was evaluated by 102 documents, which had 2722 sections, and both numbering-based and style-based definitions of document structure were used. The proposed method was found to be able to analyze both types of document structure with high accuracy (96.6%) and a high true positive rate (97.3%). It is thus accurate enough to be used for practical use.

In addition, one application for analyzing legal documents was demonstrated. Terms of Use published online on five web services were analyzed by the proposed method, a Vector Space Model with TF-IDF term weighting, and a cosine similarity score. The advantages of proposed method are (1) it estimates both numbering- and style-based document structure and (2) based on the results of document structure detection, it extracts article title and its test description, information that is useful for analyzing diverse documents.

As a result, we conclude the proposed method can preprocess diverse documents without a common structure for text analytics.

References

1. Dejean, H.: Numbered sequence detection in documents. In: Proceedings of SPIE - The International Society for Optical Engineering, vol. 7534, p. 753405 (2010). doi:10.1117/12.839494
2. Hatsutori, Y., Yoshikawa, K.: A method to estimate document structure from text document and its application to law articles. In: The Proceedings of the Ninth International Workshop on Juris-informatics (JURISIN 2015), Kanagawa, Japan, pp. 69–82 (2015)
3. Igari, H., Shimazu, A., Ochimizu, K.: Document structure analysis with syntactic model and parsers: application to legal judgments. In: Okumura, M., Bekki, D., Satoh, K. (eds.) JSAI-isAI 2011. LNCS, vol. 7258, pp. 126–140. Springer, Heidelberg (2012). doi:10.1007/978-3-642-32090-3_12
4. Iwai, I., doi, M., Yamaguchi, K., Fukui, M., Takebayashi, Y.: A document layout system using automatic document architecture extraction. SIGCHI Bull. 20(SI), 369–374 (1989). http://doi.acm.org/10.1145/67450.67520
5. Kohavi, R., Provost, F.: Glossary of terms. Mach. Learn. - Special Issue Appl. Mach. Learn. Knowl. Discov. Process 30(2–3), 271–274 (1998). http://dl.acm.org/citation.cfm?id=288808.288815
6. Mao, S., Rosenfeld, A., Kanungo, T.: Document structure analysis algorithms: a literature survey. In: Proceedings of SPIE - The International Society for Optical Engineering, vol. 5010, pp. 197–207 (2003). doi:10.1117/12.476326
7. Pasetto, D., Franke, H., Qian, W., Guo, Z., Guo, H., Duan, D., Ni, Y., Pan, Y., Bao, S., Cao, F., Su, Z.: Rts - an integrated analytic solution for managing regulation changes and their impact on business compliance. In: Proceedings of the ACM International Conference on Computing Frontiers, CF 2013, NY, USA, pp. 24:1–24:8 (2013). http://doi.acm.org/10.1145/2482767.2482798
8. Salton, G., Wong, A., Yang, C.S.: A vector space model for automatic indexing. Commun. ACM 18(11), 613–620 (1975). http://doi.acm.org/10.1145/361219.361220
9. Tata, S., Patel, J.M.: Estimating the selectivity of tf-idf based cosine similarity predicates. ACM SIGMOD Rec. 36(2), 7–12 (2007). http://doi.acm.org/10.1145/1328854.1328855
10. Yang, Y., Liu, X.: A re-examination of text categorization methods. In: Proceedings of the 22nd Annual International ACM SIGIR Conference on Research and Development in Information Retrieval, SIGIR 1999, NY, USA, pp. 42–49 (1999). http://doi.acm.org/10.1145/312624.312647

Legal Yes/No Question Answering System Using Case-Role Analysis

Ryosuke Taniguchi and Yoshinobu Kano(⊠)

Faculty of Informatics, Shizuoka University, Hamamatsu, Japan
rtaniguchi@kanolab.net, kano@inf.shizuoka.ac.jp

Abstract. A central issue of yes/no question answering is the usage of knowledge source given a question. While yes/no question answering has been studied for a long time, legal yes/no question answering largely differs from other domains. The most distinguishing characteristic is that legal issues require precise analysis of roles and relationships of agents named in sentences. We have developed a yes/no question answering system for answering questions about a statute legal domain. Our system uses case-role analysis, in order to find correspondences of roles and relationships between given problem sentences and knowledge source sentences. We applied our system to the JURISIN's COLIEE (Competition on Legal Information Extraction/Entailment) 2016 task. Our system performance was better than systems of previous task participants and shared first place in current year's task in Phase Two. This result shows the importance of the points described above, while revealing opportunities to continue further work on improving our system's accuracy.

Keywords: COLIEE · Question answering · Legal bar exam · Legal Information Extraction

1 Introduction

Automatic question answering is attracting more interests recently. Due to the increasing expectation to the Artificial Intelligence (AI) technologies, people tend to regard question answering systems as a brand new technology emerged today. However, most successful systems employ rather traditional techniques of question answering which have decades of history [1–7], including series of shared tasks such as TREC [8], NTCIR [9] and CLEF [10]. This paper describes our challenge to the COLIEE 2016 legal bar exam, which asks participants to answer true or not based on the Civil Law Articles, given text drawn from the Japanese legal bar exam.

A variety of algorithms and systems has been proposed for question answering. Typically, these question answering systems used *big data* for answering questions [11–14]. For example, Dumais et al. [15] focused on the redundancy available in large corpora as an important resource. They used this redundancy to simplify their algorithm and to support answer mining from returned snippets. Their system performed quite well given the simplicity of the techniques being utilized.

The now widely known IBM Watson system [16] would be considered as a typical example of such a question answering system of the big data approach. The IBM

© Springer International Publishing AG 2017
S. Kurahashi et al. (Eds.): JSAI-isAI 2016, LNAI 10247, pp. 284–298, 2017.
DOI: 10.1007/978-3-319-61572-1_19

Watson system won in the Jeoperdy! Quiz TV program competing with human quiz winners. The core Watson system employed a couple of open source libraries, including the traditionally well-designed DeepQA system [17] as its skeleton of question answering processing. Because their target domain, the Jeopardy! Quiz, could ask broad range of questions, they collected a huge amount of knowledge sources from the Internet, etc., extracting relevant knowledge by combining a couple of different natural language processing (NLP) techniques.

Answering university examinations is another example. The Todai Robot project [18] is a challenge to solve Japanese university examinations, focusing towards attaining a high score in the National Center Test for University Admissions by 2016, and passing the entrance exam of the University of Tokyo (Todai) in 2021 [19]. Although the Todai Robot project tries to achieve higher scores, their aim is rather to reveal the current performance and limitation of the existing AI technologies, using the examinations as its benchmark, similar to the COLIEE's legal bar exam task. In contrast to the COLIEE task, the challenge of Todai Robot project includes variety of subjects including Mathematics, English, Japanese, Physics, History, etc. all written in Japanese language. While solving any problem of these subjects could be considered as question answering, some problems require special technologies. For example, Mathematics and Physics require to process formula; Japanese requires to infer emotions of story characters. Solving the History subjects might be considered as rather an extension of the existing question answering issues. The Todai Robot project achieved better scores than the average of the real human applicants in their Mock Exam challenges.

Recognition of textual entailments (RTE or RITE) is another related issue. RTE has been intensively studied for recent days, including shared tasks such as RTE tasks of PASCAL [20, 21], SemEval-2012 Cross-lingual Textual Entailment (CLTE) [22], NTCIR RITE tasks [23–25], etc. In the third PASCAL RTE-3 task, contradiction relations are included in addition to entailment relations [21]. In the RTE-6 task, given a corpus and a set of candidate sentences retrieved by a search engine from that corpus, systems are required to identify all the sentences from among the candidate sentences that entail a given hypothesis. NTCIR-9 RITE, NTCIR-10 RITE2, and NTCIR-11 RITEVal Exam Search tasks [25] required participants to find an evidence in source documents and to answer a given proposition by yes or no. Research of RTE normally tries to employ logical processing.

As described above, question answering techniques could include logic, reasoning, syntactic and semantic analysis. Many previous related works tried to employ such deeper analyses. However, required techniques more or less differ depending on a target domain.

Another issue is whether the knowledge source needs to be "big data" or not. Regarding the COLIEE's legal problems, required knowledge source can be limited. In this paper, we suggest to use small data in a precise way, rather than to use enormous amount of data as knowledge source. Due to this small data issue, supervised machine learning methods would suffer from insufficient training data. In addition, there are no "similar" problems exist for most of the legal bar exam problems. Therefore, a solver needs to "comprehend" the contents of the knowledge sources. Moreover, it is difficult to analyze why the approaches using machine learning answer so, due to their black box architecture. Rule-based methods would make analyses less difficult, and are especially effective in a limited domain like legal documents.

Based on these thoughts, we built our yes/no question answering system. Our system does not employ any machine learning. The main method of our system is a case-role based analysis, coupling with end-of-sentence expressions which has two levels of abstraction. We tried a couple of different combinations of our methods, optimized to Phase Two, while also tried Phase Three. Our system achieved more than 7 points better score than the best participant's system in the previous COLIEE 2015, shared first place in COLIEE 2016. There are still many difficult issues remained to be solved though.

We describe datasets of NTCIR RITE challenge, and datasets of previous and this COLIEE tasks in Sect. 2. Section 3 describes our design of the yes/no question answering system. Section 4 shows our experimental results for this COLIEE task and the RITE task. We discuss our achievements and limitations in Sect. 5, comparing the different tasks together with other COLIEE participants' systems, mentioning possible future works in Sect. 6. We conclude our paper in Sect. 7.

2   Related Tasks and Datasets

2.1   Exam Search Subtask in NTCIR RITEVal

While there were a couple of subtasks in the NTCIR RITE series, we describe the exam search subtask of NTCIR-11 RITEVal because the COLIEE dataset adopted the same format as the RITEVal dataset. RITEVal is an evaluation-based workshop held in 2013, aiming to recognize entailment, paraphrase, and contradiction between sentences, which is a common problem shared widely among researchers of NLP and information access [21, 26].

The entrance exam subtask attempts to emulate human's process of answering entrance exam questions. A system solves multiple-choice questions of real university entrance exams by refereeing to textual knowledge such as Wikipedia and textbooks. The Entrance Exam subtask provides two types of evaluation challenges. In this paper, we treat the RITE-2 Search Style evaluation, whose explanation is given below. This style of subtask was called FV (Fact Validation) subtask in the RITEVal task. We refer to this RITE-2 Entrance Exam Search Style (ExamSearch) subtask simply as RITEVal in this paper. We only regard Japanese version of the subtask, while there were English and Chinese subtasks.

RITEVal's dataset was developed from the past Japanese National Center Test questions for University Admissions (Center Test). The Center Test asks students multiple-choice style questions. The RITEVal dataset consists of three types of questions, "select the correct choice" type, "select the wrong choice" type, and "combination" type.

In the RITEVal task, the original multiple-choices were not given as a whole, but given one by one. In "select the correct choice" type questions, given a choice, RITEVal participant systems are asked to return a confidence value for that choice. Evaluation is performed by comparing confidence values for each original multiple-choices, regarding the largest value as the participant system's answer (smallest in case of "select wrong choice" type questions). In the "combination" type questions, the system is required to label Y or N for each choice and evaluated by a combination of these Y/N w.r.t the original multiple-choice question. In this paper, we focus on the "select correct/wrong choice" type questions.

Figure 1 is an example of the COLIEE dataset but can also be regarded as an example set of choices in the RITEVal dataset. In this example, one of the four choices is the correct one.

---

t1: （留置権の行使と債権の消滅時効）

第三百条　留置権の行使は、債権の消滅時効の進行を妨げない。

(Exercise of Rights of Retention and Extinctive Prescription of Claims)Article 300

The exercise of a right of retention shall not preclude the running of extinctive prescription of claims.

t2: 留置権者が留置物の占有を継続している間であっても，その被担保債権についての消滅時効は進行する。

Even while the holder of a right to retention continues the possession of the retained property, extinctive prescription runs for its secured claim.

Fig. 1. An example of COLIEE legal bar problem which asks to answer t1 entails t2 or not. The correct answer is "yes" in this example.

2.2 JURISIN COLIEE Datasets

The COLIEE shared task series is held in association with the JURISIN (Juris-informatics) workshop. The first one was the COLIEE 2014 shared task [27], and the second one was the COLIEE 2015 shared task [28]. This paper mainly describes our participation to the COLIEE 2016 shared task [28]. We call COLIEE 2016 simply as COLIEE in this paper.

The COLIEE shared task consists of three phases.

Phase One of this legal question answering task involves reading a legal bar exam question, and extracting a subset of Japanese Civil Code Articles.

Phase Two of the legal question answering task involves the identification of an entailment relationship. Given a question (t2) and a relevant article (t1), a participant's system has to determine if the relevant articles entail the question or not by answering yes or no.

Phase Three is combination of Phase One and Phase Two. Phase Three requires both of the legal information retrieval system and textual entailment system. Given a set of legal yes/no questions, a participant's system will retrieve relevant Civil Law articles. Then answer yes/no entailment relationship between input yes/no question and the retrieved articles.

The corpus of legal questions is drawn from Japanese Legal Bar exams, and the relevant Japanese Civil Law articles were also provided. While there was an English translation version of the dataset provided, we only used the original Japanese version. Figure 1 shows an example of the COLIEE dataset.

3   Method

3.1   Design Concepts

Wikipedia is a typical web sourced big data. However, we decided to use only the Civil Law data, or small data, in our system. A reason is that structures of Civil Law articles are clean. That is, the Civil Law articles use only one place (snippet) for one topic. Another reason is that we need precise analyses to solve the legal issues, rather than statistically calculate rough estimate values in a surficial way. We took an unsupervised approach for the same reasons.

We assume that roles and relationships in sentences play critical role in processing legal documents. In other words, the most important element should be the case-roles and its predicates of sentences. We focused on end-of-sentence expressions which normally correspond to the predicates.

Our yes/no question answering system is based on case-role analyses. We use JUMAN [29] and KNP [30] to obtain case-role analyses. JUMAN is a Japanese morphological analyzer where we added a custom dictionary for legal technical terms based on a Japanese legal term dictionary ("有斐閣法律用語辞典第4版"). KNP is a Japanese dependency case structure analyzer, works on top of JUMAN.

Using results of these tools, we obtain a subject and an end-of-sentence expression for each sentence. Unfortunately, there is no tool that can estimate deep case roles, surficial case roles are only available. A subjective case is normally specified by particles "が" or "は" in Japanese. We regard these cases as subjective cases.

When we analyze the Civil Law articles, we removed each header part "X条 (Article X)", which includes an article name and numbers. When "ただし (only provided)" appears within a sentence, we may discard the entire sentence depending on methods we employ. This is because this word "ただし" marks conditional phrase which is additional rather than dominate in the entire sentence meaning.

3.2   One-to-One Subject and End-of-Sentence Expression (Method 1)

Figure 2 shows a conceptual figure of Method 1with an example. Method 1 takes relations into account with subjects and end-of-sentence expressions, which are extracted as described before. We describe details of Method 1 below.

First, for each sentence, we make a one-to-one pair of a subject and a predicate of an end-of-sentence, using results of the morphological analyzer and the case structure analyzer. If a predicate of the end-of-sentence has no subject extracted, we only extract the predicate as an end-of-sentence expression.

Second, we simplify the extracted information to obtain better abstraction, which helps to decide Yes/No. We remove superficial case particles of "が" and "は" from the extracted subjects. Next, we remove punctuation marks from both of subjects and predicates. Additionally, if an end-of-sentence expressions contain a possibility

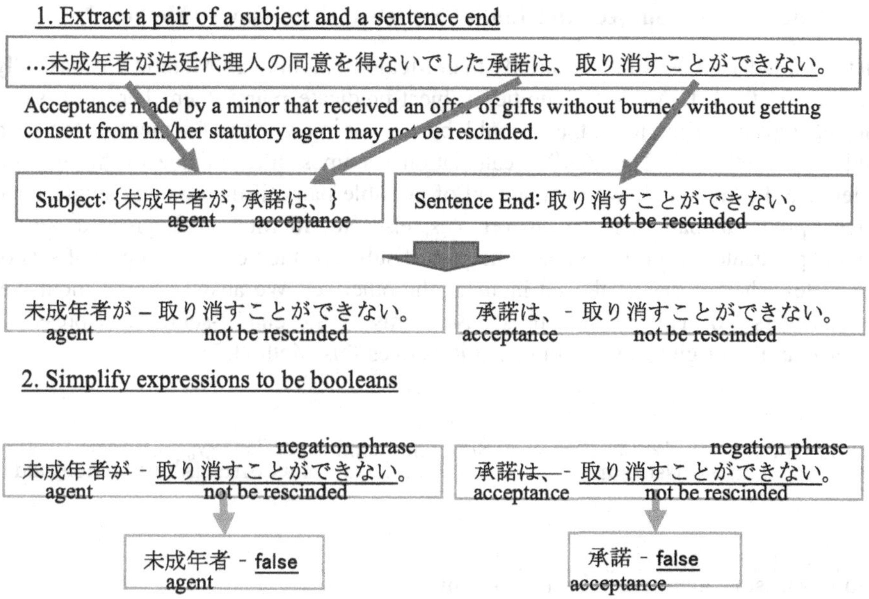

1. Extract a pair of a subject and a sentence end

...未成者が法廷代理人の同意を得ないでした承諾は、取り消すことができない。

Acceptance made by a minor that received an offer of gifts without burner without getting consent from his/her statutory agent may not be rescinded.

Subject: {未成者が, 承諾は、}
agent   acceptance

Sentence End: 取り消すことができない。
not be rescinded

未成者が－取り消すことができない。
agent   not be rescinded

承諾は、－取り消すことができない。
acceptance   not be rescinded

2. Simplify expressions to be booleans

negation phrase
未成者が－取り消すことができない。
agent   not be rescinded

negation phrase
承諾は、－取り消すことができない。
acceptance   not be rescinded

未成者－false
agent

承諾－false
acceptance

Fig. 2. A conceptual figure of Method 1.

expression, we replace the end-of-sentence expression with a boolean value. If the expression has a negation, the boolean value becomes *false*, else *true*. If an end-of-sentence expression does not contain possibility expression, we leave this end-of-sentence expression as it is without abstracting to a boolean value. The Japanese parser KNP can obtain annotations of "possibility expression (可能表現)" and "negation expression (否定表現)". We used these annotations to decide the Boolean value above.

For example, "may rescind (取り消すことができる)" is replaced with *true* because the phrase has a possibility expression. Another example, "consider as (みなす)", does not include any possibility expression is included. Therefore, we leave this predicate as it is. Possibility expressions are often used in legal documents. If there is any possibility expressions included, we can abstract different descriptions of possibility expressions like "may not be rescinded (取り消すことができない)" and "may not be rescinded (取り消せない)". However, if there is no possibility expression included, we need original forms of predicates without the abstraction above.

In Phase Two, a problem paragraph (t2) and a knowledge source (Civil Law article) paragraph (t1) are given. We apply the above process to both for t1 and t2 sentences. Then we compare results of the process for all sentence pairs between t1 and t2. When any pair of a t1 sentence and a t2 sentence has the same results, i.e. the same subject noun and end-of-sentence expression, our system answers Yes. If none of these sentence pairs matches, we determine Yes/No by matching only the end-of-sentence expression pair for each sentence pairs.

3.3   One-to-Many Subject and End-of-Sentence Expressions (Method 2)

Method 2 judges Yes/No more strictly than Method 1. Although Method 2 is basically same as Method 1, Method 2 unifies subject-predicate pairs when there is a same subject appears. Therefore, there could be one or more end-of-sentences for each subject in Method 2. The Yes/No calculation becomes different due to this method difference. In Method 1, we compare all of possible pairs; if there is any single same subject-predicate pair then we answer Yes, else No. In Method 2, given a pair of subject-predicates, we first compare the pair of subjects, then compare a pair of sets of predicates whether one of the set includes the other set. We answer Yes if inclusive, else answers No. This method uses base forms of the end-of-sentence expressions rather than the original forms. Figure 3 illustrates this Method 2.

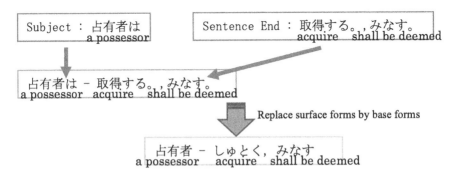

Fig. 3. An example of Method 2 process.

For example, from "A possessor in good faith shall acquire fruits derived from Thing in his/her possession. If a possessor in good faith is defeated in an action on the title, he/she shall be deemed to be a possessor in bad faith as from the time when such action was brought. (善意の占有者は、占有物から生ずる果実を取得する。善意の占有者が本権の訴えにおいて敗訴したときは、その訴えの提起の時から悪意の占有者とみなす。)", we can obtain "a possessor (占有者は)" as a subject, and "acquire (取得する)" and "shall be deemed (みなす)" as sentence ends.

4   Experiments

Experiments were conducted on the COLIEE 2016 Japanese subtask dataset. We did not use the Training/Development set except for setting the Yes/No threshold in Phase Three, because our method does not use any machine learning method i.e. unsupervised. We used these data sets to check and improve results of our system though.

Our system focused on Phase Two. Phase Three was answered using results of the Phase Two system.

4.1   Phase Two of COLIEE 2016

In Phase Two, we answered Yes or No determined by four rules as described below.

Rule 1 uses Method 1 and Method 2. Method 1 is "looser" and Method 2 is "stricter" in terms of the Yes/No judgment criteria. Therefore, in addition to Method 1/Method 2 alone, we combined Method 1 and Method 2 by using Method 1 as a base method, and overwrite its results by Method 2 only when Method 2 outputs Yes. This combination method determines whether all of t1 results are included in t2 results or not.

Rule 1 analyzes all sentences. Rule 2 is same as Rule 1 but excludes sentences from analyses when include a word "only provided (ただし)". Rule 3 and Rule 4 only use Method 1, their difference is same as Rule 1 and 2. Figure 4 summarizes these rules.

Table 1 shows results of our methods in the Phase Two training set. The H25 dataset corresponds to the previous COLIEE 2015 test dataset. Our result shows much better score (74.24) than the best participant's system in COLIEE 2015 (66.67).

Table 1. Phase Two results of COLIEE 2016 training dataset. Cells with color are the best score in corresponding year's data.

Dataset	H21	H22	H23	H24	H25
Rule1_Accuracy	66.07	63.83	65.85	67.09	72.73
Rule1_F-measure	66.06	63.76	65.67	66.44	72.32
Rule2_Accuracy	60.71	61.70	65.85	67.09	74.24
Rule2_F-measure	60.71	61.55	65.67	66.44	74.09
Rule3_Accuracy	64.29	68.09	60.98	73.42	71.21
Rule3_F-measure	64.10	68.03	60.95	73.55	70.67
Rule4_Accuracy	60.71	65.96	60.98	70.89	72.73
Rule4_F-measure	60.66	65.94	60.95	70.72	72.50

4.2   Phase Three of COLIEE 2016

In Phase Three, we used the same methods as Phase Two. We analyzed all Civil Low articles for each question regarding each article as t2 document using our Phase Two system, giving Yes/No answers. Then we counted numbers of Yes answers for each question. Finally, we set a threshold value of the Yes counts to answer Yes/No for Phase Three. In our experiments, the distribution of Yes answers became polarized to a range under and over 50. Therefore, we adopted the value of 50 as a middle value in the distribution in order to divided into two groups.

Table 2 shows results in Phase Three. The H25 result scores are quite lower than the ones in COLIEE 2015.

Table 2. Phase Three results of COLIEE 2016 training data set.

Dataset	H21	H22	H23	H24	H25
Rule1_Accuracy	48.21	48.94	56.10	58.23	43.94
Rule1_F-measure	48.20	45.98	56.07	58.20	43.93
Rule2_Accuracy	48.21	48.94	56.10	58.23	43.94
Rule2_F-measure	48.20	45.98	56.07	58.20	43.93
Rule3_Accuracy	48.21	46.81	58.54	46.84	43.94
Rule3_F-measure	39.74	41.63	53.06	39.80	36.19
Rule4_Accuracy	48.21	46.81	58.54	46.84	43.94
Rule4_F-measure	39.74	41.63	53.06	39.80	36.19

4.3   Formal Run of COLIEE 2016

In the formal run, we used the same methods for Phase Two and Phase Three. Tables 3 and 4 shows results in Phase Two and Phase Three in the formal run. The results of KIS-1 and KIS-2, which use Method 1 and Method 2, are higher than the others in Phase Two. In Phase Three, the results of KIS-3 and KIS-4 are higher than KIS-1 and KIS-2, where only Method 1 was used.

Table 3. Results of COLIEE2016 test dataset in Phase Two [31]. (Cells with colors are best scores in Phase Two and Three. KIS-1 uses Rule1, KIS-2 uses Rule2, and so on.)

Run	Accuracy	Run	Accuracy
JNLN1 [33]	40.00	iLis7 [35]	53.68
KIS-1	51.58	JNLN3 [33]	47.37
KIS-2	51.58	UofA [36]	46.32
KIS-3	52.63	UofA [36]	54.74
KIS-4	52.63	UofA [36]	55.79

Our methods shared first place with iLis7 in Phase Two, third place in Phase Three for the formal run of COLIEE 2016.

5   Discussion

In Phase Two, we obtained better results than last year's COLIEE 2015 (H25 data set). Let us discuss our results in detail for a couple of points below.

Firstly, we used a customized legal technical term dictionary for morphological analyses. Table 5 shows results when we did not use the legal dictionary but only used the default one. When comparing Table 1 with Table 5, Table 1 has better results than Table 5 entirely in most cells while sometimes slightly worse.

Table 4. Results of COLIEE2016 test dataset in Phase Three [31]. (Cells with colors are best scores in Phase Two and Three. KIS-1 uses Rule1, KIS-2 uses Rule2, and so on.)

Run	Accuracy	Run	Accuracy
JNLN1 [33]	52.86	UofA [36]	55.71
KIS-1	62.86	N01-1 [32]	54.29
KIS-2	62.86	N01-2 [32]	54.29
KIS-3	58.57	N01-3 [32]	48.57
KIS-4	58.57	N01-4 [32]	48.57
iLis7 [35]	62.86	N01-5 [32]	57.14
JNLN3 [33]	48.57		

Table 5. Results of COLIEE2016 training dataset in Phase Two without our custom dictionary. Cells with blue color are worse than corresponding cells in Table 1. Cells with red color are better than corresponding cells in Table 1.

Dataset	H21	H22	H23	H24	H25
Rule1_Accuracy	66.07	63.83	65.85	65.82	74.24
Rule1_F-measure	66.06	63.76	65.67	65.47	73.95
Rule2_Accuracy	60.71	61.70	65.85	67.09	75.76
Rule2_F-measure	60.71	61.55	65.67	66.66	75.67
Rule3_Accuracy	64.29	63.83	58.54	72.15	71.21
Rule3_F-measure	64.10	63.76	58.54	72.11	70.67
Rule4_Accuracy	60.71	61.70	58.54	70.89	72.73
Rule4_F-measure	60.66	61.68	58.54	70.72	72.50

In order to analyze this effect in detail, we focus on results in H22 (H22-9-I and H22-23-O) as shown in Fig. 5. In this figure, Rule 3 with the legal technical term dictionary is referred to as Rule 3A, Rule 3 without the legal technical term dictionary is referred to as Rule 3B. H22-9-1, the morphological analysis failed without the custom dictionary; "joined (付合)" was divided into "付" and "合", "付" has matched with other fragment of morphemes, leading its yes/no answer to wrong "yes". H22-23-O shows an opposite case. "lessor (貸借人)" was divided into "貸借" and "人" without the custom dictionary. Because "person (人)" is a common word, it matches with other morpheme fragments frequently. In this case, this leads its yes/no answer to wrong "no".

As a whole, these results suggest that our legal technical term dictionary was effective in our system.

Secondly, we discuss effect of the word "only provided (ただし)". In Phase Three, no difference is observed between Rule 1 and Rule 2, Rule 3 and Rule 4, as shown in Table 2. Differences of Rule 1 and Rule 2, Rule 3 and Rule 4, are whether we eliminate

> Rule1 : Using Method1 and Method2 with "ただし"
> Rule2 : Using Method1 and Method2 without "ただし"
> Rule3 : Using Method1 with "ただし"
> Rule4 : Using Method1 without "ただし"

Fig. 4. Summary of the rules employed in our methods. "ただし" means "only provided".

phrases which include "only provided". On the other hand, we got better results when retaining phrases of "only provided" (Rule1 and Rule3) except for H25 in Phase Two. Because this word "only provided" could lead to the opposite result, there should have a positive effect when excluding phrases of "only provided", while the results imply not. Further consideration will be needed to examine the effect of this word.

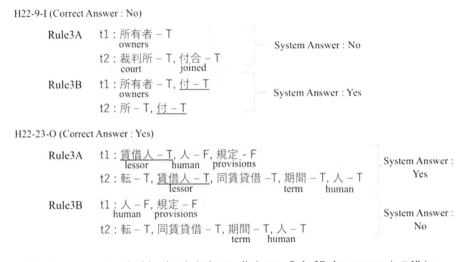

Rule 3A uses a customized legal technical term dictionary, Rule 3B does not use it. "t1" is a result from the Civil Law articles, and "t2" is from the problem sentences. T = True, F = False

Fig. 5. Examples of failed process without the custom legal technical term dictionary.

Thirdly, results of Phase Two seem not correlate with results of Phase Three, while we used methods and results of Phase Two directly to Phase Three. This might be because our methods were optimized for Phase Two, not for Phase Three. Deeper analyses of threshold and parameter effects in Phase Three would be useful.

Fourthly, our system depends on the accuracy of the case structure analyzer, KNP.

Because the analyzer was mainly trained by newswire texts, its accuracy in legal texts would not be sufficient. Analysis of errors in the case structure analyzer would be required.

iLis7, which shared the first place with our team in Phase Two, used majority vote with decision tree, linear SVM, and CNN using various linguistic features, which are lexical, semantic, and syntactic. iLis7 and other teams used machine learning methods. On the other hand, we used simple but essential heuristic rules. The optimization of machine learning could improve the overall performance in terms of parameter tuning, while it could not capture the essential features to solve problems of legal documents where complex underlying structures and implicit knowledge are required. The training data is insufficient for machine learning methods, and they would not have captured deeper features to reveal relationship between the civil law and bar exams, like sentence structures and anaphora.

Finally, we observe a range of different scores among years from H21 to H25 and the test dataset. When comparing Table 3 with Table 1 in our methods, Table 3 has worse results than Table 1 entirely. The test dataset is more difficult than the training dataset to solve. There could be different tendency among the datasets and the test datasets of these years. Precise analysis of each sentence would be needed to find causes of this score variation.

6 Future Work

A large difference with other past COLIEE systems is that we used Japanese while others used English. While the English dataset is a direct translation of the original Japanese dataset, this difference could result in different issues when solving the problems. For example, Japanese text requires explicit tokenization process. When this tokenization fails, the final result could also be failed. As far as we observed, our morphological analyses were fine with the legal technical term dictionary, but there were still some failures observed. Comparison with English version will help.

The technical terms in the COLIEE dataset tend to include logical relations implicitly. This sort of logical relation extraction would be still very difficult to perform in sufficiently high accuracy considering the current NLP technologies. Furthermore, there are sometimes "instantiations" in the Civil Law articles like "when A claims...". This sort of instantiations requires higher level of abstraction process.

The document structure of the given knowledge source, the Japanese Civil Law articles, is special. The Civil Law articles have references; logic and conditions are described in a single sentence, or scattered across snippets.

The Civil Law articles include specific negation and acronym expressions which are critical in solving the problems. While our system handles these expressions to some extent, there may be some expressions missing.

These results in COLIEE's task lead to the conclusion that our system has reasonable efficacy. However, we currently only check superficial cases which can be obtained from JUMAN and KNP analysis results. Deep case theory could make a breakthrough to reveal importance of semantics in the legal domain.

Future work would include solutions to the above issues, in addition to those described in the discussion section. The most difficult issue to solve would be the logic and abstraction.

7   Conclusion

Legal document processing requires different issues to solved than other domains. A large difference in legal yes/no question answering is that legal issues require roles and relationships of agents in sentences to be precisely analyzed. Based on this observation, we developed a yes/no question answering system for legal domain. Our system uses case-role analysis and end-of-sentence expressions, in order to find correspondences of roles and relationships between given problem sentences and knowledge source sentences. We applied our system to COLIEE 2016 Japanese task. Our system performance was 7 points better than the best system of previous COLIEE task participants in Phase Two, and shared first place in Phase Two in the COLIEE 2016 test set. While this result shows the importance of the points we employed in our system, there are still a couple of issues to be resolved as future work.

Acknowledgements. This work was partially supported by MEXT Kakenhi, National Institute of Informatics, and JST CREST.

References

1. Lin, D., Pantel, P.: Discovery of inference rules for question-answering. Nat. Lang. Eng. **7** (4), 343–360 (2001)
2. Ravichandran, D., Hovy, E.: Learning surface text patterns for a question answering system. In: Proceedings of the 40th Annual Meeting on Association for Computational Linguistics, pp. 41–47 (2002)
3. Yu, H., Hatzivassiloglou, V.: Towards answering opinion questions: separating facts from opinions and identifying the polarity of opinion sentences. In: Proceedings of the 2003 Conference on Empirical Methods in Natural Language Processing, pp. 129–136 (2003)
4. Pinto, D., McCallum, A., Wei, X., Croft, W.B.: Table extraction using conditional random fields. In: Proceedings of the 26th Annual International ACM SIGIR Conference on Research and Development in Informaion Retrieval, pp. 235–242 (2003)
5. Cui, H., Sun, R., Li, K., Kan, M.-Y., Chua, T.-S.: Question answering passage retrieval using dependency relations. In: Proceedings of the 28th Annual International ACM SIGIR Conference on Research and Development in Information Retrieval, pp. 400–407 (2005)
6. Xue, X., Jeon, J., Croft, W.B.: Retrieval models for question and answer archives. In: Proceedings of the 31st Annual International ACM SIGIR Conference on Research and Development in Information Retrieval, pp. 475–482 (2008)
7. Bian, J., Liu, Y., Agichtein, E., Zha, H.: Finding the right facts in the crowd: factoid question answering over social media. In: Proceedings of the 17th International Conference on World Wide Web, pp. 467–476 (2008)
8. Voorhees, E.M., Harman, D.K.: TREC: Experiment and Evaluation in Information Retrieval (Digital Libraries and Electronic Publishing). The MIT Press, Cambridge (2005)
9. Kando, N., Kuriyama, K., Nozue, T.: NACSIS test collection workshop (NTCIR-1) (poster abstract). In: Proceedings of the 22nd Annual International ACM SIGIR Conference on Research and Development in Information Retrieval, pp. 299–300 (1999)
10. Braschler, M.: CLEF 2000—overview of results. In: Peters, C. (ed.) CLEF 2000. LNCS, vol. 2069, pp. 89–101. Springer, Heidelberg (2001). doi:10.1007/3-540-44645-1_9

11. Kwok, C.C.T., Etzioni, O., Weld, D.S.: Scaling question answering to the web. In: Proceedings of the 10th International Conference on World Wide Web, pp. 150–161 (2001)
12. Etzioni, O., Cafarella, M., Downey, D., Kok, S., Popescu, A.-M., Shaked, T., Soderland, S., Weld, D.S., Yates, A.: Web-scale information extraction in knowitall: (preliminary results). In: Proceedings of the 13th International Conference on World Wide Web, pp. 100–110 (2004)
13. Jeon, J., Croft, W.B., Lee, J.H.: Finding similar questions in large question and answer archives. In: Proceedings of the 14th ACM International Conference on Information and Knowledge Management, pp. 84–90 (2005)
14. Kanayama, H., Miyao, Y., Prager, J.: Answering yes/no questions via question inversion. In: The 24th International Conference on Computational Linguistics (COLING 2012), pp. 1377–1391 (2012)
15. Dumais, S., Banko, M., Brill, E., Lin, J., Ng, A.: Web question answering: is more always better? In: Proceedings of the 25th Annual International ACM SIGIR Conference on Research and Development in Information Retrieval, pp. 291–298 (2002)
16. Ferrucci, D.: Introduction to 'This is Watson'. IBM J. Res. Dev. **56**(3.4), 1:1–1:15 (2012)
17. Ferrucci, D.A.: IBM's Watson/DeepQA. SIGARCH Comput. Arch. News **39**(3) (2011)
18. Arai, N.H.: The impact of AI—can a robot get into the university of Tokyo? Natl. Sci. Rev. **2** (2), 135–136 (2015)
19. The Todai Robot project. http://21robot.org/
20. Dagan, I., Glickman, O., Magnini, B.: The PASCAL recognising textual entailment challenge. In: Quiñonero-Candela, J., Dagan, I., Magnini, B., d'Alché-Buc, F. (eds.) MLCW 2005. LNCS, vol. 3944, pp. 177–190. Springer, Heidelberg (2006). doi:10.1007/11736790_9
21. Giampiccolo, D., Magnini, B., Dagan, I., Dolan, B.: The third PASCAL recognizing textual entailment challenge. In: Proceedings of the ACL-PASCAL Workshop on Textual Entailment and Paraphrasing, pp. 1–9 (2007)
22. Negri, M., Marchetti, A., Mehdad, Y., Bentivogli, L., Giampiccolo, D.: Semeval-2012 task 8: cross-lingual textual entailment for content synchronization. In: Proceedings of the First Joint Conference on Lexical and Computational Semantics - Volume 1: Proceedings of the Main Conference and the Shared Task, and Volume 2: Proceedings of the Sixth International Workshop on Semantic Evaluation, pp. 399–407 (2012)
23. Shima, H., Kanayama, H., Lee, C., Lin, C., Mitamura, T., Miyao, Y., Shi, S., Takeda, K.: Overview of NTCIR-9 RITE: recognizing inference in text. In: NTCIR-9 Workshop, pp. 291–301 (2011)
24. Watanabe, Y., Miyao, Y., Mizuno, J., Shibata, T., Kanayama, H., Lee, C.-W., Lin, C.-J., Shi, S., Mitamura, T., Kando, N., Shima, H., Takeda, K.: Overview of the recognizing inference in text (RITE-2) at NTCIR-10. In: The NTCIR-10 Workshop, pp. 385–404 (2013)
25. Matsuyoshi, S., Miyao, Y., Shibata, T., Lin, C.-J., Shih, C.-W., Watanabe, Y., Mitamura, T.: Overview of the NTCIR-11 recognizing inference in text and validation (RITE-VAL) task. In: The 11th NTCIR (NII Testbeds and Community for Information Access Research) Workshop, pp. 223–232 (2014)
26. Dagan, I., Glickman, O., Magnini, B.: The PASCAL recognising textual entailment challenge. In: Proceedings of the PASCAL Challenges Workshop on Recognising Textual Entailment (2005)
27. Competition on Legal Information Extraction/Entailment (COLIEE-14), Workshop on Juris-informatics (JURISIN) 2014 (2014). http://webdocs.cs.ualberta.ca/~miyoung2/jurisin_task/index.html
28. Kim, M.-Y., Goebel, R., Ken, S.: COLIEE-2015: evaluation of legal question answering. In: Ninth International Workshop on Juris-informatics (JURISIN 2015) (2015)

29. JUMAN (a User-Extensible Morphological Analyzer for Japanese). http://nlp.ist.i.kyoto-u.ac.jp/EN/index.php?JUMAN
30. Japanese Dependency and Case Structure Analyzer KNP. http://nlp.ist.i.kyoto-u.ac.jp/EN/index.php?KNP
31. Kim, M.-Y., Goebel, R., Kano, Y., Ken, S.: COLIEE-2016: evaluation of the competition on legal information extraction/entailment. In: Tenth International Workshop on Juris-informatics (JURISIN 2016) (2016)
32. John, A.K., Di Caro, L., Boella, G., Bartolini, C.: Team-Normas' participation at the Coliee2016 bar legal exam competition. In: Tenth International Workshop on Juris-Informatics (JURISIN 2016) (2016)
33. Carvalho, D.S., Tran, V.D., Van Tran, K., Lai, V.D., Nguyen, M.-L.: Lexical to discourse-level corpus modeling for legal question answering. In: Tenth International Workshop on Juris-Informatics (JURISIN 2016) (2016)
34. Do, P.K., Nguyen, H.-T., Tran, C.-X., Nguyen, M.-T., Nguyen, M.L.: Legal question answering using ranking SVM and deep convolutional neural network. In: Tenth International Workshop on Juris-Informatics (JURISIN 2016) (2016)
35. Kim, K., Heo, S., Jung, S., Hong, K., Rhim, Y.-Y.: An ensemble based legal information retrieval and entailment system. In: Tenth International Workshop on Juris-Informatics (JURISIN 2016) (2016)
36. Kim, M.-Y., Xu, Y., Lu, Y., Goebel, R.: Legal question answering using paraphrasing and entailment analysis. In: Tenth International Workshop on Juris-Informatics (JURISIN 2016) (2016)

Question Answering of Bar Exams
by Paraphrasing and Legal Text Analysis

Mi-Young Kim[3]([⊠]), Ying Xu[1], Yao Lu[2], and Randy Goebel[1]

[1] Alberta Machine Intelligence Institute, University of Alberta,
Edmonton, AB, Canada
{yx2, rgoebel}@ualberta.ca
[2] iLab Tongji, School of Software Engineering, Tongji University,
Shanghai, China
95luyao@tongji.edu.cn
[3] Department of Science, University of Alberta,
Augustana Campus, Camrose, AB, Canada
miyoung2@ualberta.ca

Abstract. Our legal question answering system combines legal information retrieval and textual entailment, and exploits paraphrasing and sentence-level analysis of queries and legal statutes. We have evaluated our system using the training data from the competition on legal information extraction/entailment (COLIEE)-2016. The competition focuses on the legal information processing required to answer yes/no questions from Japanese legal bar exams, and it consists of three phases: legal ad-hoc information retrieval (Phase 1), textual entailment (Phase 2), and a combination of information retrieval and textual entailment (Phase 3). Phase 1 requires the identification of Japan civil law articles relevant to a legal bar exam query. For this phase, we have used an information retrieval approach using TF-IDF and a Ranking SVM. Phase 2 requires decision on yes/no answer for previously unseen queries, which we approach by comparing the approximate meanings of queries with relevant articles. Our meaning extraction process uses a selection of features based on a kind of paraphrase, coupled with a condition/conclusion/exception analysis of articles and queries. We also identify synonym relations using word embedding, and detect negation patterns from the articles. Our heuristic selection of attributes is used to build an SVM model, which provides the basis for ranking a decision on the yes/no questions. Experimental evaluation show that our method outperforms previous methods. Our result ranked highest in the Phase 3 in the COLIEE-2016 competition.

Keywords: Legal question answering · Recognizing textual entailment · Information retrieval · Paraphrasing

1 Task Description and Summary of Our Approach

Our approach to legal question answering combines information retrieval and textual entailment. We achieve this combination with a number of intermediate steps. For instance, consider the question "Is it true that a special provision that releases warranty can be made, but in that situation, when there are rights that the seller establishes on

© Springer International Publishing AG 2017
S. Kurahashi et al. (Eds.): JSAI-isAI 2016, LNAI 10247, pp. 299–313, 2017.
DOI: 10.1007/978-3-319-61572-1_20

his/her own for a third party, the seller is not released of warranty." A system must first identify and retrieve relevant documents (typically legal statutes), and subsequently, identify a most relevant sentence. Finally it must extract and compare semantic connections between the question and the relevant sentences, and confirm a threshold of evidence about whether an entailment relation holds.

The Competition on Legal Information Extraction/Entailment (COLIEE) 2016[1] focuses on two aspects of legal information processing related to answering yes/no questions from legal bar exams: legal document retrieval (Phase 1), and whether there is a textual entailment relation between a query and relevant legal documents (Phase 2). In addition, Phase 3 is about combing them for the whole task.

We treat Phase 1 as an ad-hoc information retrieval (IR) task. The goal is to retrieve relevant Japan civil law statutes or articles that are related to a question in legal bar exams, from which we can confirm a yes or no answer based on deciding if there is an entailment relation between the question (or the negation of the question) and the relevant statutes.

We approach the information retrieval part of this problem (Phase 1) with two models based on statistical information. One is the TF-IDF model [1], i.e., term frequency-inverse document frequency. The idea is that relevance between a query and a document depends on their intersecting word set. The importance of words is measured with a function of term frequency and document frequency as parameters. Our terms are lemmatized words, which mean verbs like "attending," "attends," and "attended" are lemmatized as the same form "attend."

Another popular model for text retrieval is a Ranking SVM model [2]. We use that model to re-rank documents that are retrieved by the TF-IDF model. The model's features are lexical words, dependency path bigrams and TF-IDF scores. The intuition is that the supervised model can learn weights or priority of words based on training data in addition to, or as an alternative to TF-IDF.

The goal of Phase 2 is to construct yes/no question answering systems for legal queries, by heuristically confirming entailment of a query (or its negation) from relevant articles. The answer to a question is typically determined by measuring some kind of semantic similarity between question and answer. Because the legal bar exam query and relevant articles are complex and varied, we need to carefully determine what kind of information is needed for confirming textual entailment. Here we exploit a kind of paraphrasing based on term expansion and word embedding for semantic analysis, coupled with condition/conclusion/exception analysis on the query and relevant articles. After constructing a set of pre-trained semantic word embeddings using *word2-vec*[2], we train the system to learn models for semantic matching between question and corresponding articles. These feature extraction methods are coupled with negation analysis, then used to construct an SVM model to provide the required yes/no answers.

---

[1] https://webdocs.cs.ualberta.ca/∼miyoung2/COLIEE2016/.

[2] https://code.google.com/p/word2vec.

2 Phase 1: Legal Information Retrieval

2.1 IR Models

Our information retrieval model is a combination of the term frequency–inverse document frequency (tf-idf) model and a support vector machine (SVM) re-ranking model. We will describe the two components in the following.

2.1.1 The TF-IDF Model

One of our baseline models is a tf-idf model implemented in Lucene, an open source IR system[3].

The simplified version of Lucene's similarity score of an article to a query is:

$$tf\text{-}idf(Q,A) = \sum_{t \in Q \cup A} \left\{ \sqrt{tf(t,A)} \times [1 + \log(idf(t))]^2 \right\} \tag{1}$$

The score $tf\text{-}idf(Q,A)$ is a measure which estimates the relevance between a query Q and an article A. First, for every term t in query A, we compute $tf(t,A)$, and $idf(t)$. The score $tf(t,A)$ is the term frequency of t in the article A, and $idf(t)$ is the inverse document frequency of the term t, which is the number of articles that contain t. The final score is the sum of the scores of terms in both the article and the query. The bigger $tf\text{-}idf(Q,A)$, the more relevance between the query Q and the article A.

The choice of terms in documents is as important as choosing the score functions. Instead of using the original words in a text, we lemmatize the text with the Stanford NLP tool [15]. After lemmatization, words such as *steal, stole, and steals* become *steal*. In this way, if there is *steal* in the question, but *stole* in the article, we can still retrieve the article as a match.

2.1.2 The Ranking SVM Model

Previous tf-idf models rank the articles based on frequency information. However, other features, such as the matched phrases between the article and the queries, are useful too. We use an SVM Ranking model to learn the importance of such features and then re-estimate the score of each retrieved article from the tf-idf output.

The ranking SVM model was proposed by [2]. That model ranks documents based on user's click through data; in our case, the correct articles in the training data. Given the articles retrieved from the *tf-idf* model, the ranking SVM will learn to rank correct articles higher than incorrect ones. More precisely, given the feature vector of a training instance, i.e. a retrieved article set given a query, denoted by $\Phi(Q,A_i)$, the model tries to find a ranking that satisfies constraints:

$$\Phi(Q, A_i) > \Phi(Q, A_j) \tag{2}$$

where A_i is a relevant article for the query Q, while A_j is less relevant.

---

[3] Lucene can be downloaded from http://lucene.apache.org/core/.

To use this ranking SVM, we incorporate the following types of features:

- Lexical words: the lemmatized normal form of surface structure of words in both the retrieved article and the query. In the conversion to the SVM's instance representation, this feature is converted into binary features whose values are one or zero, i.e. if a word exists in the intersection word set or not.
- Dependency pairs: word pairs that are linked by a dependency link, arising from a dependency parsing. The intuition is that, compared with the bag of words information, syntactic information should improve the capture of salient semantic content. Dependency parse features have been used in many NLP tasks, and improved IR performance [3]. This feature type is also converted into binary values.
- TF-IDF score (Sect. 2.1.1).

2.2   Experiments

The COLIEE legal IR task has several sets of queries with the Japan civil law articles as documents (1044 articles in total). Here follows one example of the query and a corresponding relevant article.

Question: A person who made a manifestation of intention which was induced by duress emanated from a third party may rescind such manifestation of intention on the basis of duress, only if the other party knew or was negligent of such fact.

Related Article: (Fraud or Duress) Article 96(1) Manifestation of intention which is induced by any fraud or duress may be rescinded. (2) In cases any third party commits any fraud inducing any person to make a manifestation of intention to the other party, such manifestation of intention may be rescinded only if the other party knew such fact. (3) The rescission of the manifestation of intention induced by the fraud pursuant to the provision of the preceding two paragraphs may not be asserted against a third party without knowledge.

Before the final test set was released, we received 8 sets of queries for a dry run. The 8 sets of data include 412 queries. We used the corresponding 8-fold leave-one-out cross validation evaluation. The metric for measuring our IR models is Mean Average Precision (MAP):

$$MAP(Q) = \frac{1}{|Q|} \sum_{q \in Q} \frac{1}{m} \sum_{k \in (1,m)} precision\ (R_k) \tag{3}$$

where Q is the set of queries, and m is the number of retrieved articles. Rk is the set of ranked retrieval results from the first until the kth article. In the following experiments, we set m as 3 for all queries, corresponding to the column MAP@3 in Table 1. The SVM's parameters are set according to the 8-fold cross validation IR performance. Given the top 20 articles returned by the tf-idf model, the SVM model extracts features for every article and trains according to the order that relevant articles are ranked higher than irrelevant ones.

Table 1 presents the results of using the different models. The result shows that the ranking SVM with all three features achieves the highest performance. We also show

Table 1. IR results on dry run data with different models.

Id	Models	MAP@3 (%)	Standard deviation (%)	Smallest (%)	Largest (%)	Average F-score @1 (%)
1	tf-idf with lemma	39.8	7.0	23.8	45.5	53.4
2	SVM-ranking with lemma	39.8	6.5	26.1	46.8	60.0
3	SVM-ranking with lemma and dependency pair	41.2	5.4	30.1	48.4	56.7
4	Model 3 plus tf-idf score	43.1	7.1	27.2	49.0	60.0

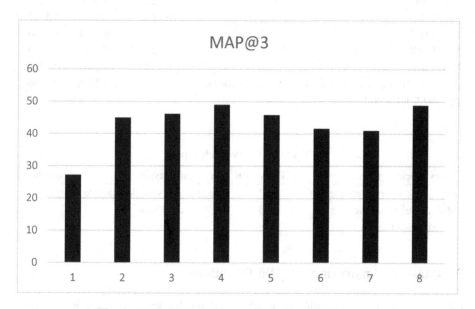

Fig. 1. MAP@3 for the 8 cross-validation set of Model 4.

the standard deviation of the cross-validation, the smallest and largest MAP@3 for 8 folds to show the effect of small training data. It seems the tf-idf model causes larger deviation than the SVM models. In the last column of Table 1, we show the F-score results of different models with the top first answers for every query. The F-score is used as the metric of the competition. We can observe that the SVM models are better than the tf-idf model. However, no difference is observed between the second model and the fourth model.

Figure 1 shows the MAP@3 values for every training fold for Model 4 in Table 1. It shows the model achieves a MAP@3 value larger than 40% for most of the folds.

Table 2 shows our IR result of the final SVM model on the test data and other systems' results. iLis7 [19] system with majority vote of decision tree, linear SVM, and CNN achieved the best result, but in Sect. 3, we will show that our method outperformed iLis7 [19] and showed the best performance in answering yes/no questions when it is combined with our textual entailment methods.

Table 2. Our IR results on test data vs. other systems' results

Systems	Precision	Recall	F-score
JNLN1 [17]	0.6105	0.4427	0.5133
HUKB-1 [21]	0.6154	0.4886	0.5447
HUKB-2 [21]	0.6250	0.4962	0.5532
HUKB-3 [21]	0.6316	0.4580	0.5310
HUKB-4 [21]	0.6316	0.4580	0.5310
JNLN2 [22]	0.6211	0.4504	0.5221
iLis7 [19]	0.7272	0.5496	0.6261
JNLN3 [20]	0.6526	0.4733	0.5487
N01-1 [23]	0.3053	0.2214	0.2566
N01-2 [23]	0.4211	0.3053	0.3540
N01-3 [23]	0.4000	0.2901	0.3363
Our system (SVM-ranking with lemma and dependency pair and tf-idf score)	0.5895	0.4275	0.4956

Table 3. Query-article types

Query-article type	Proportion	Query-article type	Proportion
One article refers to another article	0.182	Question is a specific example	0.092
Multiple relevant articles	0.388	Multiple conditions	0.731
Exceptional case	0.148		

3 Phase 2: Answering Yes/No Questions

Our system uses a combination of word embedding for semantic analysis and paraphrasing for term expansion to predict textual entailment. Here we describe the entailment types and the extraction of features from sentences.

3.1 Entailment Types

We identify a variety of types of entailment as shown in Table 3. By classifying a yes/no problem as one of these types, we can determine what kind of further information is required to provide a decision on entailment.

Table 3 shows our list of query-article types. Note that one article can refer to another, such as *"If there is any latent defect in the subject matter of a sale, the provisions of Article 566 shall apply mutatis mutandis."* This makes textual entailment more complex because we also need to analyze the meaning of the referred article.

In another case, one query can have multiple relevant articles, so we have to combine the multiple articles' meanings, or choose one as most relevant for determining entailment.

Note also that many statutes have exceptional cases, so we need to recognize if the query is included in the exceptional case or not. In addition, a query may be one example of the article case. There are also cases where some articles have multiple conditions for one conclusion, so we must then confirm if each condition is satisfied in the query. Overall, many query-article types also require the identification of negation and synonym/antonym relations to confirm the correct entailment.

The overall description of our procedure of textual entailment is as follows:

1. Find the most relevant article for a given query
2. Divide a query and the corresponding article into "Condition(s)," "Conclusion," and "Exception-condition(s)."
3. Term expansion using Paraphrasing
4. Negation and synonym detection
5. Extract features and perform learning using the features

In the following subsections, we explain each step in detail.

3.2 Finding the Most Relevant Article/Sentences

In case that there are multiple relevant sentences, we choose the article with the most overlapping words with the query. In the selected article, if there exist multiple regulations, we also choose the one regulation that has most overlapping words with a query.

3.3 Negation and Synonym Detection

We exploit a process for managing negation and antonyms as described in Kim et al. [10]. In addition, we approximate word semantic similarity by converting words to vector representations using the *word2vec* tool. The output of the semantic similarity is vector similarity. We used 1,044 legal law articles to train the *word embedding* by setting the *word2vec* vector dimension to 50 which has been most commonly chosen as the vector dimension in previous work.

3.4 Condition/Conclusion/Exception Detection

From our analysis of the structure of statutes we extract components based on the following rules:

$$conclusion := segment_{last}(sentence, keyword),$$

$$condition := \sum_{i \neq last} segment_i(sentence, keyword),$$

$$condition := condition \text{ [or] } condition$$

$$condition := sub\_condition \text{ [and] } sub\_condition$$

$$exception\_conclusion := segment_{last}(sentence, exception\_keyword),$$

$$exception\_condition := \sum_{i \neq last} segment_i(sentence, exception\_keyword),$$

$$exception\_condition := exception\_condition \text{ [or] } exception\_condition$$

$$exception\_condition := sub\_exception\_condition \text{ [and] } sub\_exception\_condition$$

So from keywords of a condition, we segment sentences. The keywords of the condition are as follows: "in case(s)," "if," "unless," "with respect to," "when," and "(comma)." After this segmentation, the last segment is considered to be a conclusion, and the rest of the sentence is considered as a condition. (We used the symbol $\sum$ to denote the concatenation of the segments.) We also distinguish segments which denote exceptional cases. Currently, we take the *exception_keyword* indication as "… this shall not apply, if (unless)."

The original bar law examinations in the COLIEE data are provided in Japanese and English, and our initial implementation used a Korean translation, provided by the Excite translation tool[4]. We chose Korean because we have a team member whose native language is Korean, and the characteristics of Korean and Japanese language are similar. In addition, the translation quality between two languages ensures relatively stable performance. Because our study team includes a Korean researcher, we can easily analyze the errors and intermediate rules in Korean. Therefore, the above rules may not be appropriate for all English sentences, because the segment order can differ.

The following is an example of condition and conclusion detection:

<Civil law example> A person who employs others for a certain business, shall be liable for damages inflicted on a third party by his/her employees with respect to the execution of that business; Provided, however, that this shall not apply, if the employer exercised reasonable care in appointing the employee or in supervising the business, or if the damages could not have been avoided even if he/she had exercised reasonable care.

---

[4] http://excite.translation.jp/world/.

(1) Conclusion => shall be liable for damages inflicted on a third party by his/her employees with respect to the execution of that business.
(2) Condition => A person who employs others for a certain business
(3) Exception
      Conclusion => this shall not apply (opposite of main conclusion)
      Condition
        Condition =>
          Condition => if the employer exercised reasonable care in appointing
                the employee
          Condition (OR) => in supervising the business
        Condition(OR) => if the damages could not have been avoided even if
          he/she had exercised reasonable care.

3.5   Term Expansion Using Paraphrasing

There are many words with similar meanings but different lexical forms (e.g., 'obligor' vs. 'debtor', 'rescind' vs. 'cancel', 'lien' vs. 'privilege', etc.). To resolve these diverse terms, we use language translation-based paraphrasing. The idea of translation-based paraphrase is that translating from one language to another and then back, will often produce semantically similar but lexically distinct outputs. If we assume that the language translations preserve semantics, more or less, then lexically distinct terms can be considered as paraphrases. In our application of this idea, we translate the original English query/document into German, and then back-translate the German sentences into English. We then can detect pairs of words/phrases which can be considered as semantically related: the original English sentence and double-translated English sentence. We used Google translate[5], and chose German as the pivot language, which is a closely related to English, which we hope reduces the number of translation errors.

We performed double translation with 100 article laws in the Japanese Civil Code. We used the monolingual alignment tool of Sultan et al. [12] to create automatic word alignments in English. Table 4 shows examples of detected paraphrases using language translation. We can see that it also detects plural forms and past tense forms, in addition to words with similar meanings. We extract the top 100 paraphrases, and manually extracted corresponding Korean words in the Korean-translated Query-Article text.

Table 4. Examples of detected paraphrases

Original word	Paraphrased word	Original word	Paraphrased word
Year	Years	Establishes	Sets
Makes	Made	Purpose	Aim
Warranties	Guarantees	Matter	Area
Released	Relieved	Pledge	Commitment
Assigned	Transferred	Demand	Claim
Respect	Relation	Referred	Designated

---

[5] https://translate.google.com/.

3.6   Supervised Learning with SVM

Since we cannot anticipate the impact of each linguistic attribute, we use a machine learning algorithm that learns what information is relevant in the text to achieve our goal. We have compared our method with SVM, as a kind of supervised learning model. Using the SVM tool included in the Weka [4] software library[6], we performed cross-validation for the 412 questions. We used a linear kernel SVM because it is popular for real-time applications as they enjoy both faster training and classification speeds. Even though our system does not require much time for training, we chose a linear kernel to see the training performance for this simplest kernel. We used the following features:

(a) Word Lemma
(b) Lexical semantic features
(c) Negation feature
(d) Sentence analysis feature (condition, conclusion, and exception).

For concept features, we have exploited word embedding using *word2vec*. When we use word embedding, we assume the concepts of two words are the same if their cosine similarity in vector space is larger than 0.8.

The detailed features that we use are as follows:

Feature 1: *If* $\exists_{i,j}\{(\text{concept}(w_i)\,,\,\text{Query}_{condition})\cap(\text{concept}(w_j)\,,\,\text{Article}_{condition})\}$

Feature 2: *If* $\exists_{i,j}\{(\text{concept}(w_i)\,,\,\text{Query}_{conclusion})\cap(\text{concept}(w_j)\,,\,\text{Article}_{conclusion})\}$

Feature 3: *If* $\exists Article_{sub\_condition}\cap$

$\qquad\qquad \exists_{i,j,k}\{(\text{concept}(w_i)\,,\,\text{Query}_{condition})\cap(\text{concept}(w_j)\,,\,\text{Article}_{sub\_condition_k})=\varnothing\}$

Feature 4: *If* $\exists_{i,j}\{(\text{concept}(w_i)\,,\,\text{Query}_{condition})\cap(\text{concept}(w_j)\,,\,\text{Article}_{exception\_condition})\}$

Feature 5: *If* $\exists Article_{sub\_exception\_condition}\cap$

$\qquad\qquad \exists_{i,j,k}\{(\text{concept}(w_i)\,,\,\text{Query}_{condition})\cap(\text{concept}(w_j)\,,\,\text{Article}_{sub\_exception\_condition_k})=\varnothing\}$

Feature 6 : *If* $neg\_level(\text{Query}_{condition}) = neg\_level(\text{Article}_{condition})$

Feature 7 : *If* $neg\_level(\text{Query}_{conclusion}) = neg\_level(\text{Article}_{conclusion})$

Feature 8 : *If* $neg\_level(\text{Query}_{condition}) = neg\_level(\text{Article}_{exception\_condition})$

Features 1 and 2 check if there are overlapping concepts between a query condition (conclusion) and its relevant article condition (conclusion). Feature 3 checks if there is an overlapping word between a query condition and its relevant article sub-condition. Because the article sub-condition is connected with other sub-condition(s), using "*and*" as a connector, the query should include the meanings of all the article sub-conditions. Feature 4 checks if there are overlapping concepts between a query condition and its article exception-condition. We want to check if the query is included in the exceptional case using the feature. Feature 5 confirms that there is no overlapping word between a query condition and its relevant article sub-exception-condition. Features 6,

---

[6] The SVM function in Weka is provided by libsvm https://www.csie.ntu.edu.tw/~cjlin/libsvm/, and the linear kernal is from liblinear https://www.csie.ntu.edu.tw/~cjlin/liblinear/.

7, and 8 check the negation levels between the query condition, article condition, query conclusion, article conclusion, and article exception-condition. The negation level (*neg_level(segment)*) is computed as following: if [negation + antonym] occurs an odd number of times in the segment, its negation level is 1. Otherwise if the [negation + antonym] occurs an even number of times, including zero, its negation level is 0.

4 Phase 2: Experimental Results

4.1 Comparison of Our System's Performance with Others

In the general formulation of the textual entailment problem, given an input text sentence and a hypothesis sentence, the task is to make predictions about whether or not the hypothesis is entailed by the input sentence. We report the accuracy of our method in answering yes/no questions of legal bar exams by predicting whether the questions are entailed by the relevant civil law articles.

There is a balanced positive-negative sample distribution in the dataset (51.70% yes, and 48.30% no) for a dry run of COLIEE 2016 dataset, so we consider the baseline for true/false evaluation is the accuracy when always returning "yes," which is 51.70%. Our total data for the dry run has 412 questions.

Table 5 shows the experimental results. An SVM-based model showed accuracy of 62.14% when we did not use word embedding but used the lexical form of each word; the method of Kim et al. [10] showed 60.92% and that of Kim et al. [11] showed 61.65%. Our SVM augmented system outperformed Kim et al. [10, 11]. The differences were significant using the Wilcoxon Signed Rank Test at the level of significance of 0.05. We guess the reasons that our current system shows better results than the previous systems [10, 11] are as follows: (1) we analyzed queries in more detail and detected multiple conditions such as "and/or" connections, and then performed entailment based on the "and/or" logics. (2) We did paraphrasing as term expansion.

Table 5 also shows the experimental results arising when we adjust some of the features in our method. For example, the accuracy was reduced by 1.70% when we removed paraphrasing, and the accuracy was reduced by 1.47% when we used word

Table 5. Experimental results on dry run data for Phase 2

Method	Accu. (%)
(a) Baseline	51.70
(b) Our method using cross-validation with Supervised learning (SVM) not using word embedding but using lexical word itself	62.14
(c) Our method using cross-validation with Supervised learning (SVM) using word embedding	60.67
(d) Cross-validation using Kim et al. [10]	60.92
(e) Cross-validation using Kim et al. [11]	61.65
Without term expansion using paraphrasing from (b)	60.44
Without neg_level() from (b)	49.27

embedding. This suggests that word embedding does not help capture the semantics better than the lexical word by itself. We can guess that it may be because of the small training data for *word2vec* training. This suggests that we need to construct a higher volume of legal text data for *word2vec* training, and then check the performance of word embedding. When we did not use the negation feature, the accuracy became lower by 12.87%, which demonstrates the importance of the negation feature.

Table 6 shows the experimental results on the COLIEE-2016 test data. The test data size is 70 queries for Phase 2 (extracted from the bar exam of 2015), and 95 queries for Phase 3 (extracted from the bar exam of 2014) which are the same with the test data for Phase 1. Our accuracy on test data is 55.71% for Phase 2, and 55.79% for Phase 3. As shown in Table 7, our system showed best performance when two phases are combined (Phase 3), even though our Phase 1 and Phase 2 systems were not the best in the COLIEE 2015 competition [16]. Our system also performed paraphrasing, and detected condition-conclusion-exceptions for the query/article; our system extracted the article segment for which the query is semantically related. In contrast to other systems (except for Carvalho et al. [17]) that recognized textual entailment from the whole article to the query, our system compared the approximate semantics from a specific article segment to the approximated semantics of the query.

Table 6. Experimental results on formal run data

Method	Accu. (%)
Phase 2 baseline when 'yes' labels are all chosen	52.86
Phase 2 system (entailment)	55.71
Phase 3 system (1) (TF-IDF and entailment)	46.32
Phase 3 system (2) (ranking SVM lemma and dependency bigram as features (a) and entailment)	54.74
Phase 3 system (3) (adding IR score as features into (a) and entailment)	55.79

4.2 Error Analysis

From unsuccessful instances, we manually classified the error types as shown in Table 8. The biggest error arises, of course, from the semantic similarity error, and we believe our word embedding is not sufficient for estimating semantic similarity. In the future, we will try to include the bar exam text in the training data for the word embedding. The second biggest error is because of complex constraints in conditions. As with the other error types, there are cases where a question is an example case of the corresponding article, and the corresponding article embeds another article. We also found cases that indicate the need to do more extensive temporal analysis.

It will be interesting if we compare our performance using Korean-translated sentences with that using original Japanese sentences. We would expect the system using original sentences to show improved performance, because there would be no translation errors. As future work, we will construct a Japanese system using paraphrase/synonym/antonym dictionaries for Japanese, and then analyze how the translation affects performance.

Table 7. IR+Entailment results (Phase 3) on the formal run data in the COLIEE-2016

Run	Accu.	Run	Accu.
JNLN1 [17]	0.4000	iLis7 [19]	0.5368
KIS-1 [18]	0.5158	JNLN3 [20]	0.4737
KIS-2 [18]	0.5158	Our system (1)	0.4632
KIS-3 [18]	0.5263	Our system (2)	0.5474
KIS-4 [18]	0.5263	Our system (3)	**0.5579**

Table 8. Error types

Error type	Accuracy (%)	Error type	Accu. (%)
Specific example case	9.62	Semantic similarity error	28.85
Incorrect detection of the most similar article sentence	10.90	Constraints in condition	25.00
Incorrect detection of condition, conclusion, and mismatch	11.54	Etc.	14.10

5 Related Work

A previous textual entailment method from Bdour and Gharaibeh [5] provided the basis for a yes/no Arabic question answering system. They used a kind of logical representation, and compared the logical representation between queries and documents. This method may be appropriate for the task where queries and documents have similar logical representations so it is easier to confirm entailment from one logical representation to another. However, our task's entailment type is more complex, so we take an approach that approximates the logical content of queries and documents, rather than attempt any complete transformation to a logical form.

Nielsen et al. [6] extracted features from dependency paths, and combined them with word-alignment features in a mixture of an expert-based classifier. Zanzotto et al. [7] proposed a syntactic cross-pair similarity measure for RTE. Harmeling [8] took a similar classification-based approach with transformation sequence features. Marsi et al. [9] described a system using dependency-based paraphrasing techniques.

Many methods have been proposed for paraphrasing. One of the methods is the idea of semantic parsing via paraphrasing [13]. They transform a sentence into a logical form, and then convert logical forms to canonical form using the Freebase database. Subsequently, they obtain an association between the original sentence and canonical forms. However, hundreds of logical/canonical forms have been generated per sentence in their method, and the method does not show how to choose the best amongst them.

The method of Zhang et al. [14] also uses a pivot language for paraphrasing. Like us, they translate one language to another, then re-translate from the translated language into the original language. They then obtain a paraphrasing set between the original utterance and double-translated utterance. They showed improved performance in paraphrase detection using the pivot language translation, so we also employ the language translation-based paraphrasing. But instead of their use of the GIZA++ alignment, we used the monolingual alignment tool of Sultan et al. [12], because GIZA++, which is for alignment between two different languages, did not show good performance for our dataset.

6   Conclusion

We have described our most recent implementation for the Competition on Legal Information Extraction/Entailment (COLIEE)-2016 Task.

For Phase 1, legal information retrieval, we implemented a Ranking-SVM model for the legal information retrieval task. By incorporating features such as lexical words, dependency links, and tf-idf score, our model shows better mean average precision than tf-idf.

For Phase 2, we have proposed a method to answer yes/no questions from legal bar exams related to civil law. We used an SVM model using paraphrasing and pre-trained word embedding and query/article condition/conclusion/exception analysis. We show improved performance over a previous system, and paraphrasing and negation detection contributed to the performance. In the COLIEE 2016 competition, our system combining the Phase 1 and Phase 2 ranked highest in the accuracy of answering yes/no questions. As future work, we will train word2vec by larger texts not by articles to get the benefit of word embedding, and also try different kernels for SVM training to check if the kernel selection can increase the entailment performance.

Acknowledgements. This research was supported by the Alberta Machine Intelligence Institute (www.amii.ca). We are indebted to Ken Satoh of the National Institute for Informatics, who had the vision to create the COLIEE competition.

References

1. Jones, K.S.: A statistical interpretation of term specicity and its application in retrieval. In: Willett, P. (ed.) Document Retrieval Systems, pp. 132–142. Taylor Graham Publishing, London (1988)
2. Joachims, T.: Optimizing search engines using clickthrough data. In: Proceedings of 8th ACM SIGKDD International Conference on Knowledge Discovery and Data Mining, KDD 2002, pp. 133–142. ACM, New York (2002)
3. Maxwell, K.T., Oberlander, J., Croft, W.B.: Feature-based selection of dependency paths in ad hoc information retrieval. In: Proceedings of 51st Annual Meeting of the Association for Computational Linguistics, (vol. 1: Long Papers), pp. 507–516. Association for Computational Linguistics, Sofia, August 2013
4. Hall, M., Frank, E., Holmes, G., Pfahringer, B., Reutemann, P., Witten, I.H.: The WEKA data mining software: an update. SIGKDD Explor. **11**(1), 10–18 (2009)
5. Bdour, W.N., Gharaibeh, N.K.: Development of yes/no Arabic question answering system. Int. J. Artif. Intell. Appl. **4**(1), 51–63 (2013)
6. Nielsen, R.D., Ward, W., Martin, J.H.: Toward dependency path based entailment. In: Proceedings of 2nd PASCAL Challenges Workshop on RTE (2006)

7. Zanzotto, F.M., Moschitti, A., Pennacchiotti, M., Pazienza, M.T.: Learning textual entailment from examples. In: Proceedings of 2nd PASCAL Challenges Workshop on RTE (2006)

8. Harmeling, S.: An extensible probabilistic transformation-based approach to the third recognizing textual entailment challenge. In: Proceedings of ACL PASCAL Workshop on Textual Entailment and Paraphrasing (2007)

9. Marsi, E., Krahmer, E., Bosma, W.: Dependency-based paraphrasing for recognizing textual entailment. In: Proceedings of ACL PASCAL Workshop on Textual Entailment and Paraphrasing (2007)

10. Kim, M.-Y., Xu, Y., Goebel, R.: Alberta-KXG: legal question answering using ranking SVM and syntactic/semantic similarity. In: 8th International Workshop on Juris-Informatics (JURISIN), 2014

11. Kim, M.-Y., Xu, Y., Goebel, R.: A convolutional neural network in legal question answering. In: JURISIN Workshop (2015)

12. Sultan, M.A., Bethard, S., Sumner, T.: Back to basics for monolingual alignment: exploiting word similarity and contextual evidence. Trans. Assoc. Comput. Linguist. 2, 219–230 (2014)

13. Berant, J., Percy, L.: Semantic parsing via paraphrasing. In: Proceedings of Conference of the Association for Computational Linguistics (ACL), pp. 1415–1425 (2014)

14. Zhang, W., Ming, Z., Zhang, Y., Liu, T., Chua, T.S.: Exploring key concept paraphrasing based on pivot language translation for question retrieval. In: AAAI, pp. 410–416 (2015)

15. Manning, C.D., Surdeanu, M., Bauer, J., Finkel, J., Bethard, S.J., McClosky, D.: The Stanford CoreNLP natural language processing toolkit. In: Proceedings of 52nd Annual Meeting of the Association for Computational Linguistics: System Demonstrations, pp. 55–60 (2014)

16. Kim, M.-Y., Goebel, R., Kano, Y., Satoh, K.: COLIEE-2016: evaluation of the competition on legal information extraction and entailment. In: Tenth International Workshop on Juris-Informatics (JURISIN) (2016)

17. Carvalho, D.S., Tran, V.D., Tran, K.V., Lai, V.D., Nguyen, M.-L.: Lexical to discourse-level corpus modeling for legal question answering. In: Tenth International Workshop on Juris-Informatics (JURISIN) (2016). (Submission ID: JNLN1)

18. Taniguchi, R., Kano, Y.: Legal yes/no question answering system using case-role analysis. In: Tenth International Workshop on Juris-Informatics (JURISIN) (2016). (Submission ID: KIS)

19. Kim, K., Heo, S., Jung, S., Hong, K., Rhim, Y.-Y.: An ensemble based legal information retrieval and entailment system. In: Tenth International Workshop on Juris-Informatics (JURISIN) (2016). (Submission ID: iLis7)

20. Do, P.-K., Nguyen, H.-T., Tran, C.-X., Nguyen, M.-T., Minh, N.L.: Legal question answering using ranking SVM and deep convolutional neural network. In: Tenth International Workshop on Juris-Informatics (JURISIN) (2016). (Submission ID: JNLN3)

21. Onodera, D., Yoshioka, M.: Civil code article information retrieval system based on legal terminology and civil code article structure. In: Tenth International Workshop on Juris-Informatics (JURISIN) (2016). (Submission ID: HUKB)

22. Nguyen, T.-S., Phan, V.-A., Nguyen, T.-H., Trieu, H.-L., Chau, N.-P., Pham, T.-T., Nguyen, L.-M.: Legal information extraction/entailment using SVM-ranking and tree-based convolutional neural network. In: Tenth International Workshop on Juris-Informatics (JURISIN) (2016). (Submission ID: JNLN2)

23. John, A.K., Di Caro, L., Boella, G., Bartolini, C.: Team-normas' participation at the COLIEE 2016 bar legal exam competition. In: Tenth International Workshop on Juris-Informatics (JURISIN) (2016). (Submission ID: N01)

SKL 2016

3rd International Workshop on Skill Science

Tsutomu Fujinami

Japan Advanced Institute of Science and Technology,
1-1 Asahidai, Nomi, Ishikawa 923-1292, Japan

1 Aims and Scope

Human skills involve well-attuned perception and fine motor control, often accompanied by thoughtful planning. The involvement of body, environment, and tools mediating them makes the study of skills unique among researches of human intelligence. The symposium invited researchers who investigate human skill. The study of skills requires various disciplines to collaborate with each other because the meaning of skills is not determined solely by efficiency, but also by considering quality. Quality resides in person and often needs to be transferred through the master-apprentice relationship. The procedure of validation is strict, but can be more complex than scientific activities, where everything needs to be described by referring to evidences. We are keen to discussing the theoretical foundations of skill science as well as practical and engineering issues in the study.

2 Topics

We invited wide ranges of investigation into human skills, from science and engineering to sports, art, craftsmanship, and whatever concerns cultivating human possibilities. Fourteen pieces of work were presented at the workshop, including one invited lecture. Two selected pieces of work are included in the issue from our workshop.

The article titled "A Basic Study of Gaze Behavior Measurement Methodology for Drivers in Autonomous Vehicles", written by Rie Osawa and her co-authors, proposes a way for tracking drivers' gaze on car. Evaluating driver's skill is an important topic given that future cars may work semi-automatic by collecting data of the driver.

The other article titled "Toward a mechanistic account for imitation learning: an analysis of pendulum swing-up" by Takuma Torii and Shohei Hidaka proposes a model of imitation, where the observer hypothesizes the goal to be achieved by a particular action. Goal recognition has been one of the important issues in communication. The article extends the area of research to human movements and their unique approach should invite further experiments among researchers.

The workshop organizer is honored to present the two reports, which deal with varieties of issues ranging from practical to theoretical. He hopes that the reader will find them interesting and will be stimulated to look into the field of Skill Science.

A Basic Study of Gaze Behavior Measurement Methodology for Drivers in Autonomous Vehicles

Rie Osawa[(✉)], Shota Imafuku, and Susumu Shirayama

Department of System Innovation, School of Engineering,
The University of Tokyo, Tokyo, Japan
rie-u@nakl.t.u-tokyo.ac.jp, shota.imafuku@gmail.com,
sirayama@sys.t.u-tokyo.ac.jp

Abstract. Research and development of autonomous driving technology is accelerating in the automotive industry. Currently, drivers of such vehicles are considered to pay less attention to environmental conditions while being driven, due to potential overestimations of autonomous driving functionality and its reliability. In this paper, methods to quantitatively measure the driver's gaze behavior are proposed, followed by investigation methodology and results on how auditory warning signals influence the behavior, where the difference between novice and experienced drivers is also compared.

1 Introduction

Research and development of autonomous driving technology is accelerating in the automotive industry, with fatality reduction being one of the main driving forces. In future autonomous vehicles, passengers are not likely to pay attention to forward or lateral environmental conditions, because they are assumed to overestimate autonomous driving functions and its reliability.

When driving, auditory interfaces can lessen the actual driving workload compared to visual interfaces in the vehicle [1]. Also, spearcons (time-compressed speech sounds) are known to significantly reduced total glance time toward vehicle monitors [2]. Although there are several researches to investigate the relationship between gaze behavior and audio stimuli, detailed information of the gaze behavior such as scanpath or pupil diameter have not been analyzed.

In our research, we first propose methods to quantitatively measure passengers' gaze behavior in a simulated experimental autonomous vehicle apparatus. The level of autonomous driving is assumed to be level 3 (conditionally autonomous) as defined by SAE [3]. We then measure the gaze behavior of driving experts, and novices to examine the difference.

S. Kurahashi et al. (Eds.): JSAI-isAI 2016, LNAI 10247, pp. 317–326, 2017.
DOI: 10.1007/978-3-319-61572-1_21

2 Proposed Methodology

2.1 Overview of Experimental Apparatus

From the safety viewpoint, there are many constraints to have a meaningful experiment involving actual driving. Therefore we have built an in-lab simulator which recreates the vehicle cabin and the driving experience.

Figure 1 illustrates the overview of the experimental apparatus, devised to measure the passengers' gaze behavior. As the first step towards establishing a quantitative measurement method, we have prepared a simple experimental apparatus utilizing a single screen, one projector and one pair of headphones (Fig. 2). A head-mounted eye-tracking device was used to capture what exactly the examinees were looking at, and their pupil diameters.

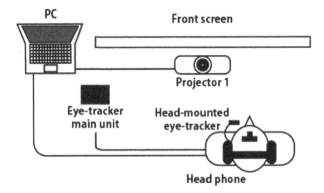

Fig. 1. Overview of the experimental apparatus

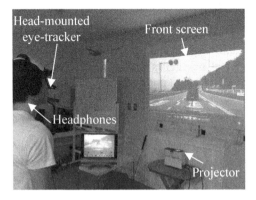

Fig. 2. Experimental apparatus of eye-tracking

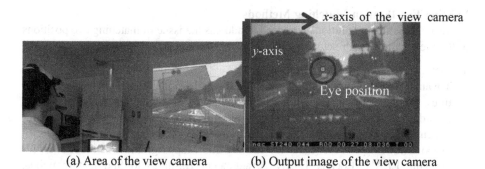

(a) Area of the view camera   (b) Output image of the view camera

Fig. 3. View of head camera

In psychological research, it is common for the subjects' heads to be fixed, in order to obtain accurate eye movement measurements. However, in our experiment, we have decided not to restrict the subjects' head motion, since humans are known to move their heads, consciously or unconsiously, to help localize audio sources, according to Wallach [4]. Head motion represents one of the major human physiological behaviors, and is essential in daily life, which is why the decision against any motion restriction was made.

For the eye-tracking device, a head-mounted type has been selected. However, there is one problem specific to such devices: the output eye position data is affected by head movement. Therefore a method to detect the exact eye position through excluding the effect of head movement has been developed.

2.2   Eye-Tracking

Eye-Tracking Methodology
We have selected NAC Image Technology's EMR-9 as the eye-tracking device used to record data of the eye position and pupil diameter, which includes a view camera attached to the subject's forehead for video recording. The eye position is indicated by the x-y coordinates in the area recorded by the view camera (Fig. 3). Therefore, even if the eye position is fixed on a specific item, head movement will cause shifting of the view camera area and x-y coordinates, leading to difficulties in identifying the target object, as seen in Fig. 4.

Fig. 4. Variations of view and axis caused by head movement

Existing Eye Position Matching Methods

There are three major solutions proposed to address the issue of matching eye positions in the view camera with actual eye positions on the screen.

1. Methods based on features in images

   Toyama et al. [5] propose utilizing SIFT(Scale-Invariant Feature Transform) features. Points which have high contrast characteristics, or points at the corner are regarded as key points with highly noticeable features. These are suitable for matching because they are not affected by rotation or scaling. Takemura et al. [6] propose utilizing PTAM (Parallel Tracking and Mapping) and Chanijani et al. [7] applying LLAH (Locally Likely Arrangement Hashing) to find feature points. However, these methods require a sufficient number of feature points in the image for accuracy, which may or may not be present depending on the contents of the view camera image.

2. Methods with markers

   NAC Image Technology offers a method utilizing AR markers to create artificial feature points [8], where AR markers captured in the view camera are matched with spacial coordinates. Tomi and Rambli [9] also proposes to utilize eye-tracker with AR application into calibration of head-mounted display. Huang and Tan [10] use circular patterns as markers. However, rather large markers could have an influence on eye movements due to their size and appearance. Kocejko et al. [11] propose the algorithm to compensate head movement with three cameras (to obtain eye, scene and head angle) and LED markers. However, objects of view are limited in the monitor and movement of subjects is limited.

3. Method with infrared data communication

   Tobii Technology offered a solution which utilizes infrared data communication markers. Eight such markers (approximately 30 mm-cubic) are required for position detection, where each marker communicates with the eye-tracking device and matches the image of the view camera with respective spacial coodinates. However, the size of such markers could also have a significant impact on eye movement as well. Note that currently this device is not available.

New Method of Utilizing Artificial Feature Points with Infrared LED Markers

In this paper, anew eye-tracking method is proposed, through creating artificial feature points made of invisible NIR (near-infrared)-LED markers. NIR-LEDs are basically invisible to the naked human eye, therefore reducing the effect to eye-tracking despite their presence. At the same time, NIR-LEDs are actually visible through IR filters, as seen in Fig. 5. In the robot technology domain, it is popular to utilize NIR-LEDs for robots to detect location, or to follow target objects [12]. However, to the authors' best knowledge, there have been no applications of NIR-LEDs used for eye-tracking, which has the potential of enabling eye movement detection even with head movement.

The view camera with IR filters captures feature points of NIR-LEDs installed on the projection screen. This image can be used to first verify the eye position relative to the NIR-LED feature points, which can then be used to calculate what exactly the subject is looking at on the screen. We shall call these invisible NIR-LED markers "IR markers" hereafter.

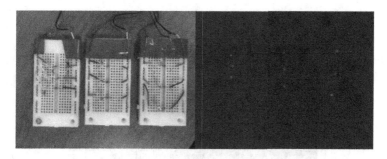

Fig. 5. IR markers with the naked eye (left) and through the filter (right)

Image processing is another question which requires attention. SIFT features could be potential options. However, these methods are not adequate for images of IR markers received through the IR filter, because single NIR-LED IR markers are homogeneous and less characteristic, as shown on the right side image of Fig. 5. As a countermeasure, several patterns composed of multiple NIR-LEDs have been developed as templates for matching, as described below.

Patterns of IR Markers

IR Marker patterns have been created, taking into account the four following conditions.

1. Patterns should have a sufficient amount of features
2. Patterns should be composed of the least number of markers possible
3. Patterns should have sufficient differentiation between one another
4. Patterns should be easily produced

To decide on the exact patterns, the similarity among patterns which illustrate filtered IR markers schematically (Fig. 6) have been calculated. Taking condition 2 into consideration, a three-point pattern was selected from a 5*5 dot matrix for each pattern, which was the best balance to ensure noticeable differentiation. Table 1 shows the result of similarity calculation, using Hu invariant moment [13]. Template images are on the table head, searched images are on the table side, where a lower score of matching evaluation means higher similarity, and are represented with red cells. Hu invariant moment allows both rotational and scale invariance check, hence relevant combinations of patterns with high similarity scores can be calculated.

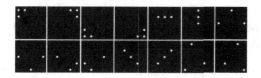

Fig. 6. Patterns used for the experiment

Table 1. The result of matching experiment

	1	2	3	4	5	6	7	8	9	10	11	12	13	14
1		0.002	0.005	0.039	0.077	0.079	0.415	0.416	0.308	0.301	0.203	0.205	0.452	0.453
2			0.003	0.041	0.035	0.038	0.307	0.308	0.212	0.203	0.119	0.120	0.367	0.367
3				0.043	0.037	0.040	0.310	0.311	0.214	0.206	0.121	0.122	0.370	0.370
4					0.006	0.003	0.267	0.268	0.171	0.162	0.078	0.079	0.326	0.327
5						0.004	0.338	0.340	0.231	0.225	0.126	0.128	0.375	0.377
6							0.336	0.337	0.229	0.222	0.124	0.126	0.373	0.374
7								0.002	0.148	0.150	0.212	0.210	0.107	0.105
8									0.150	0.152	0.213	0.212	0.106	0.105
9										0.016	0.105	0.103	0.232	0.231
10											0.098	0.097	0.238	0.237
11												0.002	0.249	0.250
12													0.247	0.249
13														0.001
14														

3 Implementation and Experiment

3.1 Implementation of the Screen for Eye-Tracking

Based on the findings in the previous section, several patterns were chosen and created with IR markers. Specifically, NIR-LEDs and resistors were attached to a solder-less breadboard, which were then mounted onto a polystyrene board. The chosen width of the polystyrene board was 1600 mm, which is representative of the width of an actual automobile's front view. Subjects of the experiments were placed at a distance from the screen representative of that between a front windshield and driver seat. Twelve patterns were found to be required to be on the board for at least three patterns to be within the view camera at a given time, to ensure high accuracy of the template matching. The layout of the IR markers was decided based on the results of similarity seen in Table 1, where the actual implementation can be seen in Fig. 7. Each pattern of these invisible IR markers shall be expressed as IRn (n = 1 to 12), where n is the identifier or each pattern, and each NIR-LED shall be represented in IRn_i, where i is the identifier of each LED (i = 1 to 3).

A preliminary experiment was conducted, in order to examine the proposed method's correlation between the eye position as seen through the view camera and the actual projected image. An image captured from a driving video as seen in Fig. 8 was first shown to the subject. View camera images were then taken through the IR filter as seen in Fig. 9, which show IR filtered marker locations, and the eye-tracker x-y coordinates of the eye position. Through image processing, these images were aligned by superimposing the IR marker patterns, in order to have a corrected eye position on the source video. Detailed procedures are as follows.

1. Apply template matching on the images of view camera in order to detect the IDs of IR markers and their coordinates.
2. Calculate the coordinates of the eye position on the entire screen of screen, based on the eye position coordinates relative to the IR markers (Fig. 7 right side)

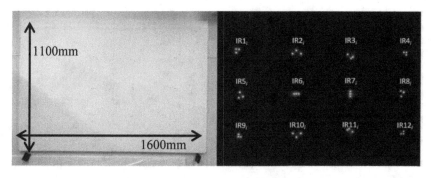

Fig. 7. IR marker-embedded screen with the naked eye (left) and through the filter (right)

Fig. 8. Projected image on the screen

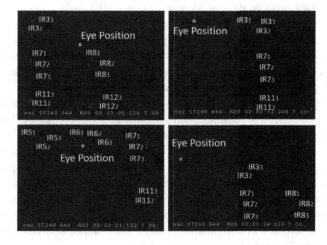

Fig. 9. Captured images of the head camera and IR marker IDs

Fig. 10. The result of the eye position matching

3. Map the corrected eye coordinates on the image projected on the screen (Fig. 8).
4. Output the movie with eye positions mapped.

Results of the above can be seen in Fig. 10.

3.2   Results and Discussion

The proposed methodology showed promise towards proceeding with further data acquisition and analysis. Following this, several steps have already been taken for further experiment design.

As a first step to evaluate how close the apparatus was to an actual autonomously driven environment, subjects were asked to perform simple tasks, such as watching video footage of driving on roads, and offering feedback on how realistic it felt. There were no secondary tasks or devices to operate.

The gaze behavior of three novice drivers and two expert drivers were measured separately, in order to confirm the feasibility of the proposed methodology, as well as to see if there were any differences in the simulated driving experience. Novice drivers were all males in their twenties who rarely drive, whereas experts were males who have more than 30 years of driving experience, and drive on a weekly basis.

The driving video footage was taken from the inside of a vehicle while driving, and was projected on the IR marker embedded screen with a short projector. In addition to the visual experience, three audio cues intended to simulate warning sounds were included, consisting of (1) beep sound of a digital watch, (2) notification sound of an airplane, and (3) horn sound of a vehicle. These cues were directionally placed on the right side and left side, utilizing Adobe Premiere, played back through a pair of Bose Quiet Comfort 25 headphones worn by the subjects. As previously mentioned NAC Image Technology EMR-9 was used for the eye-tracking. The distance between the subjects and the screen was designed to replicate the distance between a driver and a front windshield of a general passenger vehicle.

Fig. 11. An example of corrected eye position mapped on the video footage (corrected eye positions circled)

Figure 11 shows the corrected eye positions mapped on the source movie after taking into account the head movement, by following procedures proposed in the Sect. 3.1.

Through comparing the filtered images and the frame images of the video footage, the proposed method was confirmed to work well, correcting the effects of head movements as intended, and successfully outputting adjusted eye positions on the video screen. One interesting observation made was that there were no significant differences in gaze behavior between novice and expert drivers. Inexperienced novices did indeed pay necessary attention to traffic signs, center lines on roads, and oncoming vehicles.

In terms of subject feedback, although there is a need for many more subjects to analyze statistical significance, early feedback is as follows. The novices said there was a lacking sense of reality, or a sense of tension for driving, which actually allowed them to pay more attention to the environmental conditions. However, they were not fully confident in being able to pay as much attention when actually driving. On the other hand, the experts seemed to feel as if they were sitting in the passenger seat, and noticed a significant gap between the experiment and actual driving conditions. With regard to the directional audio sources, though one novice did show slightly bigger eye movements, there were no significant changes observed with the feature addition. One novice did mention how he felt urged to look in the direction of the audio source, however felt awkward due to the lack of side mirrors. One expert mentioned how he could not understand the meaning of the directional audio source, hence was not able to do anything consciously. Again, more data points are necessary for further studies.

4 Conclusion and Remarks

A new method to measure vehicle driver gaze behavior has been developed, taking into account corrective measures which utilize invisible markers. This will enable higher eye position detection accuracy, which has been a problem specific to head-mounted eye-tracking devices. However, there is need to increase the number of subjects for statistical reasons, and there is also room to improve the accuracy of template matching, leading to better eye position recognition. It is also necessary to quantitatively analyze eye-tracking data and pupil diameter data, known to reflect the subject's psychological status, which also requires a significant number of subjects to conduct a meaningful statistical analysis. Furthermore, the autonomous driving vehicle simulator requires further studies on recreating a realistic environment. There is much work ahead, but considering the result of this experiment so far, there is great potential for future applications through the development of (1) an actual vehicle-like environment (e.g. view to the left/right, steering wheel, visual instruments such as meters), (2) operational tasks (e.g. smartphones, navigation, switches), and (3) effective directional audio sources, which will all benefit from the proposed measurement methodology.

References

1. Sodnik, J., Dicke, C., Tomazic, S., Billinghurst, M.: A user study of auditory versus visual interfaces for use while driving. Int. J. Hum.-Comput. Stud. **66**, 318–332 (2008)
2. Larsson, P., Niemand, M.: Using sound to reduce visual distraction from in-vehicle human-machine interfaces. Traffic Inj. Prev. **16**(1), S25–S30 (2015)
3. SAE International: "Automated Driving Levels of Driving Automation are Defined in New SAE International Standard J3016"
4. Wallach, H.: On sound localization. J. Acoust. Soc. Am. **10**, 270–274 (1939)
5. Toyama, T., Kieninger, T., Shafait, F., Dengel, A.: Gaze guided object recognition using a head-mounted eye tracker. In: Proceedings of 7th ACM Symposium on Eye Tracking Research & Applications (ETRA2012), pp. 91–98 (2012)
6. Takemura, K., Kohashi, Y., Suenaga, T., Takamatsu, J., Ogasawara, T.: Estimating 3D point-of-regard and visualizing gaze trajectories under natural head movements. In: Proceedings of 6th ACM Symposium on Eye Tracking Research & Applications (ETRA2010), pp. 157–160 (2012)
7. Chanijani, S.S.M., Al-Naser, M., Bukhari, S.S., Borth, D., Allen, S.E.M., Dengel, A.: An eye movement study on scientific papers using wearable eye tracking technology. In: 9th International Conference on Mobile Computing and Ubiquitous Networking (ICMU) (2016)
8. NAC Image Technology, EMR-dStream. http://www.eyemark.jp/product/emr_dstream/
9. Tomi, A.B., Rambli, D.R.A.: Automated calibration for optical see-through head mounted display using display screen space based eye tracking. In: 3rd International Conference on Computer and Information Science (ICCOINS), pp. 448–453 (2016)
10. Huang, C.W., Tan, W.C.: An approach of head movement compensation when using a head mounted eye tracker. In: International Conference of Consumer Electronics-Taiwan (2016)
11. Kocejko, T., Bujnowski, A., Ruminski, J., Bylinska, E., Wtorek, J.: Head movement compensation algorithm in multi-display communication by gaze. In: 7th International Conference on Human System Interactions (HSI), pp. 88–94 (2014)
12. Sohn, B., Lee, J., Chae, H., Yu, W.: Localization system for mobile robot using wireless communication with IR landmark. In: Proceedings of the 1st International Conference on Robot Communication and Coordination, pp. 1–6 (2007)
13. Hu, M.K.: Visual pattern recognition by moment invariants. IRE Trans. Inf. Theory **8**(2), 179–187 (1962)

Toward a Mechanistic Account for Imitation Learning: An Analysis of Pendulum Swing-Up

Takuma Torii[(✉)] and Shohei Hidaka

Japan Advanced Institute of Science and Technology,
1-1 Asahidai, Nomi, Ishikawa, Japan
{tak.torii,shhidaka}@jaist.ac.jp

Abstract. Learning an action from others require to infer their underlying goals, and recent psychological studies have reported behavioral evidences that young children do infer others' underlying goals by observing their actions. The goal of the present study is to propose a mechanistic account for how this goal inference is possible by observing others' actions. For this purpose, we performed a series of simulations in which two agents control pendulums toward different goals, and analyzed with which types of features it is possible to infer their different latent goals and control schemes. Our analysis showed that pointwise dimension, a type of fractal dimension, of the pendulum movements is sufficiently informative to classify the types of agents. With respect to its invariant nature, this result suggests that the fine-grained movement patterns such as the fractal dimension reflect the structure of the underlying control schemes and goals.

Keywords: Imitation learning · Goal inference · Dynamical systems · Pendulum swing-up

1 Introduction

It is crucial for human being as a social being to learn actions by observing others' actions. We refer to a sequence of movements with a certain goal or plan behind it as *action* [2]. Thus, learning an action includes inference of the underlying goal or plan as well as production of bodily movements to achieve the goal inferred. In this study, we exclusively use the term *action* in the sense of Bernstein [2], which is a movement controlled to solve a certain problem. In this sense, the actions should be classified by their goals, although movements in general can be classified by their physical appearance. Beyond mere replication of movements or mimicking on the basis of apparent similarity, learners need to infer goals behind actions. Thus, our main question here is how this is possible with less knowledge on the goals or the actions.

Importantly, inference of the goal of an action does not require complete knowledge of bodily movements. According to accumulating pieces of empirical

© Springer International Publishing AG 2017
S. Kurahashi et al. (Eds.): JSAI-isAI 2016, LNAI 10247, pp. 327–343, 2017.
DOI: 10.1007/978-3-319-61572-1_22

evidence, children as young as 18 month old can infer the goal without observing the intended outcome of action, with which the goal becomes evident [13,14]. In their experiments, the psychologists presented to each infant an incomplete accidental and/or intentional action performed by an adult and observed response of the infant. In an experimental condition, children observe a person trying to get an out-of-reach object he *accidentally* dropped on the floor – that is an act repeating the unfulfilled actions. On the other hand, in the control condition, the person *intentionally* threw the object on the floor. For their purpose, the experimenters acted carefully so as not to make apparent difference between the movements (i.e., "drop" or "throw") for 18 month old. The psychologists found that only in the former condition, infants show helping behavior to complete the unfulfilled action of the person (i.e., the children handed him the object). This suggests that children can predict what movements follow the adult's action and thus discriminate the difference in intentions behind seemingly similar movements.

This goal inference underlying actions is a key step toward social learning from others' behavior. In this study, we seek for a mechanistic account for goal or plan inference by observation of unfulfilled actions. There are, however, two major problems to address the mechanism of the goal inference of actions. First, the learner may have a different body from the instructor, and thus observation is never complete and the learner needs to specify missing variables. In theoretical studies, this problem is often formulated as an *inverse problem* [8,9]. In this formulation, the unobservables are inferred using a generative model (or forward model) and an assumed optimality principle, and this inferrence goes backward to the generating process.

Second, a sequence of bodily movements is typically just one of possible means to achieve a given goal, and thus seemingly different movements may be similar in terms of a certain type of goals, and vice versa – two very similar movements may be performed for two different goals. This problem requires to solve another type of inverse problem, in which the one needs to identify and differentiate bodily movements according to the goals.

To solve the inverse problem, it is often assumed that the learner knows an appropriate class of forward models (a generator of bodily movements), which defines the generative process, from the goal to movements, and allows to estimate a likely model for given observations [8,9]. Thus, this approach is limited in learning actions, especially to goal inference from incomplete actions, as in such cases no or little observations is available for the learner to successfully infer the hidden goal (e.g., [3,10,11]).

To seek for a new theoretical account for goal inference on unfulfilled actions, as instantiated in the empirical studies [13,14], we analyze seemingly similar movements generated to follow different intentions to accomplish a task. In a prior stage of learning actions, without knowing the goals, learners have to classify given movements by supposed intentions. Here by the term "intention" we refer to as a motor control scheme that outputs the motor action for a given bodily state as the input, by which a sequence of bodily movements is generated for a given initial state. The motor control scheme is constructed by near-optimally fulfilling a given goal for the actor.

As a first step toward understanding the mechanism how children recognize the intention behind the other's actions, we ask the two basic questions:

1. How can we identify multiple seemingly different actions generated by the same motor control scheme?
2. How can we differentiate multiple seemingly similar actions generated by two different motor control schemes?

To address these questions, we resorted a computer simulations of the simplest possible physical body—the classical pendulum swing-up task of one degree-of-freedom, that can produce seemingly similar movements with different intentions. By identifying the motor control scheme underlying the unfulfilled actions in a simple physical model, we address the goal inference found in the psychological studies [13,14].

2   Simulation

2.1   Rationale

In this study, we idealize and simplify the psychological experimental paradigm of goal inference reported by [13,14]. Their goal inference experiments can be summarized by the two conditions, goal-achieved and/or goal-failed demonstration. In the goal-achieved demonstration (i.e., intentionally "throw" on the floor), the demonstration is successful: the demonstrator's behavior does meet his/her goal. In the goal-failed demonstration (i.e., accidentally "drop" on the floor), the demonstration is unsuccessful: the demonstrator's behavior does not meet his/her goal, but the resulting movements of the unsuccessful demonstration (accidentally "drop") is quite similar to those of the successful demonstration (intentionally "throw"). At a coarse-grain observation, the movements of the goal-achieved and goal-failed demonstration look similar (due to the experimenters' careful acts in front of children [13,14]), but the intended goals are actually different.

To capture difference in intended goals and similarity in movements of physical body, our simulation has two agents, goal-achieved (GA) and goal-failed (GF) agent; each agent goes through two phases, a learning and demonstration phase. In the learning phase, each agent forms the motor control scheme by learning to meet a goal given its physical body structure. In the demonstration phase, each agent shows movements according to the learned motor control scheme. In our simulation, the types of agents, goal-achieved and goal-failed, are defined by the consistency of the goals between the learning and demonstration phase. We set two different goals in the learning phase. The GA agent demonstrates movements under the same condition as the agent learned his/her motor control. Thus, the GA agent's movements in the demonstration phase are supposed to be the best or closely-optimized to be intended for the given goal in the learning phase. In contrast, the GF agent demonstrates movements under a different condition from the one the agent learned his/her motor control. Thus, the GF agent's movements in the demonstration phase are supposed to be not the best or sub-optimal that count as failure or unintended movements. Due to the same constraint for both GA and GF agents, they show apparently similar movements with different intentions in the demonstration phase.

How can we discriminate the GA and GF agents, who both show similar movements, by only observing their movements in the demonstration phase? Specifically, we choose the pendulum swing-up task, which is one of the simplest motor control task and has been analyzed in the past studies. The pendulum model has the control of one degree-of-freedom and two dimensional state space, that is a minimally sufficient setting we can use to answer our questions. In the following two sections, we briefly introduce the basic physical model of the pendulum, and next describe the learning/demonstration phases and the GA/GF agents in this framework.

2.2   Pendulum Swing-Up Task

Figure 1 (left) depicts a mathematical model of simple pendulum. A simple pendulum is composed of the rod of length $l = 1.0$ and the ball of mass $m = 1.0$. A state of this pendulum at an instant of time is determined by the rod angle θ and velocity $\dot{\theta}$. The equation of motion [5] is given by

$$ml^2\ddot{\theta} - mgl\sin\theta = -b\dot{\theta} + u \tag{1}$$

where $g = 9.8$ is the gravity constant, $b = 0.01$ is dumping force (torque), and $u \in [-5, 5]$ is control input (torque) from a control scheme.

The pendulum swing-up task [5,11] is originally to design a control scheme that can swing the pendulum up and hold it at the inverted position given by $\theta = 0$. We will introduce a modified version of this task in the next section.

In the learning phase, the learner (the demonstrator) is required to learn a control scheme from experience, without knowing the mechanics of the pendulum. The control scheme is learned using a reinforcement learning technique, which is described in-depth in Appendix A. A control scheme for this task is defined by the function $u = g(\theta, \dot{\theta})$, which outputs torque u for a given state

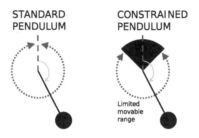

Fig. 1. The standard (left) and constrained pendulum (right).

$(\theta, \dot{\theta})$. The task for the agent in the learning is to construct the function g, which maximize the reward $\sum_t \cos(\theta_t)$ under the condition that the control input is restricted by $u \in \pm 5$, which prevents the agent from getting the maximum reward easily. For each trial in the learning phase, the initial position of the pendulum is set to randomly within $\theta \in \pm[\pi/8, \pi)$ and $\dot{\theta} = 0$.

A characteristic of the simple pendulum is the mechanistic energy that is the sum of the kinetic energy and the potential energy:

$$E(\theta, \dot{\theta}) = \frac{1}{2}ml^2\dot{\theta}^2 + mgl(\cos\theta - 1) \tag{2}$$

For the pendulum swing-up task, the goal state, holding about the inverted position given by $\theta = 0$ and $\dot{\theta} = 0$, is characterized by $E(\theta, \dot{\theta}) = E(0, 0) = 0$. Thus, if the learner knows the pendulum model of his/her task, the control problem is reduced to adjust the energy E at the moment to be closer to zero for any $(\theta, \dot{\theta})$ [1]. In the simulated learning, instead of applying this knowledge directly, the agent interatively update their control scheme g by maximizing the reward $\sum_t \cos(\theta_t)$, which is specifically implemented by the reinforcement learning framework.

2.3   Goal-Achieved and Goal-Failed Agent

To construct the goal-achieved and goal-failed agent, we consider two different conditions for the learning phase and one condition for the demonstration phase, which are summarized in Table 1 and depicted in Fig. 2. Besides the standard pendulum introduced in the previous section, we additionally introduce the *constrained* pendulum, Fig. 1 (right). The constrained pendulum is the exactly same as the standard pendulum except for the "wall" (see Fig. 1 (right)), that is, its limitation in the movable angular range ($\theta \in \pm\pi/8$) to which the agent cannot move the pendulum. In addition, the constrained pendulum is limited to have $u = 0$ in the range $\theta \in \pm\pi/6$.

Table 1. Correspondence to the goal-inference experiment in psychology

	Goal-achieved	Goal-failed
Reward	$\sum_t \cos\theta_t$	$\sum_t \cos\theta_t$
Pendulum in learning	Constrained	Standard
Pendulum in demonstration	Constrained	Constrained
Human experiment	Intended action	Unfulfilled action

By introducing the constrained pendulum, we made up a clear dissociation between the learning and demonstration phase of the GF agent, and design both GA and GF agents move similarly with different their own control schemes.

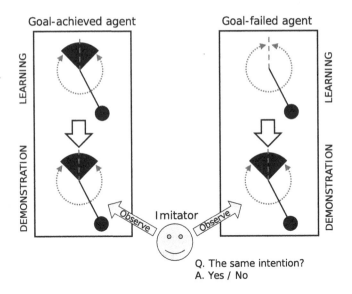

Fig. 2. Simulation design: the goal-achieved (GA) agent learns and demonstrates with the constrained pendulum (left), the goal-failed (GF) agent learns with the standard pendulum but demonstrates with the constrained one (right).

Specifically, the GA agent learns the motor control scheme with the constrained pendulum, and demonstrates movements with it (the left panel of Fig. 2). The GF agent learns the motor control scheme with the standard pendulum, but demonstrates movements with the constrained pendulum (the right panel of Fig. 2). Both GA and GF agent learn to maximize the same goal indicated by the reward $\sum_t \cos \theta_t$, but they control the different pendulums. Thus, their motor control schemes are different – the GF agent tries to swing the standard pendulum up to the top most position, which is allowed in his/her learning phase, but the GA agent tries to swing the constrained pendulum up to the topmost of the feasible region (Fig. 1). In the demonstration with the constrained pendulum, the GF agent cannot reach the top-most position which was reachable in only his/her learning phase, while the GA agent can reach the top most position reachable as well in his/her learning phase. We treat the consistency in the GA agent and inconsistency in the GF agent in their learned motor control schemes and physical body in the demonstration phase as "intended" and "unfulfilled" action in the human experiment reported by [13,14] (i.e., intentionally "throw" and accidentally "drop").

2.4   Reinforcement Learning

To construct motor control scheme g for each agent, we used a reinforcement learning framework. Reinforcement learning [12] is a framework rooted in behavioral psychology and control theory. The key idea is that in the task environment in state s, the learner takes an action a, and next, the learner observes the environment in a new state s' and receives a reward r from the environment. The learner continues the task from the new state s'. The goal of learning is to acquire a control scheme $g(a|s)$ that maximizes the cumulative reward. See Appendix A for details.

For this pendulum swing-up task, the state transition of the environment is given by the motion of the pendulum. The learner can partially modify the future movements of the pendulum by supplying control inputs (i.e., actions taken by the learner). The reward function for this task must characterize the angular error from the inverted position. From [5], we adopted $\cos(\theta)$, which gives the highest reward $\cos(\theta) = 1$ at the inverted position $\theta = 0$ and the lowest reward $\cos(\theta) = -1$ at the hanging-down position $\theta = \pm\pi$. The pendulum swing-up task is solved successfully after 5000 trials, each consists of 10000 time steps (about 100 s).

3   Typical Pendulum Movements

We generated datasets for the imitator's task (the demonstration phase in Fig. 2) using motor controls learned in the way of Sect. 2.4 (the learning phase in Fig. 2). We systematically sampled thirty different initial conditions from the ranges $\theta = \pm[\pi/8, \pi)$ with the same interval. Below we first show the typical movements of the standard pendulum and constrained pendulum controlled by the GA and GF agents. In the next section, we analyze the movements (i.e., datasets for the imitator) in place of the imitator to answer the questions.

3.1   Standard Pendulum

A typical successful movement of the standard pendulum controlled was started from a random initial position, and the position of the pendulum was eventually reached to the inverted position $\theta = 0$ (or $\theta = 2\pi$) by gradually increasing the amplitude. The mechanical energy of the pendulum increases together with its amplitude, until it got charged a sufficient energy to swing up to the inverted position. Since no constraints here, the agent learned the control scheme with the standard pendulum can swing up the standard pendulum.

3.2   Constrained Pendulums

Figure 3 show typical movements of the constrained pendulums by the goal-achieved (GA) and goal-failed (GF) agent, respectively. The initial condition is the same for both the cases in the figure. Unlike the standard pendulum, the constrained pendulum cannot be swung up to the inverted position, and the mechanical energy cannot be exactly zero. For the reason of the constraints added to the pendulum, the mechanical energy reduces near the limits of the movable range. Recall that the task of the imitator is to decide whether or not the two movements exposed were generated by the same control scheme. From Fig. 3, it seems that it is uneasy to differentiate the GF agent from the GA agent behind the movements observable.

Constrained pendulum by the goal-achieved (GA) agent

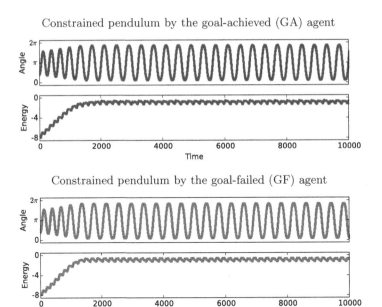

Fig. 3. Constrained pendulum swing-up movements from the same initial condition. Each figure contains the pendulum angle $\theta \in [0, 2\pi)$ and the mechanical energy. Note that $\theta = 0$ or $\theta = 2\pi$ is the inverted position. (Color figure online)

4   Analysis: Discrimination of Motor Control Schemes

Given the simulated movements of the two different agents, here we analyze them from two perspectives. One is an observer with little knowledge on the internal parameters of the pendulum, and the other is that with full knowledge on it. The former ignorant observer only accesses the angle and angular velocity $(\theta, \dot{\theta})$, but the latter knowledgeable observer knows the other physical parameters, such as mass, length, and friction of the pendulum, as well as energy of the system. A typical imitator, with respect to young children in the goal inference experiment [13,14], is supposed an ignorant observer. Thus, to understand the mechanism to recognize intended or unfulfilled action, the GA and GF agents should be discriminable from the perspective of the ignorant observer with a minimal possible access to the actor-specific parameters.

To test whether the minimal accessible feature, the angle and angular velocity $(\theta, \dot{\theta})$, is sufficient to discriminate the type of agents, GA or GF, we performed classification analysis of the agent types using the features of the demonstrated movements. Here we consider angle, angular velocity, mechanical energy, frequency spectrum of the angle, and a type of fractal dimensions of the pendulum position as candidate features for the classification analysis. The mechanical energy requires knowledge of the physical model of the pendulum, and thus it is accessible only by the knowledgeable observer.

4.1   Hypothesis

Our working hypothesis is that intention behind the movements would result in the structural complexity of the motor-controlling system as a whole. In our simulation, the task in the learning phase is defined by the reward and the physical constraints on the pendulum. The pair of the reward (objective function) and the physical constraints (the domain or state space to find the maximum of the objective function) together forms the motor control scheme through the reinforcement learning. Thus, the motor control formed as the result of reinforcement learning is supposed to reduce unnecessary movements according to the task. In the other words, the motor control scheme formed for the standard pendulum is generally not the best choice for the constrained pendulum to maximize the reward. This sub-optimality in the motor control scheme is expected to increase unnecessary movements which does not directly gain the reward.

Specifically, let us consider the GA agent in our simulation. The GA agent is readily given the wall (inadmissible region in the state space) in the learning phase, and thus it should avoid hitting the wall but try to stay longer in time near the wall, at which the agent gains the largest reward value. Namely, the ideal learning leads a pendulum movement free from the wall (as it never hits the wall). Thus, we expect that the dynamical system of the GA agent can be analyzed by the pendulum movements without consideration to the wall. On the other hand, the GF agent learns the control scheme with the standard pendulum. As the GF agent finds the best rewarding region of the standard pendulum in the inadmissible region of the constrained pendulum, its ideally learned motor control swings up through the potential wall in the constrained pendulum. In its demonstrating with the constrained pendulum, it is unavoidable for the GF agent to hit the wall (unintentionally), and the wall comes as one of key factor in the dynamical system of the GF constrained pendulum. Therefore, this observation naturally leads the idea that the dynamical system of the GA agent has lower dimension than that of the GF agent, as the GF dynamical system has the additional factor, the unexpected wall in the way swinging up, playing a substantial role forming the dynamics.

4.2   Pointwise Dimension

To characterize this type of complexity in the pendulum movements, we analyze the attractor dimension of the movements treating it as dynamical systems. Specifically, we exploited a sort of fractal dimension called *pointwise dimension* for the classification analysis.

Here, we briefly introduce the basic nature of pointwise dimension. The pointwise dimension is a type of dimension, which is defined for a small open set or measure on it including a point in a given set (for the formal definition, see [4,15]). It is invariant under arbitrary smooth transformation. As it is associated for each point, we can analyze distribution of pointwise dimension across points in a set by estimating the pointwise dimension for a set of data points.

Informally speaking, pointwise dimension of a point characterize "degree of freedom" around the point. With this nature of the pointwise dimension, we expect it to capture the differences in the motor control scheme reflected by some local differences in the pendulum movements. We have developed a statistical technique to estimate the pointwise dimension for a set of data points [7]. Applying this technique, each point in the dataset is assigned with a positive value of pointwise dimension, with which the topological nature around each point is characterized.

Figure 3 shows a representative time series of pendulum angle θ and the mechanical energy E for the GA and GF agent in the demonstration. Around 2000 time step, the movements of the pendulums reached to the limit of angular range $\theta \in \pm\pi/8$ with the maximum amplitude, and the mechanical energy reaches at its maximum. We call the time interval after the mechanical energy reaches at its maximum *stationary period*, and the time interval before it *transient period*.

Figure 4 shows the time series of pointwise dimension for the same dataset above. The pointwise dimension is estimated for the time-delay coordinate

$$(x_t, x_{t+1}, \ldots, x_{t+9}, y_t, y_{t+1}, \ldots, y_{t+9})$$

corresponding time series of the positions $(x_t, y_t) = (\sin\theta_t, \cos\theta_t)$.

In the transient period, the pointwise dimension of both GA and GF agent stay nearly constant (Fig. 4). In the stationary period, the GF agent tended to show larger pointwise dimensions than those in the transient period. In contrast, the GA agent tended to show smaller pointwise dimensions that those in the transient period. As theoretically predicted in the previous section, this trends suggest that the GF pendulum would be described as a dynamical system with higher dimension than the GA one.

For more detailed inspection of the pointwise dimension, Fig. 5 shows a representative set of the pointwise dimensions as a function of the vertical positions, $\cos\theta - 1$, for each of the GA and GF pendulum in the stationary period. Figure 5 shows that the GA and GF agent can be discriminable by the pointwise dimensions, especially in the time interval with the higher position. As the high-position period corresponds with the impact with the wall in the constrained pendulum, this result suggests that the differences between the GA and GF agent in the pointwise dimension reflect the dynamical irregularity involving with the wall.

Hitting the wall generally decreases the mechanical energy, and the agent needs to regain the energy for the next swing. Thus, this energy loss and regain is expected to form a periodic dynamics. To visualize this energy oscillation, we plot the mechanical energy $E = U + V$, which is the sum of the kinetic U and the potential energy V, difference between the kinetic and potential energy $\Delta E = U - V$, and the pointwise dimension for each of the GA and GF pendulum (Fig. 6). This figure shows the limit cycle on the $(E, \Delta E)$ plane, and the peak of the pointwise dimension in this oscillation corresponds to the period of the higher mechanical energy E (hitting the wall). We found large changes in mechanical

Constrained pendulum by the goal-achieved (GA) agent

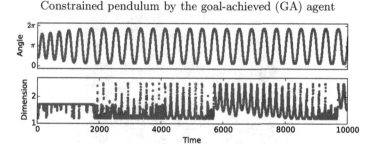

Constrained pendulum by the goal-failed (GF) agent

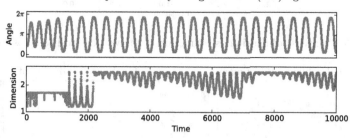

Fig. 4. Time series of pointwise dimensions in the demonstration phase.

energy E when ΔE is close to zero and the pendulum's hitting the wall, and these changes are coincident with the higher pointwise dimensions. This result means that the patterns in pointwise dimension capture the critical point in the energy dynamics of the pendulums of the GA and GF agent.

4.3 Fourier Analysis

Next, we analyzed the power spectrum of the angle time series as an alternative measure compared with the pointwise dimension. Figure 7 shows the power spectrums of the GA (green) and GF agent (red) corresponding to the angle time series shown in Fig. 3. This figure shows that both have the peaks at the nearly same frequency (Hz) with similar magnitude. We also analyzed the other 29 pairs of GA and GF samples with different initial positions, and confirmed the similar trend with the trend in Fig. 7. This result suggests that the types of agents, the GA or GF, would be difficult to discriminate with the power spectrum of the angle time series. In the next section, we tested this suggestion quantitatively (see the next Sect. 4.4).

4.4 Classification of Agent Types

Lastly, we performed classification analyses of agent types based on each of features, which can be derived from the pendulum movements. The performance in this classification analysis is considered as an indicator how informative for

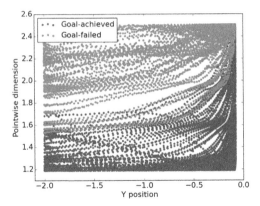

Fig. 5. Pointwise dimension as a function of the vertical position, $\cos\theta - 1$

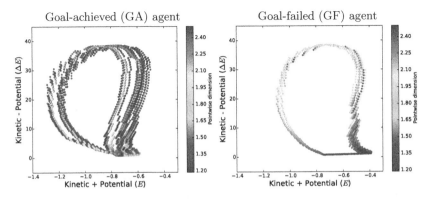

Fig. 6. Pointwise dimension (color map) as a function of $E = U + V$ and $\Delta E = U - V$, where U is the kinetic and V is the potential energy. The large change in mechanical energy E near $\Delta E = 0$ indicates hitting the wall.

the imitator to identify the seemingly different movements with different initial positions and discriminate the GF agent from the GA agent, which can produce seemingly similar movements with the same initial positions.

Specifically, we performed a two-class classification of the types of agents, the GA or GF, for each data point using one of features, angle, angle velocity, mechanical energy, the maximum power of the short interval of time series, and pointwise dimension. Suppose that the imitator is given a training dataset, consisting of a subset of all 30 time series of one out of these features, in which each data point is labeled which type of agents, the GA or GF. As our goal is to numerically test whether the pointwise dimension has significantly high information to classify the two types of agents, we chose one of well known classifiers, the Gaussian mixture model. This choice was motivated by its simplicity in computation, rather than its classification performance.

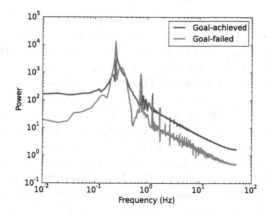

Fig. 7. Power spectrum of the angular time series in the GA and GF pendulum.

Given the training data points, the imitator constructs sample probability functions of a variable x for the GA agent $p_{GA}(x)$ and for the GF agent $p_{GF}(x)$. Using the sample probability functions, the imitator asserts that a given new sample x is of the GF agent, if $p_{GF}(x) > p_{GA}(x)$, otherwise the imitator asserts that is of the GA agent in the test. The performance is evaluated by the correct ratio of the imitator's prediction. Each of probability distributions $P_{GA}(x)$ and $P_{GF}(x)$ is represented by a Gaussian mixture model with the minimum Bayesian information criterion for one to 30 components. Gaussian mixture models represent a probability distribution of data points as a mixture of multiple Gaussian (normal) distributions. In this analysis, we found the optimal Gaussian mixture models each consists of 5 to 18 components for angle, 8 to 23 components for velocity, 5 to 19 components for power, 4 to 22 components for energy, and 4 to 11 components for pointwise dimension.

We include either or both transient or/and stationary phases in our training and test dataset, as the patterns in these phases are different qualitatively. We divided each time series, using the mechanical energy as an indicator of convergence, into the transient part (before convergence) $E(\theta, \dot{\theta}) < -1.3$ and stationary part (after convergence) $E(\theta, \dot{\theta}) \geq -1.3$.

The results of the classification analysis are summarized in Fig. 8. As both training and test dataset consists of equally balanced number of sampled labeled GA and GF, the chance level of this binary classification was 50%. The performance of the classification with the angle, angle velocity, and power spectrum were at or barely above the chance level. The correct ratio of angle was 51.0% for the transient dataset and 51.2% for the stationary one. Similarly, that of angular velocity and the maximum power was at the chance level: about 50% for both datasets.

The classification of the mechanical energy was significantly higher than the chance level: 70.9% for the stationary data and 52.0% for the transient data. Note that, however, the mechanical energy is not accessible by the ignorant imitator, as calculating the mechanical energy requires the full knowledge of the physical model such as gravity constant or friction coefficient. Moreover, the mechanical energy is the direct measure of the task, as both the GA and GF agent aim to maximize the mechanical energy, which is identical to maximization of the reward in the learning phase. Thus, the classification performance of the mechanical energy is supposed to give an upper bound, if the imitator have the complete knowledge on the pendulum model.

Lastly, the performances with the pointwise dimension are as good as these bounds of the mechanical energy : 63.9% for the stationary and 50.4% for the transient data. Note that this classification accesses externally observeable measures, as the pointwise dimension can be derived by only angle or position time series. Thus, an ignorant imitator can reach this level of performance, if he or she is asked to classify the types of agents using the pointwise dimension or other comparable measures. This implies that pointwise dimension could be a potential characteristics to identify and discriminate the motor control scheme underlying the observed movements, with little prior knowledge on the system of interest.

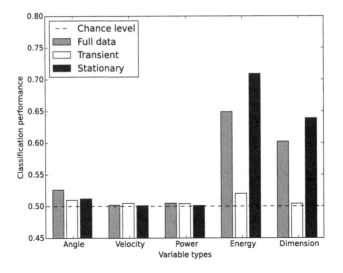

Fig. 8. Classification performance for each features

5   Discussion

In the present study, we explore a mechanistic account for intention inference from other's actions, which is crucial step toward understanding the goal-level imitation. In order to tackle this problem specifically, we consider a classical

pendulum swing-up task with constraints as a idealization of the intention inference experimental paradigm by [13,14]. By defining intention as motor control
scheme in our computational framework, intention inference can be viewed as
identification of the dynamical system with latent variables by its observeable
variables. Our analysis on the two types of agents demonstrated that the latent
motor control scheme can be identified with the observeable movements, if the
imitator computes pointwise dimension of the trajectories in state space of the
pendulum.

In past literature, identification of the motor control scheme is supposed
to require the prior knowledge on the class of systems to be estimated [8,9]. As
expected by this theory and confirmed in our analysis, the model-specific measure
such as the mechanical energy has sufficient information on the motor control
system. This approach, however, needs explicit modeling of the demonstrator,
which is hard to be accessible by an ignorant imitator. In contrast, our approach
based on pointwise dimension requires little prior knowledge on the system,
but yet it performed as good as the alternative in identification of it. Thus it
gives a possible mechanistic account for ignorant imitator to infer the imitation
underlying other's actions.

Our next step on this research program is to build an autonomous imitator
which can generates actions to meet the hidden goal identified by observing the
other's actions. Empirical evidence for the use of pointwise dimension by human
subjects is another line of our future research.

Acknowledgment. This study is supported by the JSPS KAKENHI Grant-in-Aid
for Young Scientists JP 16H05860.

A   Reinforcement learning

Reinforcement learning [12] is a framework rooted in behavioral psychology and
control theory. In the task environment in state s, the learner takes an action a
and receives a reward r from the environment in response to the action. Next,
the learner faces with the environment in a new state $s' = Q(s'|s)$, where Q is
a transition function. The goal of learning is to acquire a control scheme $g(a|s)$
that maximizes the cumulative reward.

The pendulum swing-up task is a classic control problem with continuous
space and time [5,11]. There are many researches to solve this task (e.g., [6] for
recent updates). The simple and basic algorithm for this task is so-called actor-
critic architecture [6,12]. It is composed of two, the actor and critic components.
The actor represents the control scheme $g(a|s)$. On the other hand, the critic
represents the value function $V(s)$, that tells the learner the discounted expected
reward of state s.

Since the task is in continuous space and time, it involves several engineering
problems. The typical approach is discretization of the continuous space and
time. For continuous time, we used discretized time steps for Eular integration
(step size $dt = 0.01$) and we sampled per 3 time steps. For continuous state

space, we adopted a discretized representation (tile coding [12]) in which the continuous state space $(\theta, \dot{\theta}) \in [-\pi, \pi] \times [-2\pi, 2\pi]$ is equally divided into the grid of size 40×40 (one of them is called here state s).

Learning proceeds with estimation of state values, and that shapes $g(a|s)$ to navigate to more rewarding states. Suppose the pendulum is now in state s (one of the grid) at time t. The learner supplies a control input u sampled from $g(u|s) = N(\mu_s, \sigma_s)$, a normal distribution of mean μ_s and variance σ_s^2. In response, the learner observes the pendulum in a new state s' and a reward r. Then the learner increments his value function $V(s)$ for every s by

$$\Delta V(s) = \alpha_c \left[r + \gamma_c V(s') - V(s) \right] E_c(s) \tag{3}$$

where $\alpha_c = 0.1$ is a learning rate, $\gamma_c = 0.97$ is a discount rate, and E_c is an eligibility trace with exponential decay (given by $\lambda_c = 0.65$) that is a device for continuous tasks that assigns higher weights for recently visited states.

For continuous control inputs, the control scheme $g(\cdot)$ is expressed by a collection of normal distributions for each state s. So the learner has to determine the mean μ and variance σ^2 for each state s. Formally, the learner modifies his control scheme μ_s and σ_s for every s by

$$\Delta \mu_s = \alpha_a \left[r + \gamma_a V(s') - V(s) \right] \frac{\partial N(\mu_s, \sigma_s)}{\partial \mu_s}(u) E_a(s) \tag{4}$$

$$\Delta \sigma_s = \alpha_a \left[r + \gamma_a V(s') - V(s) \right] \frac{\partial N(\mu_s, \sigma_s)}{\partial \sigma_s}(u) E_a(s) \tag{5}$$

where $\alpha_a = 0.001$ is a learning rate, $\gamma_a = 0.65$ is a discount rate, and E_a is an eligibility trace with decay (given by $\lambda_a = 0.0$).

The reward function $r = f(s, a)$ must be designed carefully. From [5], we set the reward function $r = \cos(\theta)$ for this task. It only depends on angle θ. Remark that $\cos(\theta)$ characterizes the goal of this task, because the inverted position $\theta = 0$ gives the highest reward $\cos(0) = 1$, and the hanging-down position $\theta = \pm\pi$ gives the lowest reward $\cos(\pi) = -1$.

References

1. Astrom, K.J., Furuta, K.: Swinging up a pendulum by energy control. Automatica **36**(2), 287–295 (2000)
2. Bernstein, N.A.: Dexterity and Its Development. Psychology Press, Abingdon (1996)
3. Breazeal, C., Scassellati, B.: Robots that imitate humans. TRENDS Cogn. Sci. **6**(11), 481–487 (2002)
4. Cutler, C.D.: A review of the theory and estimation of fractal dimension. In: Tong, H. (ed.) Dimension Estimation and Models, pp. 1–107. World Scientific (1993)
5. Doya, K.: Reinforcement learning in continuous time and space. Neural Comput. **12**, 243–269 (1999)
6. Grondman, I., Vaandrager, M., Busoniu, L., Babuska, R., Schuitema, E.: Efficient model learning methods for actor-critic control. IEEE Trans. Syst. Man Cybern. Part B (Cybern.) **42**(3), 591–602 (2012)

7. Hidaka, S., Kashyap, N.: On the estimation of pointwise dimension. arXiv:1312.2298 (2013)
8. Kawato, M.: Computational Theory of Brain. Sangyo Tosho, Tokyo (1996). (in Japanese)
9. Marr, D.: Vision. MIT Press, Cambridge (1982)
10. Ng, A., Russell, S.J.: Algorithms for inverse reinforcement learning. In: Proceedings of the Seventeenth International Conference on Machine Learning (ICML 2000), pp. 663–670 (2000)
11. Schaal, S.: Learning from demonstration. In: Mozer, M., Jordan, M., Petsche, T. (eds.) Advances in Neural Information Processing Systems, vol. 9, pp. 1040–1046. MIT Press, Cambridge (1997)
12. Sutton, R.S., Barto, A.G.: Reinforcement Learning: An Introduction. MIT Press, Cambridge (1998)
13. Warneken, F., Tomasello, M.: Altruistic helping in human infants and young chimpanzees. Science **311**, 1301–1303 (2006)
14. Warneken, F., Tomasello, M.: The roots of human altruism. Br. J. Psychol. **100**, 455–471 (2009)
15. Young, L.S.: Dimension, entropy, and Lyapunov exponents. Ergodic Theory Dyn. Syst. **2**(1), 109–124 (1982)

Author Index

Arisaka, Ryuta 241

Bekki, Daisuke 19, 123
Boonkwan, Prachya 253
Bowker, Mark 45

Chakraborty, Goutam 176
Chang, Shuang 204
Cooper, Robin 5

Deguchi, Hiroshi 204

Fukuda, Ken 143

Goebel, Randy 299

Hashimoto, Morito 220
Hatsutori, Yoichi 270
Hidaka, Shohei 327
Hung, Nguyen Duy 253

Imafuku, Shota 317
Imai, Haruki 270

Jafari Songhori, Mohsen 188

Kano, Yoshinobu 284
Kikuchi, Takamasa 188
Kim, Mi-Young 299
Kinoshita, Eriko 19
Kiselyov, Oleg 33
Kurahashi, Setsuya 159, 220

Liang, Huizhi 143
Liefke, Kristina 45

Lin, Meng-Ru 176
Lu, Yao 299

Matsusaka, Youichi 62
McCready, Elin 74
Mery, Bruno 90
Mineshima, Koji 19, 123

Nishimura, Satoshi 143
Nishimura, Takuichi 143

Osawa, Rie 317

Racharak, Teeradaj 253
Rieser, Lukas 108

Satoh, Ken 241
Shirayama, Susumu 317
Su, Jing 188

Tanaka, Ribeka 123
Taniguchi, Ryosuke 284
Terano, Takao 188
Tojo, Satoshi 253
Torii, Takuma 327
Toriyama, Masahiro 188

Ueda, Keiichi 159

Winterstein, Grégoire 74

Xu, Ying 299

Yang, Wei 204
Yoshikawa, Katsumasa 270

Printed in the United States
by Baker & Taylor Publisher Services